Application of Advanced Oxidation Processes

Application of Advanced Oxidation Processes

Editors

Ciro Bustillo-Lecompte
Jose Colina-Márquez
Lars Rehmann

MDPI • Basel • Beijing • Wuhan • Barcelona • Belgrade • Manchester • Tokyo • Cluj • Tianjin

MDPI

Editors
Ciro Bustillo-Lecompte
Ryerson University
Canada

Jose Colina-Márquez
Universidad de Cartagena
Colombia

Lars Rehmann
Western University
Canada

Editorial Office
MDPI
St. Alban-Anlage 66
4052 Basel, Switzerland

This is a reprint of articles from the Special Issue published online in the open access journal *Processes* (ISSN 2227-9717) (available at: https://www.mdpi.com/journal/processes/special_issues/Advanced_Oxidation).

For citation purposes, cite each article independently as indicated on the article page online and as indicated below:

LastName, A.A.; LastName, B.B.; LastName, C.C. Article Title. *Journal Name* **Year**, *Article Number*, Page Range.

ISBN 978-3-03936-888-4 (Hbk)
ISBN 978-3-03936-889-1 (PDF)

Contents

About the Editors

Ciro Bustillo-Lecompte (Ph.D.) has a multidisciplinary background in the areas of civil, environmental, and chemical engineering. He completed his Bachelor of Engineering at the University of Cartagena, Colombia, in 2008 and obtained his MASc (2012) and PhD (2016) at Ryerson University, Canada. Ciro is a certified Professional Engineer (PEng), Environmental Professional (EP), a Fraternal Member of the Canadian Institute of Public Health Inspectors (CIPHI), and a 2017–2018 Queen Elizabeth Scholar (QES). He is currently an Associate Member in the Environmental Applied Science and Management Graduate Programs and Lecturer in the School of Occupational and Public Health at Ryerson University. He has co-authored over 20 peer-reviewed scientific papers, as well as several conference proceedings, chapters and books. His research interests include advanced oxidation processes, advanced treatment of water and wastewater, waste minimization, water reuse, water, soil and air quality, energy and resource recovery, and heterogeneous catalysis.

Jose Colina-Márquez (Ph.D.) has been an Associate Professor in the Chemical Engineering Department at the University of Cartagena since 2010. He obtained his B.Sc. in Chemical Engineering at the University of Atlantico (1996) and his M.Sc. and Ph.D in Chemical Engineering at the University of Valle (2008 and 2010, respectively). Currently, he is leading the Research Group of Modeling and Applications of Advanced Oxidation Processes, that aims to solve water detoxification issues using these technologies. He is also a member of the Editorial Committee of the Revista Ingenieroa y Competitividad (University of Valle, 2012) and a member of the Editorial Committee of the Revista Ciencias e Ingenieria (University of Cartagena, 2011). He was awarded with the "Magna cum laude grade" for his PhD studies, granted by the University of Valle (2010), and "Junior Researcher of the year", granted by the Colombian Society of Catalysis (2012).

Lars Rehmann (Ph.D.) obtained a PhD in Chemical Engineering in 2007 at Queen's University after completing his undergraduate degree at the Technical University Braunschweig, Germany and his Masters work at Tottori University in Japan. He completed an NSERC post-doctoral fellowship at the University of Manchester, UK (2007–2009) holding a Canada/UK Millennium Research Award. He is an Associate Professor in the Department of Chemical and Biochemical Engineering at Western University, as well as a Visiting Professor at the RWTH Aachen, Germany and a Fellow of the Alexander von Humboldt Foundation. He is an accomplished and heavily cited researcher in the field of biochemical and environmental engineering. His research has resulted in over 70 peer-reviewed articles published in leading journals in his field, as well as three patent applications and the foundation of a spin-off company. Since his appointment at Western (May 2009) he has attracted over 4M USD in funding and has established research facilities centered around highly automated bio-processing from the micro-scale to pilot scale; with infrastructure unique in Canada. His research largely focuses on the fermentative production of platform chemicals and fuels from waste biomass, as well as recombinant protein production.

processes

MDPI

Editorial

Special Issue: Application of Advanced Oxidation Processes

Ciro Bustillo-Lecompte [1,*], Jose Colina-Marquez [2,*] and Lars Rehmann [3,*]

1 School of Occupational and Public Health, Ryerson University, 350 Victoria Street, Toronto, ON M5B 2K3, Canada

2 Department of Chemical Engineering, Universidad de Cartagena, Sede Piedra de Bolívar, Avenida del Consulado 48-152, Cartagena 130015, Colombia

3 Department of Chemical and Biochemical Engineering, Western University, Thompson Engineering Building, London, ON N6A 5B9, Canada

* Correspondence: ciro.lecompte@ryerson.ca (C.B.-L.); jcolinam@unicartagena.edu.co (J.C.-M.); lrehmann@uwo.ca (L.R.)

Received: 16 July 2020; Accepted: 16 July 2020; Published: 18 July 2020

Advanced oxidation processes (AOPs) are nowadays not only considered as a complementary treatment option but as an attractive alternative to conventional methods. AOPs are often used to improve the biodegradability of wastewaters, as well as in the inactivation of pathogenic microorganisms. This Special Issue focuses on the "Application of Advanced Oxidation Processes" and covers recent progress and trends in AOPs. A total of eleven papers constitute this Special Issue, covering different aspects of the application of AOPs. Three review papers provide comprehensive appraisals on the state-of-the-art of AOPs in industrial applications, while eight research papers cover specific technologies for wastewater treatment, dye decolorization, and nanotechnology applications.

The first review explores different photocatalytic processes used for the treatment of gold mining wastewater with an emphasis on cyanide wastewater applications [1]. In this review, Betancourt-Buitrago et al. [1] investigate the effectiveness of photocatalytic processes in the removal of free cyanide via oxidative pathways, where the need for more studies—with respect to the presence of complexed cyano-metallic compounds in photoreactors, the requirement of modifying selectivity using electron donors as scavengers of unwanted radicals, as well as the use of unconventional UV LED lamps as a growing field in photoreactors design—is highlighted.

The second review summarizes the advantages and disadvantages of different treatment methods of spent caustic wastewater from the mercaptan oxidation process in the oil industry [2]. Moreover, Pino-Cortés et al. [2] describe the characteristics of spent caustic and its environmental impacts, followed by an analysis of current neutralization processes to improve biodegradability.

The last review by Ramírez Orejel and Cano-Buendía [3] highlights the advantages of using electrolyzed water in the food industry. The authors describe the characteristics of electrolyzed water and explore the main applications of electrolyzed water in the food industry, specifically in animal products derived from poultry, cattle, swine, and fish for sanitation, cleaning and disinfection.

The first of the eight research papers presented in this Special Issue examines the effect of heating oxidation on the surface/interface features and floatability of anthracite [4]. The changes in surface and interface of coal particles were confirmed by Rong et al. [4] using scanning electron microscopy (SEM), Fourier-transform infrared spectroscopy (FTIR), and X-ray photoelectron spectroscopy (XPS).

The second research paper explores the potential of cow manure to produce highly porous activated carbon (AC) using a potassium hydroxide (KOH) activation process at different temperatures for the removal of methylene blue from an aqueous solution [5]. Based on the XPS and FTIR observations, Tsai et al. [5] concluded that the AC from cow manure has a strong interaction with the cationic compound (i.e., methylene blue) due to its large pore properties and oxygen-containing complexes on the surface.

The following two research papers are related to nanotechnology applications. Leybo et al. [6] investigated the effect of ferric and ferrous salts on the structural characteristics and physicochemical properties of nanoscale zero-valent iron samples synthesized by borohydride reduction. Samples were characterized by SEM, transmission electron microscopy, X-ray diffraction, and low-temperature nitrogen adsorption. Leybo et al. [6] showcase the synergistic action of nanoscale zero-valent iron and ferrihydrite leading to higher Pb^{2+} ion removal from water via borohydride reduction of iron from sulphate salt.

On the other hand, Hou et al. [7] synthesized mesoporous Mn-doped Fe/rGO nanocomposites by a co-precipitation method and investigated the ultrasonic-assisted removal of ethyl violet (EV) from synthetic wastewater. Artificial Intelligence (AI) models were combined with response surface methodology (RSM) to optimize the removal efficiency of Mn-doped Fe/rGO nanocomposites for EV. The Mn-doped Fe/rGO nanocomposite demonstrated its potential for the removal of dyes in water due to its facile synthesis, large adsorption capacity, high specific surface area, and excellent magnetic properties [7].

The last four research papers cover specific technologies for wastewater treatment and dye decolorization. The study by Chiliquinga et al. [8] explores the simultaneous production of the ferrate ion and free radicals on a boron-doped diamond (BDD) anode surface. Further, it examines the generation of both species in two different electrolytes with the addition of different concentrations of iron sulphate, and the effect of the current density on the degradation of a textile dye. Results showed fast degradation rates of up to 98% of colour reduction in 60 min of treatment with a COD reduction of over 60% [8].

Similarly, Lozano-Morales et al. [9] evaluated the photocatalytic degradation of nanostructured TiO_2 catalysts under UV-C radiation and the effect of the pH of paracetamol suspensions on the overall degradation efficiency. The photocatalytic oxidation degraded up to 99% of the paracetamol at a pH of 6.5, a value at which adsorption was favoured, leading to higher efficiency and exposure times below 100 min, where no traces of metabolites were detected [9].

Another study explores the use of UV-activated free chlorine and investigates the degradation efficiency of the Aqueous Polycyclic Musk Tonalide [10]. Besides, Wang and Liu [10] identified primary contributors and intermediates, described the toxicity profile of the Aqueous Polycyclic Musk Tonalide, and evaluated the influence of natural organic matter, common anions and cations in the degradation.

Finally, the study by Sun et al. [11] compared UV/O_3, US/O_3, and $UV/US/O_3$ processes based on the digestion time and conversion rate of standard N-compounds, and investigated the influence of temperature, pH, the concentration of O_3, mass flow rate, UV intensity, US frequency, and power on digestion efficiency. Results showed that the $UV/US/O_3$ process had the best digestion efficiency under optimum conditions of pH = 11, O_3 mass flow rate of 3200 mg/h, reaction temperature of 30 °C, digestion time of 25 min, 18 W UV lamp power, 20 kHz and 75 W of US frequency and power, respectively, for a wastewater conversion rate over 99% and total dissolved nitrogen in the range of 1.0–4.0 mg/L.

These diverse contributions provide sustained evidence of the large variety of applications of AOPs as well as demonstrating a promising future of the research fields of photocatalysis, UV/O_3, electrochemical oxidation, sonolysis, nanotechnology in AOPs, the development of novel catalytic material, heating oxidation, UV-activated free chlorine, electrolyzed water, industrial wastewater treatment, and resource recovery, among other applications.

To conclude, the guest editors would like to thank all the authors for their valuable contributions; their efforts and contributions are highly appreciated. State-of-the-art applications of AOPs presented in this Special Issue may serve as valuable references for future research in this ever-evolving field.

Author Contributions: Writing-review & editing, C.B.-L., J.C.-M., and L.R. All authors have read and agreed to the published version of the manuscript.

Acknowledgments: The guest editors would like to acknowledge the authors who collaborated and contributed with their research towards this Special Issue.

Conflicts of Interest: The authors declare no conflict of interest.

References

1. Betancourt-Buitrago, L.; Hernandez-Ramirez, A.; Colina-Marquez, J.; Bustillo-Lecompte, C.; Rehmann, L.; Machuca-Martinez, F. Recent Developments in the Photocatalytic Treatment of Cyanide Wastewater: An Approach to Remediation and Recovery of Metals. *Processes* **2019**, *7*, 225. [CrossRef]
2. Pino-Cortés, E.; Montalvo, S.; Huiliñir, C.; Cubillos, F.; Gacitúa, J. Characteristics and Treatment of Wastewater from the Mercaptan Oxidation Process: A Comprehensive Review. *Processes* **2020**, *8*, 425. [CrossRef]
3. Ramírez Orejel, J.; Cano-Buendía, J. Applications of Electrolyzed Water as a Sanitizer in the Food and Animal-By Products Industry. *Processes* **2020**, *8*, 534. [CrossRef]
4. Rong, G.; Xu, M.; Wang, D.; Gui, X.; Xing, Y. Effect of Heating Oxidation on the Surface/Interface Properties and Floatability of Anthracite Coal. *Processes* **2019**, *7*, 345. [CrossRef]
5. Tsai, W.; Huang, P.; Lin, Y. Reusing Cow Manure for the Production of Activated Carbon Using Potassium Hydroxide (KOH) Activation Process and Its Liquid-Phase Adsorption Performance. *Processes* **2019**, *7*, 737. [CrossRef]
6. Leybo, D.; Tagirov, M.; Arkhipov, D.; Permyakova, E.; Kolesnikov, E.; Kuznetsov, D. Effect of Initial Salt Composition on Physicochemical and Structural Characteristics of Zero-Valent Iron Nanopowders Obtained by Borohydride Reduction. *Processes* **2019**, *7*, 769. [CrossRef]
7. Hou, Y.; Qi, J.; Hu, J.; Xiang, Y.; Xin, L.; Wei, X. Mesoporous Mn-Doped Fe Nanoparticle-Modified Reduced Graphene Oxide for Ethyl Violet Elimination: Modeling and Optimization Using Artificial Intelligence. *Processes* **2020**, *8*, 488. [CrossRef]
8. Chiliquinga, M.; Espinoza-Montero, P.; Rodríguez, O.; Picos, A.; Bandala, E.; Gutiérrez-Granados, S.; Peralta-Hernández, J. Simultaneous Electrochemical Generation of Ferrate and Oxygen Radicals to Blue BR Dye Degradation. *Processes* **2020**, *8*, 753. [CrossRef]
9. Lozano-Morales, S.; Morales, G.; López Zavala, M.; Arce-Sarria, A.; Machuca-Martínez, F. Photocatalytic Treatment of Paracetamol Using TiO_2 Nanotubes: Effect of pH. *Processes* **2019**, *7*, 319. [CrossRef]
10. Wang, L.; Liu, X. Degradation of Aqueous Polycyclic Musk Tonalide by Ultraviolet-Activated Free Chlorine. *Processes* **2019**, *7*, 95. [CrossRef]
11. Sun, X.; Chen, H.; Liu, Z.; Zhou, M.; Cai, Y.; Pan, H.; Xia, L. Investigations on Ozone-Based and UV/US-Assisted Synergistic Digestion Methods for the Determination of Total Dissolved Nitrogen in Waters. *Processes* **2020**, *8*, 490. [CrossRef]

processes

MDPI

Review

Recent Developments in the Photocatalytic Treatment of Cyanide Wastewater: An Approach to Remediation and Recovery of Metals

Luis Andrés Betancourt-Buitrago [1], Aracely Hernandez-Ramirez [2], Jose Angel Colina-Marquez [3,*], Ciro Fernando Bustillo-Lecompte [4], Lars Rehmann [5] and Fiderman Machuca-Martinez [1]

1 Escuela de Ingeniería Química, Universidad del Valle, Calle 13 #100-00. Cali A.A. 25360, Colombia; betancourt.luis@correounivalle.edu.co (L.A.B.-B.); fiderman.machuca@correounivalle.edu.co (F.M.-M.)
2 Facultad de Ciencias Químicas, Universidad Autónoma de Nuevo León, San Nicolás de los Garza 64570, Mexico; aracely.hernandezrm@uanl.edu.mx
3 Department of Chemical Engineering, Universidad de Cartagena, Sede Piedra de Bolívar, Avenida del Consulado 48-152, Cartagena A.A. 130001, Colombia
4 Graduate Programs in Environmental Applied Science and Management, and School of Occupational and Public Health, Ryerson University, 350 Victoria Street, Toronto, ON M5B 2K3, Canada; ciro.lecompte@ryerson.ca
5 Department of Chemical and Biochemical Engineering, Thompson Engineering Building, Western University, London, ON N6A 5B9, Canada; lrehmann@uwo.ca
* Correspondence: jcolinam@unicartagena.edu.co

Received: 6 March 2019; Accepted: 15 April 2019; Published: 20 April 2019

Abstract: For gold extraction, the most used extraction technique is the Merrill-Crow process, which uses lixiviants as sodium or potassium cyanide for gold leaching at alkaline conditions. The cyanide ion has an affinity not only for gold and silver, but for other metals in the ores, such as Al, Fe, Cu, Ni, Zn, and other toxic metals like Hg, As, Cr, Co, Pb, Sn, and Mn. After the extraction stage, the resulting wastewater is concentrated at alkaline conditions with concentrations up to 1000 ppm of metals. Photocatalysis is an advanced oxidation process (AOP) able to generate a photoreaction in the solid surface of a semiconductor activated by light. Although it is well known that photocatalytic processes can remove metals in solution, there are no compilations about the researches on photocatalytic removal of metals in wastewater with cyanide. Hence, this review comprises the existing applications of photocatalytic processes to remove metal and in some cases recover cyanide from recalcitrant wastewater from gold extraction. The use of this process, in general, requires the addition of several scavengers in order to force the mechanism to a pathway where the electrons can be transferred to the metal-cyanide matrices, or elsewhere the entire metallic cyanocomplex can be degraded by an oxidative pathway.

Keywords: UV-LED; photoreactors; mining wastewater; cyanide; metal removal

1. Introduction

Gold has always had a high value since prehistoric times as ornaments in rituals, and it occupies an essential role in the world economy. By mid-2017, the world gold reserves were around 33,450 metric tons, with a demand of 4337 tons in 2016, destined for jewelry (47%), technology (7%), investments (36%), and central banks (9%) [1].

The gold exploitation depends on the way it is present in minerals, and its extraction can be done in the acid phase (pH < 3) with thiourea, thiocyanate, chlorine, aqua regia, ferric chloride; in neutral phase with thiosulfate, halogens, sulfuric acids, bacteria; and in alkaline phase (pH > 10) with cyanide, ammonium cyanide, ammonium, sulfur, and nitriles [2,3]. However, the practical application of these processes is limited to extraction in the alkaline phase using cyanide because of its high selectivity with respect to gold [4–6].

Latin America and the developing countries exhibit one of the primary gold and silver exploitation scenarios based on the leaching of ores with solvents, such as sodium cyanide (NaCN)—The Merrill-Crowe Process. In this extraction process, the gold-concentrated effluent is later taken to a precipitation stage with the use of zinc, called cementation [7]. The wastewater resulting from this process is rich in heavy and non-heavy metals, poor of gold, and it contains dissolved silver, which is very harmful to the environment [8]. The mining wastewater is well known to be the predominant cause of pollution problems in surface water bodies (lakes and rivers). The problems, such as death, due to poisoning, lead poisoning; cancer, due to chromium, blindness and congenital malformations, are attributed to the contamination of surface and underground water sources [9]. Furthermore, heavy metals in these waters could be bioaccumulated and present biomagnification causing serious health effects, due to their high levels of toxicity [10–12]. Besides, mining wastewaters can show problems of metal mobility and local cyanide release where they are stored; mining wastewaters are directly discharged to tailings ponds for periods of three to six months where degradation is expected by the sun (photolysis and evaporation) [13].

In large-scale operations, wastewater treatment is carried out with highly oxidizing processes, such as chlorination, sulfur dioxide, hypochlorite oxidation, electrolytic oxidation, ozonation, use of hydrogen peroxide, high thermal transformation, biological treatments, adsorption on activated carbon, among others, usually at high oxidation conditions and operation cost [14].On the other hand, advanced oxidation processes (AOPs) have the advantage of removing liquid and recalcitrant gaseous matrices by non-selective chemical species. Among these processes, heterogeneous photocatalysis (HPC) has received considerable attention as a promising technology, able to use renewable energy from the sun. It is conventionally defined as the acceleration of the rate of a chemical reaction, induced by the absorption of light by a catalyst or coexisting molecule. This definition of photocatalysis may be the most widely accepted as it encompasses all aspects of the field, including photosensitization [15].

HPC is one of the AOPs that allows the elimination of toxic compounds in a non-selective pathway, due to the generation of oxidizing species, such as the hydroxyl radical ($^{\bullet}OH$), perhydroxyl radical ($HOO^{\bullet}$), superoxide ($O_2^{\bullet-}$), and photogenerated holes (h+), transforming recalcitrant and toxic molecules into biodegradable or less harmful compounds [16]. Photocatalytic processes not only are applied to oxidize recalcitrant organic matter, but also to promote reduction reactions. Some examples are the photoreduction of benzaldehyde to benzyl alcohol, metallic ions, such as Fe^{3+}, Cr^{6+}, Hg^{2+}, Cu^{2+}, inorganic nitrogen and carbon dioxide to formic acid, simulating part of artificial photosynthesis.

The photocatalytic reduction represents an option when traditional oxidative pathways are not feasible and when the nature of semiconductor is able to transfer electrons at a high energy level on its conduction band. Although the current applications of large-scale photocatalytic processes are scarce, different assessments have been made for the contamination associated with gold mining wastewaters. In this review, different photocatalytic processes used for the elimination of synthetic and real cyanide matrices of gold extraction are explained and described.

2. Production and Characterization of Cyanide Wastewater

Cyanidation is used when gold is in the pyrite form and is not extractable by physical separation methods. This process is carried out through the use of sodium cyanide in the alkaline phase and with an excess of oxygen, as shown in Equation (1) [17]. Once Au is extracted from the ores, the gold is precipitated by adding Zn (cementation), replacing the gold of the aurocyanide ion with zinc cyanide and precipitating it in metallic form, as indicated in Equation (2). Although the efficiency of the process is in the order of 99%, the wastewater has metal cyano-complexes strong acid dissociable (SAD), such as iron, copper, and cobalt, as well as weak acid dissociable (WAD), such as nickel, silver, zinc, and arsenic [14].

Leaching or cyanidation:

$$4Au + 8NaCN + O_2 + 2H_2O \rightarrow 4Na[Au(CN)_2] + 4NaOH \quad (1)$$

Cementation:

$$2NaAu(CN)_2 + 4NaCN + 2Zn + 2H_2O \rightarrow 2Na_2Zn(CN)_4 + 2Au\downarrow + \uparrow H_2 + 2NaOH \quad (2)$$

Table 1 shows the metallic and semi-metallic cyano-composites, sorted by the logarithms of their stability constants. Thus, the most unstable compound corresponds to the hydrogen cyanide in the gas phase; the easily dissociable WAD corresponds to complexes of Cd, Zn, Ag, Ni, Cu, Cr, and the most stable SAD correspond to complexes of Fe, Au, Co. The stability of strong complexes makes necessary the use of tailings ponds for removing them by solar-evaporation [18].

Table 1. Stability of cyano-metallic complexes [19].

Group	Species	Toxicity [20]	Stability Constant (Log K_n)
Free cyanide	CN^-	High	n.a.
	$HCN_{(g)}$		9.2
Simpler compounds: Easily soluble	NaCN, KCN, $Ca(CN)_2$, $Hg(CN)_2$, $Zn(CN)_2$, CuCN, $Ni(CN)_2$, AgCN	High	n.d.
Weak complex (*WAD—Weak Acid Dissociable*)	$Cd(CN)_4^{2-}$	Intermediate	17.9
	$Cd(CN)_3^-$		n.d.
	$Zn(CN)_4^{2-}$		19.6
	$Ag(CN)_2^-$		20.5
	$Ni(CN)_4^{2-}$		30.2
	$Cu(CN)_3^{2-}$		21.6
	$Cr(CN)_6^{3-}$		n.a.
	$Cr(CN)_6^{3-}$		n.a.
Strong complexes (SAD—*Strong Acid Dissociable*)	$Fe(CN)_6^{4-}$		35.4
	$Fe(CN)_6^{3-}$	Low	43.6
	$Au(CN)_2^-$		38.3
	$Co(CN)_6^{3-}$	High	64.0
Unstable inorganic	SCN^-, CNO^-	High	n.d.
Aliphatic organic	Acetonitrile, acrylonitrile, adiponitrile, propionitrile	Intermediate	n.d.

On the other hand, weak complexes are easily hydrolyzable by changing the pH of the solution. In principle, weak complexes tend to be destroyed over three months with or without photolysis; however, strong complexes, such as $Fe(CN)_6{}^{3-}$, $Co(CN)_6{}^{3-}$ remain over time, turning these waters into recalcitrant. Additionally, degradation products, such as $NH_3/NH_4{}^+$, $NO_2{}^-$, $NO_3{}^-$, CNO^-, sulfates, and carbonates are formed by the slow rupture of cyano-metallic complexes. Thus, the resulting wastewater (concentrated by these complexes) is not suitable for being poured into surface bodies of water [13,21].

3. Existing Treatment Options

The existing treatment options of oxidative processes for the treatment of cyanide wastewater, such as natural attenuation [18,22], chemical oxidation [23], thermal treatments, precipitation, biologic oxidation [24–27], and ionic adsorption [28], are well documented [18]. Nonetheless, strictly photocatalytic treatments are scattered, and there is no clarity of existing photocatalytic processes applications for degradation of cyanide complexes.

Figure 1 shows a relational diagram constructed with VOS Viewer® using technology watch tools [29] on the main topics related to cyanide. In this figure, it is observed that the relationship with the word "photocatalysis" is not very broad. However, it appears related to "activated carbon" and "titanium dioxide" (in yellow). Likewise, other technologies appear, such as: "Biodegradation", "ozone", "hydrogen peroxide", "electroplating", and "adsorption". For this review, some applications of photocatalytic processes used in the degradation of this type of cyano-metallic wastewater are depicted.

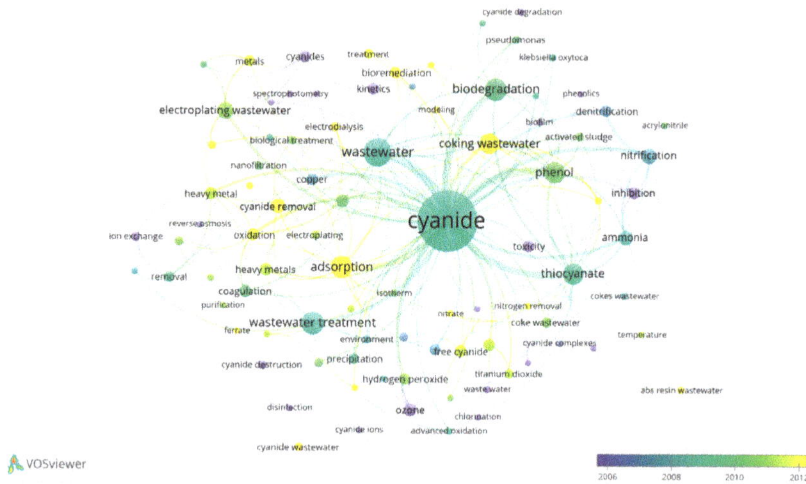

Figure 1. Keyword correlation map. Database: Web of Science. 464 documents. Date of consultation: 22 February 2018. Built with VOSViewer v1.6.7.

4. Photocatalytic Treatment Alternatives

4.1. *Classic Oxidative Photocatalysis*

AOPs are generally based on mechanisms capable of producing profound changes in the chemical structure of pollutants. Heterogeneous Photocatalysis (HPC) is a photochemical AOP where redox reactions are promoted by the interaction of a semiconductor catalyst and photons, generating active species able to degrade recalcitrant organic matter to allowable concentrations for final discharge [30].

Figure 2 shows the photocatalytic mechanism on a TiO_2 particle. The photocatalytic process initiates when a semiconductor is irradiated by photons (from solar or artificial light) whose energy is equal or greater than the band-gap energy (Eg) of the semiconductor, promoting the electrons from the balance band to the conduction band. Then, the electrons at the conduction band can react with the adsorbed oxygen on the semiconductor surface to form superoxide radicals ($O_2^{\bullet -}$), and other radicals with water, such as the perhydroxyl ($HOO^{\bullet}$) or generating reduction reactions by the direct or indirect transfer with adsorbed compounds. On the other hand, the holes in the balance band can react with the adsorbed water generating hydroxyl radicals ($^{\bullet}OH$), characteristics of the AOPs [31]. Conversely, it has also been evidenced that hydroxyl radicals are obtained by reduction of H_2O_2 and not by direct oxidation of water adsorbed on the catalyst surface [32]. Furthermore, the holes in the valence band can react directly with other adsorbed molecules as sacrificing agents oxidizing them to simpler substances [32]. This final step is the most argued mechanism in the photocatalytic treatment of recalcitrant molecules.

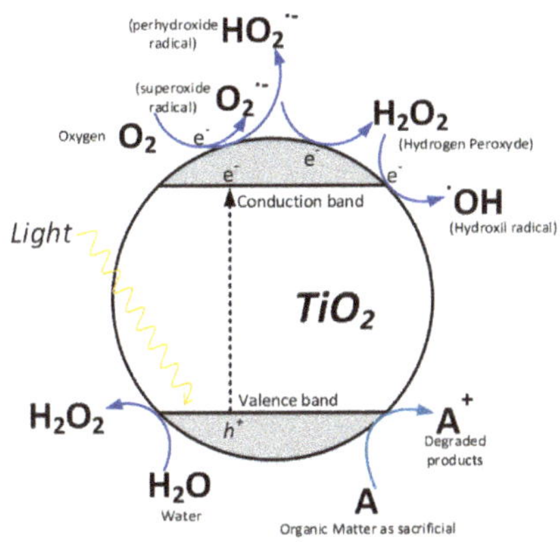

Figure 2. One electron reduction steps of oxygen to hydroxyl radical and two electron oxidation steps of water to H_2O_2. Adapted from Reference [32].

In the case of the cyanide, the use of different catalysts, such as titanium dioxide, nickel oxide, zinc oxide, platinum, zirconium, cadmium, and cobalt, have been studied for removing potassium cyanide and sodium cyanide [33–50].

In general, all the photocatalytic evaluations are carried out using the oxidative chemical pathway that works in the presence of oxygen. This process transforms the adsorbed substances in less toxic compounds (nitrates, carbonates, carbon dioxide, and nitrogen). Table 2 shows the most applied processes in matrices containing synthetic cyanide solutions of KCN and NaCN. It can be noticed that the initial concentrations oscillate around 100 ppm of CN^- (3.2 mM CN^-). Moreover, photocatalytic degradation with doped TiO_2 and ZnO has been evaluated, with the main doping agents of Ce, Ag, Zn, Pt, and Co. These photocatalysts have achieved degradations of free cyanide (over 8 and 10%) in oxic conditions and with artificial radiation using UV-C (light emitting diodes) LED light. The photocatalytic evaluations in batch and continuous reactors for degradation of organic cyanide compounds show how they have degraded and mineralized acetonitrile (CH_3CN) in the gaseous and liquid phase, obtaining better results with the gas phase [47].

Other combinations for free cyanide involve combined treatments of oxidation processes, such as photocatalysis/ozone/electrolysis/electrocoagulation that achieved degradations greater than 90% [51–59]. Although in those photocatalytic evaluations, more than 90% of the substrate was removed. Most of them were carried out with synthetic solutions of pure free cyanide, which does not address the issue related to the metal-cyanide complex remediation at laboratory scale, as shown in Table 2.

The photocatalytic degradation of free cyanide is summarized in Figure 3. As mentioned before, the oxidation is carried out by different pathways: Oxidation by holes (h^+), oxidation by superoxide ($O_2^{\bullet -}$) and oxidation by hydroxyl radicals ($^{\bullet}OH$). The reaction of CN^- with radicals transforms it into cyanate (CNO^-), ammonium (NH_4^+), nitrates (NO_2^-, NO_3^-) and carbonates (HCO_3^-, CO_3^{2-}). Although these photocatalytic treatments were used for the degradation of free cyanide, all applications were limited to synthetic solutions. Complexing agents like the metals in the ores represent the main problem of mining wastewater.

Table 2. Photocatalytic treatments applied to free cyanide matrices.

Year	Substance [C_0]	Source of Light	Wavelength	Type of Reactor	Degradation/Reaction Time	Catalyst	Main Findings
1992	KCN [100 ppm]	14 W UV Hg Low Pressure	360 nm	Compact Square batch reactor	100%/60 min	TiO_2 P25	Achieve total degradation to nitrates and cyanates. They find the CO_2 in air bubbling as harmful for the photocatalytic mechanism [33]
1999	NaCN, NaCNO [3.85 mM]	Solar light	Solar spectrum	CPC pilot scale	100%/4.1 Einstein accumulated	TiO_2 P25	Total degradation with solar light, but kinetics is only related to accumulated energy [34]
2001	NaCN [666 ppm CN^-]	450 W, 700 W Hg high-pressure lamp	UV-A	Laboratory Batch	1.5 mmol/h H_2 produced at 70 °C and 700 W	NiO/TiO_2	The process produced hydrogen and cyanate from cyanide as a photocatalytic strategy of remediation [50]
2002	Free Cyanide, phenol, atrazine, EPTC, dichloroacetic acid, and Cr(VI) among others.	Solar	Solar spectrum	Pilot-scale PSA–Solar platform of Almeria	100%/N.D.	photo Fenton and photocatalysis applications.	Several experiments applied at a solar pilot plant in Almeria with successful results [60]
2002	KCN [100 ppm]	150W Hg medium pressure lamp	>300 nm	Batch cylindrical	47%/2 h TiO_2/SBA-15	Supported TiO_2 on SBA-15 and MCM-41	Achieved geometry optimization using the support SBA. However, degradation resulted low [43]
2002	NaCN [100 ppm CN^-]	150W Hg Medium pressure	n.a.	Batch cylindrical	50%/350 min	TiO_2 Sol-gel method on four different support	Achieved a low degradation of free cyanide exploring a novel geometry configuration on the TiO_2 distribution [48]
2003	NaCN [3.85 mM]	n.a.	n.a.	n.a.	100%/420 min	TiO_2 P25	Although total degradation was achieved, authors argue the photonic efficiency is very low and radical recombination occurred. They propose a very detail degradation kinetic mechanism [61]
2004	KCN [50 ppm]	450 W High-pressure Hg lamp	>300 nm	binaural pyrex batch with intern lamp	n.a.	TPA/TiO_2, $Cs\text{-}TPA/TiO_2$	They determine the interaction of CN− with holes and electrons photogenerated. The Cs resulted in photocatalytic inhibition [62].
2005	CH_3CN (gas and liquid) [24 mM]	500 W Hg medium pressure lamp	365 nm	Annular photoreactor steady state for liquid and gas phase	21%/4 g gas phase 35%/5 g liquid phase	TiO_2 anatase for gas, and TiO_2 P25 for the liquid phase	Photocatalytic activity was low, and free cyanide ions remain in solution [47]
2007	NH_3, HCOOH, CN^- from Electric Power Plant wastewater [10 ppm CN^-, 1700 HCOOH, 150 ppm NH_3]	150 W Hg lamp	190–280 nm	Batch cylindrical	100% CN 90% NH_3 100% HCOOH/10 min	TiO_2 P25 + H_2O_2	Requires addition of H_2O_2 to enhance photocatalytic degradation [45]

Table 2. *Cont.*

Year	Substance [C_0]	Source of Light	Wavelength	Type of Reactor	Degradation/Reaction Time	Catalyst	Main Findings
2007	KCN [45 ppm CN^-]	400 W Hg UV Lamp	>300 nm	Recirculating cylindrical photoreactor	5%/100 min	Sol-gel TiO_2/SiO_2	Apply an optimization methodology to optimize the photonic efficiency of the photoreactor. However, a very low photodegradation was evidenced [63]
2007	KCN [40 ppm CN^-]	400 W Hg medium pressure lamp	>300 nm	Cylindrical with reflector	95%/60 min	Three photocatalysts were evaluated: TiO_2 P25, DBH TiO_2, nanometric TiO_2.	Evaluated the photocatalytic degradation with three photocatalysts and with the addition of O_3. A good degradation was achieved but the addition of O_3 instead O_2 resulted in photocatalytic inhibition [59]
2008	KCN [3.85 nM]	80 W and 36 W Low-pressure lamp	UV-A	Cylindrical photoreactor	n.a.	TiO_2 P25, TiO_2/SiO_2	The authors proposed an intrinsic kinetic model of cyanide degradation with an accurate fitting of experimental data. The study was more kinetic than a photocatalytic evaluation [64]
2008	NaCN, gasification plant wastewater [10 ppm CN^-]	Solar light	200 W/cm^2 of solar spectrum concentrated with a Fresnel Lens	Cylindrical photoreactor	100%/90 min	TiO_2 P25.	The evaluated the effect of solar light using a Fresnel lens to concentrate energy. They required the addition of H_2O_2 to achieve total mineralization of free cyanide [46]
2008	KCNO, Fe^{+4} [1 mM CNO^-] [1 mM Fe^{+4}]	n.a.	UV-A	Borosilicate glass cylindrical	80% cyanate degradation/120 min	TIO_2 P25	The process reduced ferrate(VI) and oxidated cyanate in a Fe(VI)-TiO_2-UV-NCO^- system. However, the role of the TiO_2 in the degradation was not specified. The possible reduction-oxidation mechanism for Fe^{+4} reduction was not clarified [65]
2008	KCN [100 ppm CN^-]	15 W Hg low-pressure lamp	UV-A	Cylindrical batch	100%/350 min	TiO_2 P25	The degradation was done using 10.5 mM EDTA as a hole scavenger. Addition of EDTA evidenced an increase in the cyanide oxidation to CNO^- [49]
2009	KCN [30 ppm CN^-]	400 W and 36 W Blacklight lamp	365 nm	Annular reactor	100%/120 min	TiO_2/SiO_2	They evaluate and compared the scaling-up process from laboratory to pilot plant, using supported TiO_2. Total elimination of cyanide was achieved in both systems. Propose a scaling up methodology for photoreactors [66]

Table 2. *Cont.*

Year	Substance [C_0]	Source of Light	Wavelength	Type of Reactor	Degradation/Reaction Time	Catalyst	Main Findings
2009	NaCN [400 ppm]	450 W Halide lamp	UV-A	Cylindrical glass batch	90%/30 min	TiO_2 nanoparticles coupled with an electrocoagulation recovery	It is proposed a recovery technique using electrocoagulation after a typical photocatalytic cyanide degradation. A study of TiO_2 reuse was also performed [54]
2010	KCN [250 ppm]	8 W Hg lamp	365 nm	Batch cylindrical	40%/100 min	Ce-ZnO sonochemical impregnation	Doping relations of 2% Ce-ZnO calcined at 500 °C. This photocatalyst works better in the visible region. There is an excess of light applied to the system, which could mix the photocatalysis with the photolysis effect on CN^- degradation [35]
2010	KCN [11 mM CN^-]	8 W Hg Lamp	365 nm	Batch cylindrical Reactor	86%/90 min	Ag-ZnO sonochemical impregnation synthesis	Ag-ZnO was found to be three times better than ZnO pure [36]
2011	KCN [10 mM]	150 W halide and 8 W Hg UV lamp	365 nm UV	Annular batch reactor	16%/150 min	ZnO-TiO_2	Photocatalytic activity was demonstrated but with important radiant field losses in the photoreactor [37]
2013	KCN [100 ppm]	150 W fluorescent lamp	450 nm	Batch cylindrical reactor	98%/60 min	Pt-TiO_2-hydroxyapatite. Prepared by Sonic method.	Hydroxyapatite enhanced the photocatalytic behavior of bare suspended TiO_2 [38]
2013	KCN [100 ppm]	150 W fluorescent lamp	450 nm	Batch annular reactor	100%/20 min	Pt/ZrO_2-SiO_2 prepared by a photo-assisted deposition method.	Evaluated the effect of catalyst load on the reactor [39]
2014	KCN [100 ppm]	150 W fluorescent blue lamp	450 nm	Batch cylindrical reactor	100%/360 min 96%/240 min	Co-TiO_2-SiO_2 prepared by a photo-assisted method and impregnation.	Obtained the best catalyst load obtained at 0.08 g/L and a decreased in the TiO_2 band-gap with the total elimination of CN^- [41]
2015	NaCN [30 ppm]	UV-LED	Not specified. UV-A UV-B UV-C	Submerged cylindrical LED photoreactor	100%/>600 min	TiO_2 P25	Demonstrated the possibility of using LED as a source of UV light in a photocatalytic treatment. The most efficient was UV-C, due to photolytic effect [21]
2015	KCN [100 ppm]	500 W Xe bulb lamp	>420nm	Pyrex reaction cell	100%/60 min	MWCNT/Au-TiO_2	They found carbon nanotubes beneficial for photocatalytic degradation in the presence of oxygen and visible light [67]

Table 2. *Cont.*

Year	Substance [C_0]	Source of Light	Wavelength	Type of Reactor	Degradation/Reaction Time	Catalyst	Main Findings
2015	KCN [100 ppm]	700 W Xenon lamp	n.a.	Pyrex reaction cell	100%/5 h	CeO_2/KLTO	CeO_2/KLTO enhanced the photocatalytic activity compared to a photolytic effect at 750W/m^2 [68]
2015	KCN [100 ppm]	150 W Blue fluorescent lamp	>400 nm	Horizontal cylinder annular batch reactor	100%/30 min	S-TiO_2	Photocatalytic activity resulted enhanced with the addition of S, being 0.3 wt %S-TiO_2 the most efficient with visible light [69]
2016	KCN [150 ppm]	150 W Blue fluorescent lamp	>400 nm	Pyrex cell reactor	100%/60 min	Ag-Sm_2O_3	The Ag was beneficial for the photocatalytic activity by 90% more than bare Sm_2O_3. The catalyst is useful up to 5 times cycles [70]
2016	CN^- [100 ppm]	150 W Blue fluorescent lamp	n.a.	Pyrex glass cylindrical	100%/70 min	Pt/Ti-Al-MCM-41	The Pt addition to Ti-Al-MCM41 resulted in 10 times more efficient the suspended TiO_2 photocatalytic activity [71]
2018	KCN [30 ppm]	25 W metal halide lamp, and UV Lamp	365 nm and 400–700 nm	Pyrex glass cylindrical	100%/350 min	ZnO-CuPc	0.5wt%Zn-CuPc enhanced cyanide degradation. However, the TiO_2 P25 still showed faster kinetic of degradation [72]
2019	NaCN [0.18 mM CN^-]	300 W Xe lamp	>400 nm	Quartz batch reactor	90%/150 min	g-$C_3N_4^-$Nanosheets	Nanotubes exhibited good photocatalytic activity, but it was the dissolved oxygen played the most important role in the oxidation of cyanide in visible light [73]
2019	KCN [10 ppm]	Xe lamp	400–800 nm	Pyrex glass cylindrical	89%/120 min	B-ZnO	B-ZnO enhanced photocatalytic activity compared to bare ZnO with visible light at low cyanide concentrations [74]

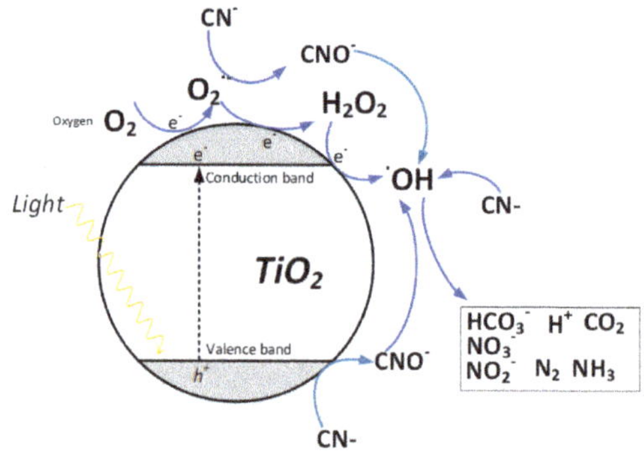

Figure 3. Photocatalytic scheme of free cyanide in a TiO_2 particle. Adapted from Reference [21].

It is well-known that the application of photocatalytic processes for chemical remediation and disinfection via hydroxyl radicals and holes oxidation has shown promising results. However, it is possible to develop other types of photodegradation without direct oxidation. Since photocatalysis is a redox process, the transformation of metallic ions and inorganic substances using the conduction band (holes) instead the valence band (electrons) can be developed. This pathway depends on the conduction and valence band energy level in the catalyst, the redox potential of the inorganic substance and the pH of the solution. In most cases, the electron transfer to a substrate is favored in the absence of oxygen in a process called reductive photocatalysis. Reductive photocatalysis has been applied to substances with oxidation states similar to CO_2, such as CCl_4, which could hardly enhance the oxidation of carbon via interaction with holes. Another example is the removal of transition metals in the solution given their multivalences when the oxidation potential is very similar to the valence band value [32].

4.2. Photoreduction of Metals

Photoreduction of metals was one of the first motivations for developing photocatalytic processes, and it was intended to be applied for precious metals exploitation. The main differences between photocatalysis with inorganic substances and photocatalysis with metals are (1) There are greater types of states excited by the participation of metal orbitals and ligands, (2) Conversion from one state to another is not always efficient, (3) Some excitations are achieved in the visible spectrum, (4) Heavy metals form strong spin pairing in the orbitals, generating stable and long-tripled states, and (5) The modular structure of the complexes does not allow radicals attacks, due to the organic substances. However, the excitation process by electron transfer can occur by *direct transfer* of the electron from the conduction band to substrates adsorbed on the surface of the catalyst; or *indirect reduction* by the formation of a radical product of the oxidation of organic molecules of low repulsion with the complex. Similarly, the presence of metals, such as Tl, Pb or Mn in solution has demonstrated their reductive ability of metals in solution [75].

Photoreduction processes require the absence of molecular oxygen for avoiding the formation of superoxide radical by transfer of the electron promoted to the conduction band. Furthermore, the efficiency of the reducing mechanism can be improved by the action of a sacrificing agent that is more selective for hydroxyl ($^{\bullet}OH$), hole (h^+), and perhydroxyl ($HOO^{\bullet}$) oxidant radicals, so that recombination is avoided and the probability of reducing other species adsorbed on the catalyst increases [76]. This reductive mechanism can decrease the oxidation state of inorganic ions and metals in solution, leading to smaller forms or their zero-valence state. This promotes the precipitation on the semiconductor surface; nevertheless, it is required that the standard potential of the level of the conduction band of the electron is sufficiently negative for generating the reduction half-reaction of the metal [77].

Figure 4 shows the different conduction and valence bands of some metal sulfides and oxides with semiconductor properties. The standard potentials (NHE) of the conduction bands (upper) and valence bands (lower) of each semiconductor are depicted. For the semiconductors with a conduction band more negative than the H^+/H_2 redox couple potential, the predominant mechanism is the reduction of adsorbed species; those are known as reductive semiconductors. In the opposite, for semiconductors with a balance band more positive than the H_2O/O_2 redox couple, are considered oxidative catalysts, generating oxidation reaction to adsorbates as its predominant mechanism.

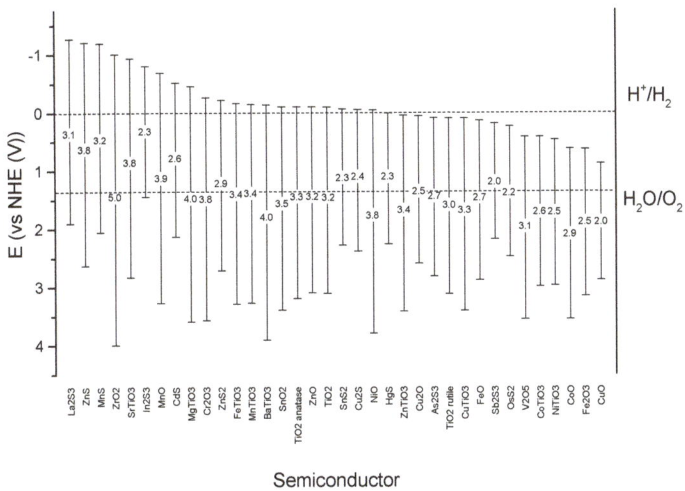

Figure 4. Relative position of the edges of the conduction and valence bands of some semiconductors. Adapted from References [78,79].

As it can be seen in Figure 4, the sulfides of La, Zn and Mn have conduction bands more negative than the H^+/H_2 redox potential, whereas metal oxides, such as Zr, Ni and $BiTiO_3$ have valence bands more positive for the H_2O/O_2 potential. Depending on the reduction potential of the metal or inorganic ion in solution, the most suitable semiconductor will be selected for a photocatalytic desired reaction. The most applied semiconductor catalyst in photocatalytic processes is the TiO_2; however, it appears in three different crystalline forms: Anatase, rutile, and brookite. From them, anatase and rutile are the most used crystalline phases in photocatalytic processes and the most commonly used in photocatalytic applications.

Few studies reported the photocatalytic reduction of metals in cyanide complexed matrices. Table 3 shows some studies related to metallic cyano-complexes using photocatalytic processes. The first evidence of solar-assisted TiO_2 photocatalysis studies with real mining wastewater by using As, Fe, Hg, Cu, and Zn complexed for precipitating metals, reported that three days were needed for free cyanide elimination and 17 days to achieve 99% of Hg and As removal [44]. Other studies were focused on the degradation of complex in synthetic samples and studied the degradation of Cu, Ni, Fe, Co, Pb, Cr, Au, and As cyano-complexes.

Table 3. Photocatalytic treatments of metallic cyano-complexes.

Year	Matrix	Light Source	Wavelength	Type of Photoreactor	Removal/Reaction Time	Catalyst	Main Finding
1995	Real mining wastewater $Cu(CN)_3^{2-}$ [22 mM] $Zn(CN)_4^{2-}$ [300 mM] $Fe(CN)_6^{4-}$ [5.2 mM] Fe [29 mM] Hg [11 mM] As [16 mM]	Solar light	Solar spectrum	Dish PVC	99% metal removals/17 days	TiO_2 P25	All metal was removed with the formation of metal-hydroxides and nitrate [44]
2002	$Fe(CN)_6^{3-}$ [0.64 mM]	150W Hg high pressure lamp	>300 nm	Pyrex batch photoreactor	50%/350 min for SBA-15/TiO_2	TiO_2 MCA-41, SBA-15	The photocatalytic activity was evaluated using two different support for TiO_2. The porous SBA-15 resulted in better degradation of $Fe(CN)_6^{3-}$ but also for the free cyanide mineralization [48].
2003	$Fe(CN)_6^{3-}$ [1 mM]	4W Hg low mercury lamp and solar light	>300 nm	Cylindrical batch	100%/1.5 h solar radiation, 77%/6 h UV Lamp	TiO_2 sol-gel	TiO_2 resulted in a better way to destroy $Fe(CN)_6^{3-}$, however resulting wastewater was rich in cyanate and incomplete oxidation was observed. Solar light exhibited better degradation rates [80].
2004	CuCN [90 ppm CN^-]	400 and 700 W halide lamp medium pressure Hg	UV	Batch cylindrical reactor	100%/180 min	TiO_2 in Raschig rings support	Evaluated four different methods and the hydrothermal was the best [81].
2004	NaCN, $Cu(CN)_3^{2-}$ [1 mM NaCN], [10 mM $Cu(CN)_3^{2-}$]	100 W high pressure Hg lamp	228–420 nm	Batch annular reactor bench scale.	100%/150 min	TiO_2 P25	The ratio Cu:CN influences photocatalytic degradation. A 10:1 ratio was the best for the process [82]
2005	$AuCN_2^-$ [75 mg/L $AuCN_2^-$]	150 W medium pressure Hg lamp	365 nm	Beaker	86%/240 min	TiO_2/L	The recovery of free cyanide is made adding methanol as $^{\bullet}OH$ acceptor. Thus, oxidation of CN^- to CNO^- is avoided. The cyano-complex $AuCN_2^-$ is the electron acceptor and Au^0 is deposited on the TiO_2 particles [83]

Table 3. *Cont.*

Year	Matrix	Light Source	Wavelength	Type of Photoreactor	Removal/Reaction Time	Catalyst	Main Finding
2005	$[Fe(CN)_6]^{3-}$ and $[Fe(CN)_6]^{4-}$ [100 ppm CN^- equivalent]	150 W Hg medium pressure	>320 nm	Beaker	70%/240 min	TiO_2 P25, TiO_2/SiO_2 prepared by sol-gel and hydrothermal method.	The maximum degradation was about 70% of the cyano-complex. It requires additional treatment. Iron complexes contaminated the semiconductor [84]
2005	KCN, $K_3(Fe(CN)_6)$, $KAu(CN)_2$ [3.85 mM KCN; 0.64 mM $K_3(Fe(CN)_6)$; 0.38 mM $KAu(CN)_2$]	150 W Hg medium pressure	365 nm	Beaker	n.d.	TiO_2/GrSiO_2, TiO_2/SBA-15.	Different methods of support were evaluated, 60% of TiO_2/SBA-15 performed better for iron-complex degradation [85]
2008	CNO^- [0.5 mM] Fe(IV) [1 mM]	Spectro line UV-A lamp	365 nm	Beaker	80%/120 min	TiO_2 P25 Degussa	There is an enhancement in the cyanate degradation related to the presence of ferrate [65]
2009	Real Wastewater from Energy Plant	UVA UVC	200–280; 320–400 nm	Pilot photoreactor	100%/15 min	$FeSO_4$, H_2O_2	Although the study demonstrates the ability of a pilot plant for cyanide degradation, it only is evaluated the degradation of free cyanide and not of its complexes [57]
2013	KCN, $Co(CN)_6{}^{3-}$, $Ni(CN)_4{}^{2-}$ [100 μM]	15 W Hg low-pressure lamp.	n.d.	Cylindrical borosilicate reactor	Ni:90%/180 min, Co: 30%/180 min	TiO_2 P25 suspension	Nickel removal was shown to be achievable by photocatalysis; however, cobalt removal is more challenging [86]
2013	KCN, Co, Pb, Cr [100 ppm CN^-, Co, Pb, Cr]	Blacklight lamp and blue light	365 nm	Annular photoreactor	100%/180 min	TiO_2/SiO_2 sol-gel.	Synthesized photocatalyst could degrade free cyanide and dissolved Co, Pb, Cr. However, the evaluation of metal photo-removal was not done in the presence of cyanide [87]
2018	$Fe(CN)_6{}^{3-}$ [100 ppm]	30 W UV-LED	300–400 nm	Mini CPC UVLED photoreactor	70%/20 min	TiO_2 P25	Using UV-LED at 30W/m^2 in a mini CPC resulted better for recovery of cyanide instead remediation [88]
2018	$Fe(CN)_6{}^{3-}$ [100 ppm]	5W UV-LED	300–400 nm	UV baffled flat plate reactor	60%/90 min	TiO_2 P25	Configuration resulted useful for light harvesting, but it is required more UV Power since the complex was not complete degraded [89]

Generally, the main characteristic of these photocatalytic treatments is the use of UV lamps with a high irradiation capacity (about 150, 400 or 700 W) and in some cases metals are recovered by reducing them to their zero valence state [37–41,46–48,50,54,59,62,63,66,81–86,90]. Nevertheless, photocatalytic processes for stable metallic cyano-complexes destruction are not yet fully effective for the treatment of mining wastewater, due to the presence of metal re-oxidation-redissolution and photocatalyst poisoning by deposition.

It is known the role of chemisorbed oxygen in photooxidation reactions. The TiO_2 chemistry depends on the O_2 coverage, temperature and the characteristics of the semiconductor crystalline phase. Those studies are performed using molecules, such as Ar, Kr, N_2, CO, CH_4 in order to understand their interaction with the adsorbed oxygen using a photon stimulated desorption. This technique has been used to understand and monitor photochemical processes occurring on the surface of photocatalyst [91]. Although these methods require specialized equipment, practical applications require more investment. A simpler method to understand the global mechanism is related to scavengers' addition to the bulk of the photocatalytic system.

In the case of metallic-cyanide matrices, the addition of scavengers increases the selective photoreduction of the metals (charge transfer efficiency) without oxidizing the free cyanide. Thus, several acceptors have been used as electron donors for hydroxyl, perhydroxyl, and holes. This selectivity enhancement was used to precipitate Ag from a solution of sodium cyano-argentate and sodium aurocyanide [83].

Table 4 shows examples of the main acceptors with which the mechanism studies on photocatalytic degradations in several matrices have been carried out. Compounds, such as NaF, have been used to inhibit the adsorption effect on the semiconductor particle and demonstrate the importance of the degradation reaction in bulk and not on its surface [92]. Moreover, the application of radical scavengers has been used to determine the main pathway that affects the photocatalytic degradation and to study the selectivity for certain radicals.

Table 4. Several radical scavenger agents used in a typical photocatalytic degradation.

Scavengers	Compound
Holes (h^+)	Glucose [93]; formic acid, sulfuric acid [9,94]; sodium oxalate [95]; ammonium oxalate [96,97]; 4-methylimidozal [98]; EDTA [97,99,100]; KI [92,100,101]; NH_4^+ [102]; oxalic acid and methylene blue [103]
Hydroxyl radical ($^{\bullet}OH$)	t-butanol [92,96,99]; isopropyl alcohol [97]; methanol [100,104]; ethanol [101]; acetonitrile [101]; KBr [105]; terephthalic acid [106]
Electrons on the conduction band (e^-)	Fe^{3+}, Cu^{2+}, Ag^+ [106,107]; $AgNO_3$ [96]; Cr^{6+} [95]; KIO_3 [102], $(S_2O_8)^{2-}$ [92]
Superoxide radical ($O_2^{\bullet -}$)	Benzoquinone [96,97]

4.3. Application of Traditional Photoreactors and LEDs in Mining Wastewater

Sunlight and UV lamps (as natural and artificial photon source, respectively) have been used directly on photocatalytic processes with several reactor geometries, such as compound parabolic concentrators (CPC), flat plates, spinning-discs, submerged lamps and microchannel among other systems, in order to harvest light energy [108]. Nevertheless, available sunlight radiation is very variable and depends on the weather condition, geography, time of the day and year. Reversely, UV lamps as a source of UV photons for photocatalysis make this process more controllable, and it is a better alternative for fine photochemistry. The development of semiconductors for being irradiated by UV LED is a trending research topic and a promising source of photons. LED lamps have been replacing traditional incandescent halide and fluorescent mercury lamps for a wide variety of applications. The advantages of LEDs are not only the small geometry, versatility, and robustness but also a more prolonged time life, a high electrical efficiency (more than 60%) and a capability for generating radiation at a particular wavelength [109–111]. This type of artificial light has been used for water purification, sterilization, protective coatings and photo sensors from the near visible, UV to the IR spectrum range [112,113]. Table 5 shows the application of this type of diodes for photocatalytic processes as an alternative to the treatment of other organic matrices. There is only a study that reported a successful application for removal of free cyanide [21]; therefore, it can be an emerging solution for the issue of availability of UV radiation from the sun by coupling with another renewable source of energy instead using direct sunlight or traditional incandescent lamps.

Table 5. LED emerging photoreactors.

Year	Description	Main Findings
2013	Phenol photodegradation using batch UV LED at 375 nm.	It is reported that UV LEDs at 800 mW are 100 times more efficient in comparison with 12 and 16 W fluorescent UV Lamps [114].
2013	Drinking water potabilization using UV LED at 365 nm	Natural organic matter and emerging pollutants were removed from drinking water. It is concluded that the photoreactor design with this type of light is more critical than the catalyst load [111].
2014	Used in the dyes photodegradation, organic matter of air and water.	Proved the capability of organic matter using this type of light, which is better in term of the photoreactor size, energy consumption [110].
2014	Oxytetracycline and 17-α-etinil estradiol as agriculture antibiotic degradation using UV LED light.	Reached a 100% degradation of total organic carbon with cumulative energy of about 12.5 kJ/L [115].
2014	Acetonitrile degradation in Green blue and red LED photoreactor with C-N TiO_2.	Degradation of about 100% was achieved in 2 h using low power (3 W) LEDs [116].
2014	Chromium photoreduction using CdS and TiO_2 with white LED photoreactor.	The removal of chromium was about 93% in 240 min of reaction [117].
2014	Methyl orange degradation modeling applying Controlled Periodic Illumination in a UV LED photoreactor.	It was found that Langmuir-Hinshelwood kinetics do not describe well the photoreactor operating at Controlled Periodic Illumination. Novel mathematic modeling is required for pulsed photoreactors [118].
2014	Selective photocatalytic reduction of nitrobenzene carried out by UV LED light.	The transformation of nitrobenzene to aniline was achieved using ethanol as the electron donor with 100% of conversion [119].
2014	Evaluation of nitro-aromatic compounds using CdS as the catalyst.	The photoreactor uses a visible LED to enhance photoreduction of amines using methanol and isopropanol as electron donors (hole scavenger). The conversion was about 90% with a selectivity of about 71% [120].
2014	A CFD simulation experimental and validation of a UV LED photoreactor for *Escherichia coli* disinfection.	The CFD established the best amount of irradiation, flowrate and photoreactor dimension in which best photo absorption is achieved for *E. coli* disinfection [121].
2014	Methyl orange degradation under Controlled Periodic Illumination with a UV LED.	The Controlled Periodic Illumination demonstrated being more critical in the photonic efficiency when the ON-OFF period is closer to the characteristic time of the reaction. Also proposes photo-reductive degradation instead of a photooxidation mechanism [122].
2014	Methyl ketone degradation using UV-vis LED with supported TiO_2 in alveolar foam.	The removal was 100% of methyl ketone in 600 min of reaction using 56 LEDs [123].

Table 5. *Cont.*

Year	Description	Main Findings
2014	Photoreactor using graphene oxide ZnC for methylene blue degradation.	Degradation of 100% of Methylene blue is achieved in 150 min using UV-A LEDs. Graphene oxide resulted in photocatalytic degradation enhancement than Degussa P25 [124].
2014	Evaluation of FeFNS-TiO_2 activated by LED in pesticides mineralization.	Degradation of 90% achieved in 100 min of reaction [125]
2015	*E. coli* disinfection modeling in a LED photoreactor	Bacteria deactivation achieved in 120 min using TiO_2 in an annular LED UV-A photoreactor [126].
2015	Direct Red 23 degradation in a continuous UV LED photoreactor assisted with $S_2O_8^{2-}$	Complete oxidation of Direct Red 23 is done in homogeneous photocatalysis and 72 UV-LED units [127].
2015	Uses a photoreactor with UV-A LED for phenol and plywood mill wastewater treatment.	Demonstrated a photocatalytic degradation of phenol about >90% in 13 min and total removal of tannic acid in plywood mill wastewater in 43 min [128].
2015	Free cyanide degradation by the oxidative pathway in synthetic wastewater.	Demonstrated the free cyanide degradation in more than 10 h, using LEDs at UVA, UVB, and UVC. The last one was the most effective in photooxidation [21].
2017	Methylene blue degradation in a mini-CPC photoreactor.	Evaluated two systems in a coupled mini CPC and a traditional beaker with external UV-A LED illumination. Demonstrated the capability of the mini CPC in harvesting LED Light in the degradation [129].
2018	CFD simulation to enhance LED light utilization and evaluation in iron cyano-metalic complexes.	Demonstrated the utilization of a baffled plat plate photoreactor is useful for UV-LED light harvesting [89].
2018	Iron cyanocomplexes degraded in anoxic conditions using a mini-CPC UV LED photoreactor	Achieved the photoreduction of iron and free cyanide liberation as a strategy of recovery instead remediation for this iron cyanocomplex [88].

5. Conclusions

In this review, the state-of-the-art in the application of photocatalytic processes for the decontamination of synthetic and real cyanide wastewaters was presented. Photocatalytic processes can be effective for removing free cyanide content via oxidative pathways. Complexed cyano-metallic compounds are less studied in photoreactors, and usually, it requires the modification of selectivity by applying electron donors as scavengers of unwanted radicals in order to enhance charge transfer to the cyano-complex. The metal removal from inorganic cyanide matrices using photocatalytic processes has been explored, and the direct metal reduction on the conduction band appears to be the main mechanism as an electron acceptor at the conduction band. The use of unconventional UV LED lamps represents a growing area for development of photoreactors. Likewise, little evidence has been found of the treatment of metallic cyano-complexes from mining activities by using this type of UV source, and the existing applications are not aimed at improving the use of photons in the illuminated area. Although the evidence shows UV-vis LED application for other types of organic compounds, the knowledge about its use for the elimination or treatment of inorganic substances is still scarce. As far as it is known, none of the studies has compared the performance of the processes between different types of radiation sources at different wavelengths for cyanide wastewater treatment using UV LED.

Author Contributions: L.A.B.-B. compiled the literature references for photocatalytic treatments of gold mining cyanide wastewater. F.M.-M., A.H.-R., J.A.C.-M., C.F.B.-L. and L.R. restructured the information and contributed to the design, analysis and edition of the manuscript. All authors discussed the results and commented on the manuscript.

Funding: This research was funded by Colciencias (GRANT No. 1106-669-45250).

Acknowledgments: The authors are grateful to Universidad del Valle and Colciencias for the financial support to produce this work (GRANT 1106-669-45250. Recuperación de oro y tratamiento de aguas residuales cianuradas en la industria aurífera de la región pacífico Colombiana).

Conflicts of Interest: The authors declare no conflict of interest.

References

1. World Gold Council Gold Demand Trends Q1. 2017. Available online: http://www.gold.org/research/gold-demand-trends (accessed on 1 June 2017).
2. Sparrow, G.J.; Woodcock, J.T. Cyanide and Other Lixiviant Leaching Systems for Gold with Some Practical Applications. *Miner. Process. Extr. Metall. Rev.* **1995**, *14*, 193–247. [CrossRef]
3. Li, J.; Miller, J.D. A review of gold leaching in acid thiourea solutions. *Miner. Process. Extr. Metall. Rev.* **2006**, *27*, 177–214. [CrossRef]
4. Konyratbekova, S.S.; Baikonurova, A.; Akcil, A. Non-cyanide Leaching Processes in Gold Hydrometallurgy and Iodine-Iodide Applications: A Review. *Miner. Process. Extr. Metall. Rev.* **2015**, *36*, 198–212. [CrossRef]
5. Haque, K.E. Gold Leaching from Refractory Ores—Literature Survey. *Miner. Process. Extr. Metall. Rev.* **1987**, *2*, 235–253. [CrossRef]
6. Brent Hiskey, J.; Atluri, V.P. Dissolution Chemistry of Gold and Silver in Different Lixiviants. *Miner. Process. Extr. Metall. Rev.* **1988**, *4*, 95–134. [CrossRef]
7. Mpinga, C.N.; Bradshaw, S.M.; Akdogan, G.; Snyders, C.A.; Eksteen, J.J. Evaluation of the Merrill-Crowe process for the simultaneous removal of platinum, palladium and gold from cyanide leach solutions. *Hydrometallurgy* **2014**, *142*, 36–46. [CrossRef]
8. Norgate, T.; Haque, N. Using life cycle assessment to evaluate some environmental impacts of gold production. *J. Clean. Prod.* **2012**, *29–30*, 53–63. [CrossRef]
9. Dozzi, M.V.; Saccomanni, A.; Selli, E. Cr(VI) photocatalytic reduction: Effects of simultaneous organics oxidation and of gold nanoparticles photodeposition on TiO_2. *J. Hazard. Mater.* **2012**, *211–212*, 188–195. [CrossRef] [PubMed]
10. Ramírez, A.V. Toxicidad del cianuro. Investigación bibliográfica de sus efectos en animales y en el hombre. *An. Fac. med.* **2010**, *71*, 54–61.

11. López-Muñoz, M.-J.; Aguado, J.; van Grieken, R.; Marugán, J. Simultaneous photocatalytic reduction of silver and oxidation of cyanide from dicyanoargentate solutions. *Appl. Catal. B Environ.* **2009**, *86*, 53–62. [CrossRef]
12. Little, E.E.; Calfee, R.D.; Theodorakos, P.; Brown, Z.A.; Johnson, C.A. Toxicity of cobalt-complexed cyanide to Oncorhynchus mykiss, Daphnia magna, and Ceriodaphnia dubia. Potentiation by ultraviolet radiation and attenuation by dissolved organic carbon and adaptive UV tolerance. *Environ. Sci. Pollut. Res. Int.* **2007**, *14*, 333–337. [CrossRef] [PubMed]
13. Logsdon, M.; Hagelstein, K.; Mudder, T. *El manejo del cianuro en la extracción de oro*; ICME—Consejo Internacional de Metales y Medio Ambiente: Canada, 2001; ISBN 1-895720-35-4.
14. Chi, G.; Fuerstenau, M.C.; Marsden, J.O. Study of Merrill-Crowe processing. Part I: S, olubility of zinc in alkaline cyanide solution. *Int. J. Miner. Process* **1997**, *49*, 171–183. [CrossRef]
15. Gaya, U.I. *Heterogeneous Photocatalysis Using Inorganic Semiconductor Solids*; Springer Science & Business Media: Dordrecht, The Netherlands, 2014; ISBN 9789400777750.
16. Domènech, X.; Jardim, W.F.; Litter, M.I. Procesos avanzados de oxidación para la eliminación de contaminantes. In *Eliminación de Contaminantes por Fotocatálisis Heterogénea*; Blesa, M.A., Ed.; CYTED: La Plata, Argentina, 2001; ISBN 987-43-3809-1.
17. Malloch, K.R.; Craw, D. Comparison of contrasting gold mine processing residues in a temperate rain forest, New Zealand. *Appl. Geochem.* **2017**, *84*, 61–75. [CrossRef]
18. Dobrosz-Gómez, I.; Ramos García, B.D.; GilPavas, E.; Gómez García, M.Á. Kinetic study on HCN volatilization in gold leaching tailing ponds. *Miner. Eng.* **2017**, *110*, 185–194. [CrossRef]
19. Adams, M.D. Impact of recycling cyanide and its reaction products on upstream unit operations. *Miner. Eng.* **2013**, *53*, 241–255. [CrossRef]
20. Johnson, C.A. The fate of cyanide in leach wastes at gold mines: An environmental perspective. *Appl. Geochem.* **2015**, *57*, 194–205. [CrossRef]
21. Kim, S.-O.S.H.; Lee, S.W.; Lee, G.M.; Lee, B.-T.; Yun, S.-T.; Kim, S.-O.S.H. Monitoring of TiO_2-catalytic UV-LED photo-oxidation of cyanide contained in mine wastewater and leachate. *Chemosphere* **2016**, *143*, 106–114. [CrossRef] [PubMed]
22. Vymazal, J. Constructed wetlands for treatment of industrial wastewaters: A review. *Ecol. Eng.* **2014**, *73*, 724–751. [CrossRef]
23. Dai, X.; Simons, A.; Breuer, P. A review of copper cyanide recovery technologies for the cyanidation of copper containing gold ores. *Miner. Eng.* **2012**, *25*, 1–13. [CrossRef]
24. Gupta, N.; Balomajumder, C.; Agarwal, V.K.K. Enzymatic mechanism and biochemistry for cyanide degradation: A review. *J. Hazard. Mater.* **2010**, *176*, 1–13. [CrossRef]
25. Luque-Almagro, V.M.; Moreno-Vivián, C.; Roldán, M.D. Biodegradation of cyanide wastes from mining and jewellery industries. *Curr. Opin. Biotechnol.* **2016**, *38*, 9–13. [CrossRef]
26. Mekuto, L.; Ntwampe, S.K.O.; Akcil, A. An integrated biological approach for treatment of cyanidation wastewater. *Sci. Total Environ.* **2016**, *571*, 711–720. [CrossRef] [PubMed]
27. Dash, R.R.; Gaur, A.; Balomajumder, C. Cyanide in industrial wastewaters and its removal: A review on biotreatment. *J. Hazard. Mater.* **2009**, *163*, 1–11. [CrossRef]
28. Al-Saydeh, S.A.; El-Naas, M.H.; Zaidi, S.J. Copper removal from industrial wastewater: A comprehensive review. *J. Ind. Eng. Chem.* **2017**, *56*, 35–44. [CrossRef]
29. Gómez-Luna, E.; Navas, D.F.; Aponte-Mayor, G.; Betancourt-Buitrago, L.A. Literature review methodology for scientific and information management, through its structuring and systematization. *Dyna* **2014**, *81*, 158–163. [CrossRef]
30. Malato, S.; Blanco, J. *Solar Detoxification*; Ilustrada; UNESCO-United Nations Educational, Scientific and Cultural Organization: Paris, Francia, 2003; ISBN 9789231039164.
31. Wen, J.; Li, X.; Liu, W.; Fang, Y.; Xie, J.; Xu, Y. Photocatalysis fundamentals and surface modification of TiO_2 nanomaterials. *Chinese J. Catal.* **2015**, *36*, 2049–2070. [CrossRef]
32. Pichat, P. *Photocatalysis and Water Purification*; Wiley-VCH Verlag GmbH & Co. KGaA: Weinheim, Germany, 2013; ISBN 9783527645404.
33. Pollema, C.H.; Hendrix, J.L.; Milosavljević, E.B.; Solujić, L.; Nelson, J.H. Photocatalytic oxidation of cyanide to nitrate at TiO_2 particles. *J. Photochem. Photobiol. A Chem.* **1992**, *66*, 235–244. [CrossRef]

34. Augugliaro, V.; Gálvez, J.B.; Vázquez, J.C.; López, E.G.; Loddo, V.; Muñoz, M.J.L.; Rodrĺguez, S.M.; Marcĺ, G.; Palmisano, L.; Schiavello, M.; et al. Photocatalytic oxidation of cyanide in aqueous TiO_2 suspensions irradiated by sunlight in mild and strong oxidant conditions. *Catal. Today* **1999**, *54*, 245–253. [CrossRef]
35. Karunakaran, C.; Gomathisankar, P.; Manikandan, G. Preparation and characterization of antimicrobial Ce-doped ZnO nanoparticles for photocatalytic detoxification of cyanide. *Mater. Chem. Phys.* **2010**, *123*, 585–594. [CrossRef]
36. Karunakaran, C.; Rajeswari, V.; Gomathisankar, P. Antibacterial and photocatalytic activities of sonochemically prepared ZnO and Ag–ZnO. *J. Alloys Compd.* **2010**, *508*, 587–591. [CrossRef]
37. Karunakaran, C.; Abiramasundari, G.; Gomathisankar, P.; Manikandan, G.; Anandi, V. Preparation and characterization of ZnO–TiO_2 nanocomposite for photocatalytic disinfection of bacteria and detoxification of cyanide under visible light. *Mater. Res. Bull.* **2011**, *46*, 1586–1592. [CrossRef]
38. Mohamed, R.M.; Baeissa, E.S. Preparation and characterisation of Pd–TiO_2–hydroxyapatite nanoparticles for the photocatalytic degradation of cyanide under visible light. *Appl. Catal. A Gen.* **2013**, *464–465*, 218–224. [CrossRef]
39. Kadi, M.W.; Mohamed, R.M. Enhanced Photocatalytic Activity of ZrO_2-SiO_2 Nanoparticles by Platinum Doping. *Int. J. Photoenergy* **2013**, *2013*, 1–7. [CrossRef]
40. Aazam, E.S.S. Environmental remediation of cyanide solutions by photocatalytic oxidation using Au/CdS nanoparticles. *J. Ind. Eng. Chem.* **2014**, *20*, 2870–2875. [CrossRef]
41. Baeissa, E.S. Photocatalytic removal of cyanide by cobalt metal doped on TiO_2-SiO_2 nanoparticles by photo-assisted deposition and impregnation methods. *J. Ind. Eng. Chem.* **2014**, *20*, 3761–3766. [CrossRef]
42. Salinas-Guzmán, R.R.; Guzmán-Mar, J.L.; Hinojosa-Reyes, L.; Peralta-Hernández, J.M.; Hernández-Ramírez, A. Enhancement of cyanide photocatalytic degradation using sol–gel ZnO sensitized with cobalt phthalocyanine. *J. Sol-Gel Sci. Technol.* **2010**, *54*, 1–7. [CrossRef]
43. Van Grieken, R.; Aguado, J.; López-Muñoz, M.J.; Marugán, J. Synthesis of size-controlled silica-supported TiO_2 photocatalysts. *J. Photochem. Photobiol. A Chem.* **2002**, *148*, 315–322. [CrossRef]
44. Rader, W.S.; Solujic, L.; Milosavljevic, E.B.; Hendrix, J.L.; Nelson, J.H. Photocatalytic detoxification of cyanide and metal cyano-species from precious-metal mill effluents. *Environ. Pollut.* **1995**, *90*, 331–334. [CrossRef]
45. Durán, A.; Monteagudo, J.M.; San Martín, I.; García-Peña, F.; Coca, P. Photocatalytic degradation of pollutants from Elcogas IGCC power station effluents. *J. Hazard. Mater.* **2007**, *144*, 132–139. [CrossRef]
46. Monteagudo, J.M.; Durán, A.; Guerra, J.; García-Peña, F.; Coca, P. Solar TiO_2-assisted photocatalytic degradation of IGCC power station effluents using a Fresnel lens. *Chemosphere* **2008**, *71*, 161–167. [CrossRef] [PubMed]
47. Addamo, M.; Augugliaro, V.; Coluccia, S.; Faga, M.; Garcia-Lopez, E.; Loddo, V.; Marci, G.; Martra, G.; Palmisano, L. Photocatalytic oxidation of acetonitrile in gas–solid and liquid–solid regimes. *J. Catal.* **2005**, *235*, 209–220. [CrossRef]
48. Aguado, J.; van Grieken, R.; López-Muñoz, M.; Marugán, J. Removal of cyanides in wastewater by supported TiO_2-based photocatalysts. *Catal. Today* **2002**, *75*, 95–102. [CrossRef]
49. Osathaphan, K.; Chucherdwatanasak, B.; Rachdawong, P.; Sharma, V.K. Photocatalytic oxidation of cyanide in aqueous titanium dioxide suspensions: Effect of ethylenediaminetetraacetate. *Sol. Energy* **2008**, *82*, 1031–1036. [CrossRef]
50. Lee, S.G.; Lee, S.G.; Lee, H.-I. Photocatalytic production of hydrogen from aqueous solution containing CN−as a hole scavenger. *Appl. Catal. A Gen.* **2001**, *207*, 173–181. [CrossRef]
51. Hernández-Alonso, M.D.; Coronado, J.M.; Javier Maira, A.; Soria, J.; Loddo, V.; Augugliaro, V. Ozone enhanced activity of aqueous titanium dioxide suspensions for photocatalytic oxidation of free cyanide ions. *Appl. Catal. B Environ.* **2002**, *39*, 257–267. [CrossRef]
52. Szpyrkowicz, L.; Ziliograndi, F.; Kaul, S.; Rigonistern, S. Electrochemical treatment of copper cyanide wastewaters using stainless steel electrodes. *Water Sci. Technol.* **1998**, *38*, 261–268. [CrossRef]
53. Shinde, S.S.; Bhosale, C.H.; Rajpure, K.Y. Photocatalytic activity of sea water using TiO_2 catalyst under solar light. *J. Photochem. Photobiol. B Biol.* **2011**, *103*, 111–117. [CrossRef]
54. Parga, J.R.; Vázquez, V.; Casillas, H.M.; Valenzuela, J.L.; VÃ¡zquez, V.; Casillas, H.M.; Valenzuela, J.L. Cyanide Detoxification of Mining Wastewaters with TiO_2 Nanoparticles and Its Recovery by Electrocoagulation. *Chem. Eng. Technol.* **2009**, *32*, 1901–1908. [CrossRef]

55. Pedraza-Avella, J.A.; Acevedo-Peña, P.; Pedraza-Rosas, J.E. Photocatalytic oxidation of cyanide on TiO_2: An electrochemical approach. *Catal. Today* **2008**, *133–135*, 611–618. [CrossRef]
56. Durán, A.; Monteagudo, J.M.; San Martín, I.; Aguirre, M. Decontamination of industrial cyanide-containing water in a solar CPC pilot plant. *Sol. Energy* **2010**, *84*, 1193–1200. [CrossRef]
57. Durán, A.; Monteagudo, J.M.; San Martín, I.; Sánchez-Romero, R. Photocatalytic treatment of IGCC power station effluents in a UV-pilot plant. *J. Hazard. Mater.* **2009**, *167*, 885–891. [CrossRef] [PubMed]
58. Mudliar, R.; Umare, S.S.; Ramteke, D.S.; Wate, S.R. Energy efficient–advanced oxidation process for treatment of cyanide containing automobile industry wastewater. *J. Hazard. Mater.* **2009**, *164*, 1474–1479. [CrossRef] [PubMed]
59. Hernández-Alonso, M.D.; Coronado, J.M.; Soria, J.; Conesa, J.C.; Loddo, V.; Addamo, M.; Augugliaro, V. EPR and kinetic investigation of free cyanide oxidation by photocatalysis and ozonation. *Res. Chem. Intermed.* **2007**, *33*, 205–224. [CrossRef]
60. Malato, S.; Blanco, J.; Vidal, A.; Richter, C. Photocatalysis with solar energy at a pilot-plant scale: an overview. *Appl. Catal. B Environ.* **2002**, *37*, 1–15. [CrossRef]
61. Chiang, K.; Amal, R.; Tran, T. Photocatalytic oxidation of cyanide: kinetic and mechanistic studies. *J. Mol. Catal. A Chem.* **2003**, *193*, 285–297. [CrossRef]
62. Kim, J.-H.; Lee, H.-I. Effect of surface hydroxyl groups of pure TiO_2 and modified TiO_2 on the photocatalytic oxidation of aqueous cyanide. *Korean J. Chem. Eng.* **2004**, *21*, 116–122. [CrossRef]
63. Marugán, J.; van Grieken, R.; Cassano, A.E.; Alfano, O.M. Quantum efficiency of cyanide photooxidation with TiO_2/SiO_2 catalysts: Multivariate analysis by experimental design. *Catal. Today* **2007**, *129*, 143–151. [CrossRef]
64. Marugan, J.; Vangrieken, R.; Cassano, A.; Alfano, O. Intrinsic kinetic modeling with explicit radiation absorption effects of the photocatalytic oxidation of cyanide with TiO2 and silica-supported TiO_2 suspensions. *Appl. Catal. B Environ.* **2008**, *85*, 48–60. [CrossRef]
65. Winkelmann, K.; Sharma, V.K.; Lin, Y.; Shreve, K.A.; Winkelmann, C.; Hoisington, L.J.; Yngard, R.A. Reduction of ferrate(VI) and oxidation of cyanate in a Fe(VI)–TiO_2–UV–NCO− system. *Chemosphere* **2008**, *72*, 1694–1699. [CrossRef]
66. Marugán, J.; van Grieken, R.; Cassano, A.E.; Alfano, O.M. Scaling-up of slurry reactors for the photocatalytic oxidation of cyanide with TiO_2 and silica-supported TiO_2 suspensions. *Catal. Today* **2009**, *144*, 87–93. [CrossRef]
67. Mohamed, R.M.; Mkhalid, I.A. Visible light photocatalytic degradation of cyanide using Au–TiO_2/multi-walled carbon nanotube nanocomposites. *J. Ind. Eng. Chem.* **2015**, *22*, 390–395. [CrossRef]
68. Pala, A.; Politi, R.R.; Kurşun, G.; Erol, M.; Bakal, F.; Öner, G.; Çelik, E. Photocatalytic degradation of cyanide in wastewater using new generated nano-thin film photocatalyst. *Surf. Coatings Technol.* **2015**, *271*, 207–216. [CrossRef]
69. Baeissa, E.S. Synthesis and characterization of sulfur-titanium dioxide nanocomposites for photocatalytic oxidation of cyanide using visible light irradiation. *Chinese J. Catal.* **2015**, *36*, 698–704. [CrossRef]
70. Barakat, M.A. Ag-Sm_2O_3 nanocomposite for environmental remediation of cyanide from aqueous solution. *J. Taiwan Inst. Chem. Eng.* **2016**, *65*, 134–139. [CrossRef]
71. Kadi, M.W.; Hameed, A.; Mohamed, R.M.; Ismail, I.M.I.; Alangari, Y.; Cheng, H.-M. The effect of Pt nanoparticles distribution on the removal of cyanide by TiO_2 coated Al-MCM-41 in blue light exposure. *Arab. J. Chem.* **2016**. [CrossRef]
72. Maya-Treviño, M.L.; Guzmán-Mar, J.L.; Hinojosa-Reyes, L.; Hernández-Ramírez, A. Synthesis and photocatalytic activity of ZnO-CuPc for methylene blue and potassium cyanide degradation. *Mater. Sci. Semicond. Process.* **2018**, *77*, 74–82. [CrossRef]
73. Guo, Y.; Wang, Y.; Zhao, S.; Liu, Z.; Chang, H.; Zhao, X. Photocatalytic oxidation of free cyanide over graphitic carbon nitride nanosheets under visible light. *Chem. Eng. J.* **2019**, *369*, 553–562. [CrossRef]
74. Núñez-Salas, R.E.; Hernández-Ramírez, A.; Hinojosa-Reyes, L.; Guzmán-Mar, J.L.; Villanueva-Rodríguez, M.; de Lourdes Maya-Treviño, M. Cyanide degradation in aqueous solution by heterogeneous photocatalysis using boron-doped zinc oxide. *Catal. Today* **2019**, *328*, 202–209. [CrossRef]
75. Weinstein, J.A. Inorganic Photochemistry. In *Applied Photochemistry*; Springer: Dordrecht, The Netherlands, 2013; pp. 105–148.

76. Kohtani, S.; Yoshioka, E.; Saito, K.; Kudo, A.; Miyabe, H. Photocatalytic hydrogenation of acetophenone derivatives and diaryl ketones on polycrystalline titanium dioxide. *Catal. Commun.* **2010**, *11*, 1049–1053. [CrossRef]
77. Guzmán-Mar, J.L.; Villanueva-Rodríguez, M.; Hinojosa-Reyes, L. Application of Semiconductor Photocatalytic Materials for the Removal of Inorganic Compounds from Wastewater. In *Photocatalytic Semiconductors*; Hernández-Ramírez, A., Medina-Ramírez, I., Eds.; Springer International Publishing: Cham, Switzerland, 2015; pp. 229–254. ISBN 978-3-319-10998-5.
78. Rodríguez, J.; Candal, R.J.; Solís, J.; Estrada, W.; Blesa, M.A. El fotocatalizador: síntesis, propiedades y limitaciones. In *Microbiologia del agua. Conceptos Básicos*; Argentina, 2005; Available online: http://www.psa.es/es/projects/solarsafewater/documents/curso/dia_14/9.%20Juan%20Rodriguez.pdf (accessed on 6 March 2019).
79. Gaya, U.I.; Abdullah, A.H. Heterogeneous photocatalytic degradation of organic contaminants over titanium dioxide: A review of fundamentals, progress and problems. *J. Photochem. Photobiol. C Photochem. Rev.* **2008**, *9*, 1–12. [CrossRef]
80. Parga, J.R.; Shukla, S.S.; Carrillo-Pedroza, F.R. Destruction of cyanide waste solutions using chlorine dioxide, ozone and titania sol. *Waste Manag.* **2003**, *23*, 183–191. [CrossRef]
81. Bozzi, A.; Guasaquillo, I.; Kiwi, J. Accelerated removal of cyanides from industrial effluents by supported TiO_2 photo-catalysts. *Appl. Catal. B Environ.* **2004**, *51*, 203–211. [CrossRef]
82. Barakat, M. Removal of toxic cyanide and Cu(II) Ions from water by illuminated TiO_2 catalyst. *Appl. Catal. B Environ.* **2004**, *53*, 13–20. [CrossRef]
83. Van Grieken, R.; Aguado, J.; López-Muñoz, M.-J.; Marugán, J. Photocatalytic gold recovery from spent cyanide plating bath solutions. *Gold Bull.* **2005**, *38*, 180–187. [CrossRef]
84. Van Grieken, R.; Aguado, J.; López-Muñoz, M.-J.; Marugán, J. Photocatalytic degradation of iron–cyanocomplexes by TiO2 based catalysts. *Appl. Catal. B Environ.* **2005**, *55*, 201–211. [CrossRef]
85. López-Muñoz, M.-J.; Van Grieken, R.; Aguado, J.; Marugán, J. Role of the support on the activity of silica-supported TiO_2 photocatalysts: Structure of the TiO_2/SBA-15 photocatalysts. *Catal. Today* **2005**, *101*, 307–314. [CrossRef]
86. Osathaphan, K.; Ruengruehan, K.; Yngard, R.A.; Sharma, V.K. Photocatalytic Degradation of Ni(II)-Cyano and Co(III)-Cyano Complexes. *Water Air Soil Pollut.* **2013**, *224*, 1647. [CrossRef]
87. Harraz, F.A.; Abdel-Salam, O.E.; Mostafa, A.A.; Mohamed, R.M.; Hanafy, M. Rapid synthesis of titania–silica nanoparticles photocatalyst by a modified sol–gel method for cyanide degradation and heavy metals removal. *J. Alloys Compd.* **2013**, *551*, 1–7. [CrossRef]
88. Betancourt-Buitrago, L.A.; Ossa-Echeverry, O.E.; Rodriguez-Vallejo, J.C.; Barraza, J.M.; Marriaga, N.; Machuca-Martínez, F. Anoxic photocatalytic treatment of synthetic mining wastewater using TiO_2 and scavengers for complexed cyanide recovery. *Photochem. Photobiol. Sci.* **2018**, *18*, 853–862. [CrossRef]
89. Devia-Orjuela, J.S.; Betancourt-Buitrago, L.A.; Machuca-martinez, F. CFD modeling of a UV-A LED baffled flat-plate photoreactor for environment applications: a mining wastewater case. *Environ. Sci. Pollut. Res.* **2019**, *26*, 4510–4520. [CrossRef]
90. Vilhunen, S.; Sillanpää, M. Recent developments in photochemical and chemical AOPs in water treatment: A mini-review. *Rev. Environ. Sci. Biotechnol.* **2010**, *9*, 323–330. [CrossRef]
91. Petrik, N.G.; Kimmel, G.A. Probing the photochemistry of chemisorbed oxygen on TiO_2 (110) with Kr and other co-adsorbates. *Phys. Chem. Chem. Phys.* **2014**, *16*, 2338–2346. [CrossRef] [PubMed]
92. Song, S.; Hong, F.; He, Z.; Cai, Q.; Chen, J. AgIO3-modified AgI/TiO2 composites for photocatalytic degradation of p-chlorophenol under visible light irradiation. *J. Colloid Interface Sci.* **2012**, *378*, 159–166. [CrossRef] [PubMed]
93. Hirayama, J.; Kamiya, Y. Combining the Photocatalyst Pt/TiO_2 and the Nonphotocatalyst SnPd/Al_2O_3 for Effective Photocatalytic Purification of Groundwater Polluted with Nitrate. *ACS Catal.* **2014**, *4*, 2207–2215. [CrossRef]
94. Chen, G.; Sun, M.; Wei, Q.; Ma, Z.; Du, B. Efficient photocatalytic reduction of aqueous Cr(VI) over $CaSb_2O_5(OH)_2$ nanocrystals under UV light illumination. *Appl. Catal. B Environ.* **2012**, *125*, 282–287. [CrossRef]

95. Zhang, L.S.; Wong, K.H.; Yip, H.Y.; Hu, C.; Yu, J.C.; Chan, C.Y.; Wong, P.K. Effective photocatalytic disinfection of E. coli K-12 using AgBr-Ag-Bi_2WO_6 nanojunction system irradiated by visible light: The role of diffusing hydroxyl radicals. *Environ. Sci. Technol.* **2010**, *44*, 1392–1398. [CrossRef]
96. Liu, S.; Zhang, N.; Tang, Z.R.; Xu, Y.J. Synthesis of one-dimensional CdS@TiO_2 core-shell nanocomposites photocatalyst for selective redox: The dual role of TiO_2 shell. *ACS Appl. Mater. Interfaces* **2012**, *4*, 6378–6385. [CrossRef]
97. Cao, J.; Li, X.; Lin, H.; Chen, S.; Fu, X. In situ preparation of novel *p–n* junction photocatalyst BiOI/$(BiO)_2CO_3$ with enhanced visible light photocatalytic activity. *J. Hazard. Mater.* **2012**, *239–240*, 316–324. [CrossRef]
98. Dai, H.; Zhang, S.; Hong, Z.; Li, X.; Xu, G.; Lin, Y.; Chen, G. Enhanced photoelectrochemical activity of a hierarchical-ordered TiO_2 mesocrystal and its sensing application on a carbon nanohorn support scaffold. *Anal. Chem.* **2014**, *86*, 6418–6424. [CrossRef]
99. Pastrana-Martínez, L.M.; Morales-Torres, S.; Kontos, A.G.; Moustakas, N.G.; Faria, J.L.; Doña-Rodríguez, J.M.; Falaras, P.; Silva, A.M.T. TiO_2, surface modified TiO_2 and graphene oxide-TiO_2 photocatalysts for degradation of water pollutants under near-UV/Vis and visible light. *Chem. Eng. J.* **2013**, *224*, 17–23. [CrossRef]
100. Yin, M.; Li, Z.; Kou, J.; Zou, Z. Mechanism investigation of visible light-induced degradation in a heterogeneous TiO_2/eosin Y/rhodamine B system. *Environ. Sci. Technol.* **2009**, *43*, 8361–8366. [CrossRef] [PubMed]
101. Chen, Y.; Yang, S.; Wang, K.; Lou, L. Role of primary active species and TiO_2 surface characteristic in UV-illuminated photodegradation of Acid Orange 7. *J. Photochem. Photobiol. A Chem.* **2005**, *172*, 47–54. [CrossRef]
102. Chen, M.; Chu, W. Degradation of antibiotic norfloxacin in aqueous solution by visible-light-mediated C-TiO_2 photocatalysis. *J. Hazard. Mater.* **2012**, *219–220*, 183–189. [CrossRef] [PubMed]
103. Pandikumar, A.; Ramaraj, R. Titanium dioxide-gold nanocomposite materials embedded in silicate sol-gel film catalyst for simultaneous photodegradation of hexavalent chromium and methylene blue. *J. Hazard. Mater.* **2012**, *203–204*, 244–250. [CrossRef] [PubMed]
104. Zhao, C.; Krall, A.; Zhao, H.; Zhang, Q.; Li, Y. Ultrasonic spray pyrolysis synthesis of Ag/TiO_2 nanocomposite photocatalysts for simultaneous H_2 production and CO_2 reduction. *Int. J. Hydrogen Energy* **2012**, *37*, 9967–9976. [CrossRef]
105. Sheng, F.; Zhu, X.; Wang, W.; Bai, H.; Liu, J.; Wang, P.; Zhang, R.; Han, L.; Mu, J. Synthesis of novel polyoxometalate $K_6ZrW_{11}O_{39}Sn_{12}H_2O$ and photocatalytic degradation aqueous azo dye solutions with solar irradiation. *J. Mol. Catal. A Chem.* **2014**, *393*, 232–239. [CrossRef]
106. Tian, L.; Ye, L.; Liu, J.; Zan, L. Solvothermal synthesis of CNTs–WO_3 hybrid nanostructures with high photocatalytic activity under visible light. *Catal. Commun.* **2012**, *17*, 99–103. [CrossRef]
107. Villa, K.; Murcia-López, S.; Andreu, T.; Morante, J.R. Mesoporous WO_3 photocatalyst for the partial oxidation of methane to methanol using electron scavengers. *Appl. Catal. B Environ.* **2015**, *163*, 150–155. [CrossRef]
108. Kowalska, E.; Rau, S. Photoreactors for Wastewater Treatment: A Review. *Recent Patents Eng.* **2010**, *4*, 242–266. [CrossRef]
109. Yeh, N.; Yeh, P.; Shih, N.; Byadgi, O.; Cheng, T.C. Applications of light-emitting diodes in researches conducted in aquatic environment. *Renew. Sustain. Energy Rev.* **2014**, *32*, 611–618. [CrossRef]
110. Jo, W.; Tayade, R.J. New Generation Energy-Efficient Light Source for Photocatalysis: LEDs for Environmental Applications. *Ind. Eng. Chem. Res.* **2014**, *53*, 2073–2084. [CrossRef]
111. Izadifard, M.; Achari, G.; Langford, C. Application of Photocatalysts and LED Light Sources in Drinking Water Treatment. *Catalysts* **2013**, *3*, 726–743. [CrossRef]
112. Chatterley, C. UV-LED irradiation technology for point-of-use water disinfection in developing communities. Master's Thesis, University of Colorado, Boulder, CO, USA, 2009.
113. Vilhunen, S.; Puton, J.; Virkutyte, J.; Sillanpää, M. Efficiency of hydroxyl radical formation and phenol decomposition using UV light emitting diodes and H_2O_2. *Environ. Technol.* **2011**, *32*, 865–872. [CrossRef]
114. Jamali, A.; Vanraes, R.; Hanselaer, P.; Van Gerven, T. A batch LED reactor for the photocatalytic degradation of phenol. *Chem. Eng. Process. Process Intensif.* **2013**, *71*, 43–50. [CrossRef]
115. Malkhasian, A.Y.S.; Izadifard, M.; Achari, G.; Langford, C.H. Photocatalytic degradation of agricultural antibiotics using a UV-LED light source. *J. Environ. Sci. Heal. Part B Pestic. Food Contam. Agric. Wastes* **2014**, *49*, 35–40. [CrossRef] [PubMed]

116. Abdollahi Kakroudi, M.; Kazemi, F.; Kaboudin, B. Highly efficient photodeoximation under green and blue LEDs catalyzed by mesoporous CN codoped nano TiO_2. *J. Mol. Catal. A Chem.* **2014**, *392*, 112–119. [CrossRef]
117. Liu, X.; Pan, L.; Lv, T.; Sun, Z. CdS sensitized TiO_2 film for photocatalytic reduction of Cr(VI) by microwave-assisted chemical bath deposition method. *J. Alloys Compd.* **2014**, *583*, 390–395. [CrossRef]
118. Tokode, O.; Prabhu, R.; Lawton, L. a.; Robertson, P.K.J. Mathematical modelling of quantum yield enhancements of methyl orange photooxidation in aqueous TiO_2 suspensions under controlled periodic UV LED illumination. *Appl. Catal. B Environ.* **2014**, *156–157*, 398–403. [CrossRef]
119. Zand, Z.; Kazemi, F.; Hosseini, S. Development of chemoselective photoreduction of nitro compounds under solar light and blue LED irradiation. *Tetrahedron Lett.* **2014**, *55*, 338–341. [CrossRef]
120. Eskandari, P.; Kazemi, F.; Zand, Z. Photocatalytic reduction of aromatic nitro compounds using CdS nanostructure under blue LED irradiation. *J. Photochem. Photobiol. A Chem.* **2014**, *274*, 7–12. [CrossRef]
121. Jenny, R.M.; Simmons, O.D.; Shatalov, M.; Ducoste, J.J. Modeling a continuous flow ultraviolet Light Emitting Diode reactor using computational fluid dynamics. *Chem. Eng. Sci.* **2014**, *116*, 524–535. [CrossRef]
122. Tokode, O.; Prabhu, R.; Lawton, L.A.; Robertson, P.K.J. The effect of pH on the photonic efficiency of the destruction of methyl orange under controlled periodic illumination with UV-LED sources. *Chem. Eng. J.* **2014**, *246*, 337–342. [CrossRef]
123. Doss, N.; Bernhardt, P.; Romero, T.; Masson, R.; Keller, V.; Keller, N. Photocatalytic degradation of butanone (methylethylketone) in a small-size TiO_2/β-SiC alveolar foam LED reactor. *Appl. Catal. B Environ.* **2014**, *154–155*, 301–308. [CrossRef]
124. Dai, K.; Lu, L.; Liang, C.; Dai, J.; Zhu, G.; Liu, Z.; Liu, Q.; Zhang, Y. Graphene oxide modified ZnO nanorods hybrid with high reusable photocatalytic activity under UV-LED irradiation. *Mater. Chem. Phys.* **2014**, *143*, 1410–1416.
125. Hossaini, H.; Moussavi, G.; Farrokhi, M. The investigation of the LED-activated FeFNS-TiO_2 nanocatalyst for photocatalytic degradation and mineralization of organophosphate pesticides in water. *Water Res.* **2014**, *59*, 130–144. [CrossRef]
126. Marugán, J.; van Grieken, R.; Pablos, C.; Satuf, M.L.; Cassano, A.E.; Alfano, O.M. Kinetic modelling of Escherichia coli inactivation in a photocatalytic wall reactor. *Catal. Today* **2015**, *240*, 9–15. [CrossRef]
127. Rasoulifard, M.H.; Fazli, M.; Eskandarian, M.R. Performance of the light-emitting-diodes in a continuous photoreactor for degradation of Direct Red 23 using UV-LED/$S_2O_{82}-$ process. *J. Ind. Eng. Chem.* **2015**, *24*, 121–126. [CrossRef]
128. Levchuk, I.; Rueda-Márquez, J.J.; Suihkonen, S.; Manzano, M. a.; Sillanpää, M. Application of UVA-LED based photocatalysis for plywood mill wastewater treatment. *Sep. Purif. Technol.* **2015**, *143*, 1–5. [CrossRef]
129. Betancourt-Buitrago, L.A.; Vásquez, C.; Veitia, L.; Ossa-Echeverry, O.; Rodriguez-Vallejo, J.; Barraza-Burgos, J.; Marriaga-Cabrales, N.; Machuca-Martínez, F. An approach to utilize the artificial high power LED UV-A radiation in photoreactors for the degradation of methylene blue. *Photochem. Photobiol. Sci.* **2017**, *16*, 79–85. [CrossRef]

processes

MDPI

Review

Characteristics and Treatment of Wastewater from the Mercaptan Oxidation Process: A Comprehensive Review

Ernesto Pino-Cortés [1,*], Silvio Montalvo [2], César Huiliñir [2], Francisco Cubillos [2] and Juan Gacitúa [2]

1 Escuela de Ingeniería Química, Pontificia Universidad Católica de Valparaíso, Av. Brasil 2162, Valparaíso 2362854, Chile

2 Departamento de Ingeniería Química, Universidad de Santiago de Chile, Av. Libertador Bernardo O'Higgins 3363, Estación Central 9170022, Chile; silvio.montalvo@usach.cl (S.M.); cesar.huilinir@usach.cl (C.H.); francisco.cubillos@usach.cl (F.C.); juan.gacitua@usach.cl (J.G.)

* Correspondence: ernesto.pino@pucv.cl; Tel.: +56-322372609

Received: 12 March 2020; Accepted: 31 March 2020; Published: 3 April 2020

Abstract: Sulfur compounds are removed from petroleum by the addition of sodium hydroxide at a very high concentration. As a result, a residue called spent soda or spent caustic is generated, being extremely aggressive to the environment. In this work, the chemical properties of this residue are described in detail. The sodium hydroxide remains that have not reacted, sulfur compounds, and organic matter are the primary pollutants reported. Additionally, the main characteristics of the methods of treatment used to reduce them are described. This review comes from comprehensive and updated research and bibliographic analysis about the investigation on the topic. The advantages and disadvantages of the different treatment methods are highlighted. We established some criteria to set out when assessing the application of each one of these treatments is considered.

Keywords: biological processes; electrochemical processes; oxidation processes; petroleum; phenols; sulfides

1. Introduction

Petroleum is around 85% carbon and 12% hydrogen, while in the remaining 3%, we find several elements that consist mainly of oxygen, nitrogen, and sulfur. The sulfur compounds limit the direct use of petroleum in any of its forms, such as liquefied petroleum gas (LPG), due to its odorous, corrosive, and environmentally harmful characteristics. Different techniques or methods are necessary for the reduction of these compounds in petroleum. Among these sulfur compounds are mercaptans and sulfides. Mercaptans are thiols containing the functional group formed by a sulfur atom and a hydrogen atom (-SH). This functional group is called the thiol or sulfhydryl group. Examples of these compounds present in LPG are methylmercaptan (H_3C-SH) and ethylmercaptan (H_5C_2-SH).

The technology used globally for mercaptan removal content in the oil industry is the mercaptan oxidation (MEROX) process [1,2]. It is a catalytic conversion process in which mercaptans react to produce disulfides. It is promoted by a catalyst which activates the oxidation at room temperature, using atmospheric oxygen according to the following reactions:

Mercaptan + NaOH to form sodium mercaptan:

$$RSH + NaOH \rightarrow RSNa + H_2O \quad (1)$$

Mercaptan oxidation to form disulfides + NaOH:

$$RSNa + \frac{1}{4}O_2 + \frac{1}{2}H_2O \rightarrow RSSR + NaOH \quad (2)$$

As sodium hydroxide depletes as it reacts, its ability to remove sulfides and mercaptans also decreases. When the disulfides accumulate to a few milligrams per liter, the OH^- content falls below 5%, whereby this depleted soda solution is purged. Other causes of sodas depletion are the accumulation of mercaptans, Na_2S, phenolic compounds, emulsified naphthalenes, thiosulfates, carbonates, and Fe^{+2} precipitates [2]. The H_2S and CO_2 that are also present in the medium to be oxidized react with the caustic soda according to the following competitive reactions [3]:

$$H_2S + 2NaOH \rightarrow Na_2S + 2H_2O \quad (3)$$

$$CO_2 + 2NaOH \rightarrow Na_2CO_3 + H_2O \quad (4)$$

These side effects can cause several problems, such as the irreversible consumption of caustic soda, low absorbing mercaptans by the presence of sodium salts, and the precipitation of solid due to the accumulation of salt in the caustic solution. That is why a caustic solution purge and a fresh soda replenishment are provided to prevent salt accumulation (Na_2S/Na_2CO_3) from restoring the appropriate NaOH concentration. Spent caustic (SC) is the name of the solution obtained once the mercaptans react with NaOH. Without adequate treatment, SC can cause environmental problems.

The main focus of this comprehensive review is to expose different techniques released and mentioned in the literature, to process the SC. The first approach is a brief description of its characteristics that explain the environmental problems caused by SC. Following, the main body of this paper has a series of processes that help neutralize sodas and allow their subsequent biological treatment. There is no review published before or paper that reports this relevant information that could help engineers to make better decisions in the petroleum industry. The analysis included information and results from reports, theses, and research articles published recently.

2. SC Characteristics

SC is classified into three types [4]: sulfidic (SSC), cresylic (SCC), and naphthenic (SNC). The composition of this type of waste is highly variable as can be seen in Table 1. SC solutions are characterized by high pH (pH > 12) [5] and high sodium concentrations up to 2%–15% (*w*/*w*). Conner et al. [6] also found that spent sulfidic caustics contained hydrosulfides (HS^-) and sulfides^{2-} greater than 2%–3% (*w*/*w*).

Table 1. Main chemical components present in spent sodas.

Parameters	SMC	Sulfidic	Cresylic	Naphtenic	Reference
pH	-	11–12.5	-	-	[7]
	-	12–13.5	-	-	[8]
	-	13.1–13.5	-	-	[9]
	7.5–13	-	-	-	[10]
	-	11.2 -13	-	-	[5]
	-	12–14	12–14	12–14	[11]
	-	13.1–13.5	-	-	[12]
	-	13–14	12–14	12–14	[13]
	-	13–14	12–14	12–14	[14]
	-	11.6–12.5	-	-	[15]

Table 1. *Cont.*

Parameters	SMC	Sulfidic	Cresylic	Naphtenic	Reference
COD ($g\ O_2/L$)	-	7.5–60	-	-	[7]
		60.3–68.1	-	-	[16]
		100–200	-	-	[8]
		20–60	-	-	[17]
	13.1–98.8	-	-	-	[10]
	-	62.7	-	-	[18]
	49.3 [1]	-	-	-	[18]
	74	-	-	-	[18]
	261	-	-	-	[19]
	-	5–90	50–100	150–240	[20]
	-	66.7–156.5	-	-	[15]
	114	-	-	-	[21]
TOC (g C/L)	-	6–20	-	-	[17]
	1.6–23.6	-	-	-	[10]
	92	-	-	-	[19]
	-	7.6	-	-	[18]
	-	1–1.6	-	-	[12]
	-	2–3	24–60	10–24	[20]
BOD_5 ($mg\ O_2/L$)	-	5000–10,000	-	-	[17]
	-	18,100	-	-	[18]
	20,100 [1]	-	-	-	[18]
Sulfides (mg/L)	-	5000–20,000	-	-	[7]
	-	80–90	-	-	[8]
		34,500	-	-	[17]
	6500–22,500	-	-	-	[10]
	-	5100–7700	-	-	[7]
	-	17,800	-	-	[18]
	8040	-	-	-	[18]
	-	15,200–17,600	-	-	[12]
	-	2000–52,000	0–63	<1	[20]
		-	0–1	0.1	[14]
		30,600–66,800	-	-	[15]
	24,000	-	-	-	[21]
Sulfides (% *w/w*)	-	1–4	-	-	[17]
	-	0.5–4	0–1	0–0.1	[11]
	1.4	-	-	-	[19]
	-	0.5–4	0–4	0–0.1	[4]
	-	0.5–4	-	-	[14]
	-	0–5	-	-	[7]
	-	2–300	-	-	[8]
	-	0–2000	-	-	[17]
	1.6–20	-	-	-	[10]
Phenols (mg/L)	1990 [1]	-	-	-	[18]
	6110	-	-	-	[18]
	-	1.8–33.8	-	-	[12]
	540	-	-	-	[21]
	-	0–30,000	-	-	[8]
Mercaptans (mg/L) (% *w/w*) [a]	-	0.1–4 [a]	-	-	[17]
	-	9800	-	-	[18]
	1800	-	-	-	[18]
	-	0–30,000	0–5400	<30	[20]

Table 1. *Cont.*

Parameters	SMC	Sulfidic	Cresylic	Naphtenic	Reference
Benzene (mg/L)	-	47–780	-	-	[7]
	-	7.8–63.1	-	-	[12]
	-	600	-	-	[15]
Toluene (mg/L)	-	0.2–7.8	-	-	[12]
	-	360	-	-	[15]
Cresylic acids (% *w/w*)	-	-	10–25	0–3	[11]
	-	0–4	2–25	0–3	[4]
	-	-	2–25	0–3	[14]
Napthenic acids (% *w/w*) (mg/L) [b]	-	-	-	2–15	[11]
	19,700 [1,b]	-	-	-	[18]
	-	-	-	2–15	[4]
	-	-	-	2–15	[14]
NaOH (% *w/w*)	-	4–5	-	-	[8]
	-	7.5	-	-	[17]
	2–2.9	-	-	-	[10]
	-	11.1	-	-	[3]
	-	2–10	10–15	1–4	[11]
	-	2–10	1–15	1–4	[4]
	-	2–10	1–15	1–4	[14]
Carbonates (% *w/w*)	2–2.9	-	-	-	[10]
	-	0–4	0–0.5	-	[11]
	-	-	0–0.5	-	[4]
	-	0–4	0–0.5	-	[14]

[1] napthenic + cresylic; [a] % *w/w*; [b] mg/L.

Often plants do not have the facilities to segregate the soda solutions used in these three types of classifications. Therefore, a spent mixed caustic (SMC) is usually produced. The latter is considered the fourth type of depleted or spent soda.

Processes of fuel gas, liquefied petroleum gas, and gasoline treatment generate sulfidic soda [13]. This depleted soda has a high concentration of sulfide and also a terrible smell. This type of waste can not only be treated but could also be a reactant for the cellulose pulping process. Nevertheless, the cost of transport generally discards this possibility [13].

The washing of the diesel and the fractions of jet fuels generate the exhausted naphthenic soda. It contains very few sulfides, being mainly naphthenic acids. Naphthenes generally do not impart unpleasant odors or toxicity to the caustic sodas, being partially soluble in sodas by increasing the total organic carbon considerably in SC. Naphtenes is a general term used for cyclic alkanes or non-aromatic hydrocarbons called cycloalkanes. Refinery naphthenes typically include cyclopentanes and cyclohexanes as compounds of this group. This type of waste can also be sold if it is kept insufficiently in pure form. However, processing at the same plant is most often the best option.

The exhausted cresylated soda (often called phenolic) is generated from the washing of gasoline fractions and is composed mainly of aromatic, acid oils, cresols, and other organic acids. Cresylates are a form of phenols containing a methyl group (CH_3), often called phenolics as a family or group of compounds. In their three forms, ortho, meta, and para cresol, they are not as toxic as phenols. However, the cresylates have relatively high concentrations, between 1% and 35%. Cresylates or cresylic acids, often referred to as acid oils, are solubilized at high pH, but readily separated at low pH. It is not a question of eliminating these compounds from petroleum since they increase the amount of octane in the final fuel; however, these are present in different concentrations in the depleted sodas. The number of phenols in spent soda is the factor to consider it as a reagent for the production of phenolic compounds. Otherwise, the low amount of oily acids drastically increases the costs of transportation and its treatment in the plant is the best option.

Due to the high content of organic matter, in addition to sulfur compounds, phenols, and mercaptans, the depleted sodas are a liquid residue tough to treat [2,17], with a series of related problems with their treatment of depleted among those highlights [2]:

- Strong odors of sulfides and mercaptans: the odor traces of these compounds are in the order of parts per billion. As is known, these compounds are highly toxic even at minimal concentrations.
- High phenol concentrations: phenol is a highly inhibiting compound of biological activity. Deficient concentrations of phenol have been shown to inhibit the biological removal of organic matter.
- High concentrations of bio-refractory material: the presence of cresylic and naphthenic acids, which are difficult to biodegrade, is another problem associated with SC. Naphthenic acids facilitate foaming formation.

Shailja Singh and Shikha [22] mentioned the primary pollutants found in the effluent from various processes in refineries. The liquid effluents generated in the oil refineries differ from one industrial plant to another due to the variation in the configuration of the plant and in particular, to the type of petroleum processed. For example, ammonia and hydrogen sulfide are the main substances present in the wastewater from the isomerization process; meanwhile, the alkylation produces hydrofluoric acid. Other processes in this industry such as fluid catalytic cracking, crude desalting, catalytic hydrocracking, coking, distillation, catalytic hydrotreating, lubricating oil manufacture, sulfur removal, catalytic reforming, and thermal cracking are summarized in the same publication.

In general, liquid petrochemical residues contain significant concentrations of suspended solids (SS), chemical oxygen demand (COD), biochemical oxygen demand (BOD), oils and fats, sulfides, ammonium, phenols, hydrocarbons, benzene, toluene, ethylbenzene, xylene, polycyclic aromatic hydrocarbons, and heavy metals [23–26].

In order to avoid or reduce the environmental problems caused by SC, a series of processes have been proposed that help neutralize sodas and allow their subsequent biological treatment.

3. Treatment Methods of SC

3.1. Initial Considerations

In general terms, the treatment of spent sodas is carried out in two types of processes in series; the first one considers the removal of sulfides, phenols, and other compounds that can inhibit the biological processes. These are later applied in order to remove the organic matter present in this residue. The spent sodas cannot be directly biotreated for different reasons:

- The presence of phenols inhibits, at a specific concentration, the healthy metabolism of the microorganisms that operate in the biological process.
- Spent sodas contain some low biodegradable compounds, such as naphthenic acids.
- The presence of naphthenic acids may result in operational issues in aerobic processes through foam formation.
- The high COD concentration in SC makes impossible their direct treatment.
- They have a high pH, which is not adequate for the development of microorganisms.

3.2. Wet Air Oxidation (WAO)

In practice, this is the most widely used chemical treatment applied to spent sodas. This process has been executed on several occasions as a pilot [27] at a plant [21,28–32] and laboratory level [33,34], in addition to the diverse industrial plants [35] that have implemented the WAO system. It is basically the oxidation of the soluble and suspended compounds that are present in this residue using oxygen or air as an oxidant [21]. The oxidation is carried out at very high temperatures and pressures, which depend on the strength of the material to be oxidized in the process and the quality required from the final effluent.

In the WAO process, several chemical reactions take place, such as:

$$\text{Organics} + O_2 \rightarrow CO_2 + H_2O + RCOOH \tag{5}$$

$$\text{Sulphur species} + O_2 \rightarrow SO_4^{2-} \tag{6}$$

$$\text{Organochlorides} + O_2 \rightarrow Cl^- + CO_2 + RCOOH \tag{7}$$

$$\text{Organic nitrogen} + O_2 \rightarrow NH_3 + CO_2 + RCOOH \tag{8}$$

$$\text{Phosphorus} + O_2 \rightarrow PO_4^{3-} \tag{9}$$

Three ranges divide the operational conditions: low, medium, and high-temperature systems [36]. Oxidation takes place at temperatures of 200 °C and 27.5 bars in low-temperature systems. This system partially oxidizes sulfides into sulfates and thiosulfates [37]. In medium temperature systems, oxidation takes place in the range of 200 °C and 27.5 bars to 260 °C and 86 bars. The ideal supply for this system is naphthenic caustics sodas. Sulfides react to sulfates, and the mercaptans are also destroyed [38]. The high-temperature systems oxidize the cresylic caustics at a range from 240 °C and 55 bars to 260 °C and 85 bars. Complete oxidation of sulfides, mercaptans, and other organic compounds can be carried out, such as cresylic acids [39]. The typical industrial WAO systems can work with spent sodas flows between 1 and 50 m^3/h and with a COD between 10,000 and 100,000 mg/L. However, for COD levels above 50,000 mg/L, dilutions with water or fresh caustic solution are required. Diluted caustic solutions help to reach different goals, among them, avoiding diluted salts concentrations below the solubility levels in order to prevent incrustations. The other reason is to guarantee that alkalinity is not consumed by oxidation when acid effluents could damage the construction materials of the system. Finally, the addition of the caustic solution can help the oxidation of contaminants according to the reactions that are described below [36].

Sulfide.

$$NaSH + 2\,O_2 + NaOH \rightarrow Na_2SO_4 + H_2O \tag{10}$$

$$NaSH + O_2 \rightarrow \frac{1}{2}\,Na_2S_2O_3 + \frac{1}{2}\,H_2O \tag{11}$$

Thiosulfate.

$$Na_2S_2O_3 + 2\,O_2 + NaOH \rightarrow 2\,Na_2SO_4 + H_2O \tag{12}$$

Mercaptan.

$$NaSR + \frac{3}{2}\,O_2 \rightarrow RSO_3 - Na \tag{13}$$

Cresylic acid.

$$C_6H_5O - Na + 7\,O_2 + 11\,NaOH \rightarrow 6\,Na_2CO_3 + 8\,H_2O \tag{14}$$

$$C_6H_5O - Na + 5\,O_2 + 8\,NaOH \rightarrow 4\,Na_2CO_3 + CH_3COO - Na + 8\,H_2O \tag{15}$$

Naphthenic acid.

$$Na - C_{12}H_{22}O_2 + \frac{67}{7}\,O_2 + 23\,NaOH \rightarrow 2\,Na_2CO_3 + \frac{45}{2}\,H_2O \tag{16}$$

$$Na - C_{12}H_{22}O_2 + \frac{59}{4}\,O_2 + 20\,NaOH \rightarrow 10\,Na_2CO_3 + CH_3COO - Na + \frac{39}{2}\,H_2O \tag{17}$$

The WAO process can treat liquid currents containing cyanide, several heterocyclic compounds, industrial sludge, and spent coal from adsorption processes [40]. Several industrial residues apply this process, such as paper manufacture, textile sludge, among others. Figure 1 shows an outline of a typical WAO process.

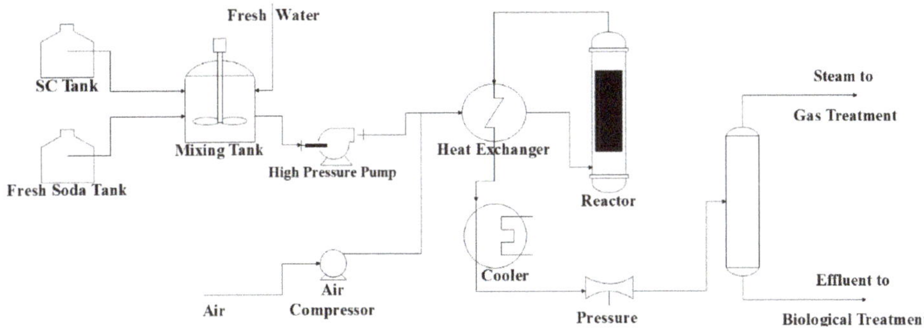

Figure 1. Typical outline of a wet air oxidation (WAO) system.

As can be seen in Figure 1, fresh soda or freshwater dilute the spent soda. The dilution ratio depends on the strength of the spent soda. Then, the diluted spent soda is pumped at 27.5–85 bars through a high-pressure pump. The addition of compressed air to the diluted spent soda is required in order to supply the needed oxygen for the oxidation reaction. A heat exchanger preheated the mixture at 200–260 °C. The reactor must operate for a sufficient residence time that allows high oxidation of the organic matter contained in the spent sodas (COD disposal). Since the reaction is exothermic, the reactor's effluent has a high temperature, so it can be used to preheat the diluted spent sodas before entering the reactor [21]. Subsequently, the effluent coming out of the heat exchanger is cooled and depressurized before being sent to a gas–liquid separator. Then, the separated liquid goes to biological treatment. The effluent gas has 5%–21% of oxygen and some volatile organic compounds. In Table 2, the values of a sample of the operational conditions of a WAO system from the Oil Refinery of Manguinhos, S. A. in Rio de Janeiro, Brazil, are presented.

Table 2. Operational conditions of the WAO system in Manguinhos, Brazil [21].

Parameters	Supply	Effluent
Residence time (h)	-	1
Oxidation temperature (K)	-	533
COD (g/L)	114	≈23
Sulfides (g/L)	≈24	<0.001
Total phenols (g/L)	≈0.45	<0.002

As can be observed, despite obtaining a removal percentage of 80% of COD, measured from organic matter, it is still very high. This concentration is required to be reduced until reaching the possible values to discharge it into a receiving current or another place for its final disposal. It could be carried out through a biological process that can be done in this case, provided the low levels of sulfides and phenols in the WAO effluent since the two chemical species were reduced in the process by almost 100%. Additionally, most of the remaining COD in the treated effluent comes from aliphatic acids of low molecular weight, which can be biologically oxidized [37]. The oxidation pressure is directly related to the oxidation temperature. The oxidation pressure is responsible for keeping the reaction in the liquid phase. In this way, as the pressure increases, the oxidation pressure rises to keep the reaction in the liquid phase. The WAO processes typically operate with hydraulic residence times of 45–120 min, the oxidation degree that takes place within the reactor affecting this parameter. Each WAO system uses different hydraulic residence times to reach the degree of COD reduction required. The selection of the construction material is critical in the WAO system, considering that the operation takes place at a high temperature and pressure. Thus, the materials used must be appropriate for these conditions. The WAO system presents some disadvantages and difficulties, among which the initial investment cost stands out. This cost depends mainly on the operating conditions, which, at the same time, are related to the strength of the residue to be treated. As this strength increases, more severe operating conditions are required, which increases the cost of the plant.

Tests and research are still carried out in order to improve this system, as well as combining this process with others. A way of reducing the severe operating conditions is the use of solid catalysts in the WAO to accelerate the reactions, and there are reports of cases where catalysts have been used in this process [8,41]. This innovative technique is referred to as catalytic wet air oxidation (CWAO). In recent years it has been studied in order to increase the efficiency to treat pollutants in industrial effluents [42]. The studies have been developed to remove individual toxic substances and to expose the best selection of metal catalyst-based support at lab scale. In general, the catalytic activity of transition metal such as Vanadium, Ferrum, Nickel, Copper and others have been chosen as the most common option.

Otherwise, researchers are still looking for better conditions of the process to treat most pollutants in spent caustic using this technology.

Nunez et al. [43] added silver to Al_2O_3-ZrO_2 mixed oxide to reduce p-cresol at 160 °C and 15 bar of O_2 pressure, showing faster oxidation associated with more selectivity of this metal compared to the test without it. Additionally, it reported that the highest concentration (20%) of ZrO_2 in the catalyst removed 50% and 96% for TOC and p-cresol, respectively.

Jagushte and Mahajani [34] studied the effect of the heterogeneous copper catalyst in the kinetics of Equation (12) mentioned above at the lab-scale. As a result, the catalyst reduced the temperature and time process to 30 °C from 150 °C and 4 min from 12 min, respectively, in a conversion of 99% of the pollutant in alkaline conditions. It is remarkable to say that the batch experiments were carried out at low temperatures and 0.69 MPa of oxygen partial pressure, showing first and 0.5 kinetic order concerning thiosulfate and oxygen, respectively. Additionally, this study explained a strategy to treat spent caustic under its results associated with the enhanced effect of phenols in the oxidation of thiosulfate.

A similar study was reported by Zermeño-Montante et al. [44]. In this case, copper catalyst in a silica support material achieved the complete sulfide oxidation at 70 °C in 20 min. It showed better results than a similar experiment with Vanadium/Clinoptilolite just for 6 min latter. Both cases enhanced the WAO process for this pollutant. This report explained some of its results from Jagushte and Mahajani [34],

Recently, Barge and Vaidya [45] experimented with ferrous sulfate to eliminate sodium sulfide. That study innovated using a cheap and abundant catalyst, destroying 94% of COD at 100 °C and 0.69 MPa of oxygen partial pressure within 1 h. It is remarkable to say those authors exposed the Langmuir–Hinshelwood as the kinetic model to oxidize the reactant. These same authors, Barge and Vaidya [46] used Graphene oxide (GO) and Ruthenium (Ru) as a catalyst in WAO for the treatment of the cresylic spent caustic, obtaining the best results at 175 °C and 0.60 MPa of O_2 pressure, with removals of 54.9%, 48.9%, and 61.2% as TOC, for 0-cresol, m-cresol, and p-cresol, respectively, in the typical WAO. Meanwhile, when the Ru/GO catalyst was used for the same operating conditions, the removals obtained were 66.4%, 53.4%, and 73.9%, as TOC, for *0*-cresol, *m*-cresol, and *p*-cresol, respectively.

3.3. Acid Neutralization

In this type of process, alkaline caustic solution traps the acid components released from acidified spent sodas [36]. This action results in sulfides and mercaptans being released as acid gases, and naphthenic acids come to form some oil layer. One of the most significant differences between acid neutralization and the WAO process is that the acid components of spent sodas are removed but not destroyed, for example, phenols. In some cases, this entails the requirement of additional treatments for the acidification effluents, for example, gas ignition or methods for sulfur recovery. In other cases, this allows for any component present in spent sodas to be captured and reused.

As phenols are highly soluble in water, they are complicated to acidify. However, acidification eliminates the trend of the formation of foam, similar to the one obtained in the WAO process at high temperatures. The cause of this is mainly the fact that the naphthenic compounds that tend to form foam move to an insoluble phase and, even though they are not eliminated, they reduce their trend to form foam. Additionally, some alkaline compounds that could form foam are acidified, reducing this trend.

In some cases, acidification as a pretreatment for oxidation processes has been studied. Sheu and Weng [47] treated the spent sodas from olefin plants containing a considerable amount of H_2S and some mercaptans, phenols, and crude oils undergo a process of acid neutralization before a process with Fenton reagent. Using sulfuric acid with a pH of up to 5, and 70 °C, a little more than 90% of the dissolved H_2S changes to the gas phase. From this pH, the conversion to dissolved sulfide (S^{2-}) is minimal [48]. These authors were the first to report this pretreatment process.

Nuñez et al. [49] showed a reduction of 71% and 82% of COD in two batch experiments using this procedure, under conditions of pH below 4. In this context, only the sulfide content was reduced by pH reduction, the phenol concentration not being affected. The removal percentages were lower, considering the analysis made by Sheu and Weng [47] and it was attributed to the use of synthetic and real samples used. The last type has other substances that could inhibit the pH reduction to oxidize the reactants.

In summary, acid neutralization or acidification reduces most of the COD and the trend to form foam removing the naphthenic acids. Sulfuric acid is the chemical agent used in most cases in these processes, due to its strength and lower price in comparison with other acids [50].

3.4. Advanced Oxidation Process (AOP)

These processes are defined as the processes that involve the formation of radical hydroxyls (OH) that oxidize both organic and inorganic contaminants present in water and wastewater [51]. These radicals are the second most potent oxidants after fluorides [52]. Unlike the physical process such as adsorption, filtration, and air stripping, the AOPs can destroy the contaminants rather than transferring them from one medium into another [53]. The AOPs can be used to treat wastewater with high COD and low biodegradability [54]. With these processes, complete oxidation of residues can be reached, converting contaminants into water, CO_2, and innocuous inorganic products. However, it is expensive and impractical to use these processes for the complete mineralization of compounds since there are intermediate products that are resistant to chemical oxidation. A practical solution is to use these processes as a pretreatment for biological treatments. The partial mineralization produced by the AOPs generates intermediates with higher biodegradability and lower toxicity, which makes them available for biological oxidation. It is very important to determine the quality of the residue before selecting any AOP process, since this factor will affect the efficiency of the process, for example, residues with high alkalinity have an excess of carbonate and bicarbonate compounds, which may interfere with the oxidation reactions of hydroxyl radicals. These radicals react with those compounds forming weaker radicals resulting in lower efficiencies of the minor oxidation processes [55]. This issue can be solved by reducing the alkalinity to make sure there is no excess of these compounds in the residues, which can be achieved in a relatively simple way by lowering the pH. Nitrates and nitrites can also affect the efficiency of the AOP processes, which use UV light to generate hydroxyl radicals. These compounds can absorb UV light and, thus, reduce the generation rate of free OH radicals. The turbidity has the same effect over the efficiency of the AOP processes as nitrates and nitrites. Turbidity reduces the production of hydroxyl radicals, since it acts as a barrier against UV light, which cannot penetrate through the residue to be treated.

The AOP reaction systems consist of a catalyst and an oxidant. The catalyst task is to form hydroxyl radicals from oxidants. There are several means to form these radicals, which can be photochemical, photo-catalyzed, ultrasonic, and chemical oxidations.

The photochemical processes include UV radiation. Some studies mention the use of this technology individually. Gurol and Vatistas [56] irradiated by UV a mixture of various phenolic compounds, using 16 low-pressure mercury lamps, each of which emitted 2.2 Watts of radiation at

254 nm. The results showed a faster initial rate remotion of phenol at pH 2.5 than p-cresol and xylenol at the same condition. The duration of the experiments was 100 min, starting in 50 mg/L of the three substances and achieving remotions in the range of 20%–30%. Neutral and basic (pH 9) conditions did not exhibit significant changes in the initial rate remotion, observing the same removal efficiency for those compounds at the same time. The authors concluded that the understanding of the reaction mechanism at low-pH irradiation requires further research.

Spent caustic from the ethene plant were irradiated with UV by Yu et al. [57]. The results showed low-efficiency COD removal (under 10%) in 180 min using an ultraviolet light source of 254 nm wavelength. The high content of several persistent pollutants in spent caustic is not affected when this technology is used by itself. That is why the UV radiation has not attracted industrial applications and also lab-scale studies, and this is the evidence of a few reports about it.

Other photochemical technologies like ozone systems (UV/O_3) and (UV/O_2) systems have been reported. Photo-catalysis includes photo-Fenton, while chemical oxidation includes O_3/H_2O_2 and H_2O_2/Fe^{3+}.

3.4.1. Ozone and UV Radiation (O_3/UV)

Ozone in water is an unstable compound, dissociating in alkaline water and forming free radicals according to the following mechanism [58]:

$$O_3 + OH^- \rightarrow O_3^- + OH^- \tag{18}$$

$$O_3^- \rightarrow O_2 + O^- \tag{19}$$

$$O^- + H^+ \rightarrow OH \tag{20}$$

The pH increase promotes the formation of hydroxyl radicals. The main disadvantage of using only ozone is that the number of radicals produced is low, so in order to increase the efficiency of the ozone, it is combined with other physical methods, such as UV rays, ultrasound or chemicals, like hydrogen peroxide and Fenton reagent.

The photolysis of O_3 starts the advanced oxidation processes with O_3 and UV according to the following reaction:

$$O_3 + H_2O \overset{h\gamma}{\rightarrow} H_2O_2 + O_2 \tag{21}$$

The hydrogen peroxide formed reacts then with the ozone to produce the free radical according to:

$$O_3 + H_2O_2 \rightarrow 2\,OH\cdot + 3\,O_2. \tag{22}$$

Ozone photolysis occurs when UV radiation (wavelength of 254 nm) is applied to ozone-saturated water [59]. According to Equations (21) and (22), in order to produce 1 mol of hydroxyl radical, the consumption of 1 mol of ozone and half a mol of hydrogen peroxide is required. In this system, contaminant degradation can occur due to the chemical oxidation of hydroxyl radical, the direct oxidation of ozone, as well as the photolysis caused by the UV.

Two essential parameters considered those systems: the UV dose and the ozone concentration. The UV dose required per ozone is responsible for the formation of radicals and it depends on the strength of the water to be treated. Other important parameters to be considered in order to guarantee efficient operations of these systems are the pH and the cleaning system of the radiation lamp [60]. The pH can influence the hydrogen peroxide generation. For a pH of less than 1.8, the photolysis requires 1 mol of ozone to react with 1 mol of hydrogen peroxide. However, as the pH increases, this ratio decreases. Cleaning the lamps is very important in order to avoid their siltation. The cleaning frequency depends on the concentration of the compounds that can produce siltation, such as iron, calcium, and magnesium.

Gurol and Vatistas [56] also studied this technology. At lower pH 2.5, selective removal of phenols, p-cresol, and xylenol were shown compared to neutral and alkaline conditions using ozone doses of 3.8 mg/L. Otherwise, a direct relation between pH and pollutant removal was reported. These three substances were completely reduced at 100 min and pH 9. Therefore, this research exposed an additive effect of the O3/UV system compared to its use individually.

Results for spent caustic from the ethene plant also were reported by Yu et al. [57]. They showed COD removal and BOD/COD ratio improvement with the increase of ozone dosage in this technology. An optimum of 3.3 g of COD removal per gram of O_3 at a low dosage (0.6 g O_3/L) was found.

The advantages of these systems are the high removal that ozone and UV radiations produce since more free radicals are produced than with other AOP systems. The main disadvantage of this system is that ozone has to be disseminated through the liquid, which may be limited by the mass transfer in the medium. The low stability of ozone in water can result in a higher energy demand, which increases the capital and operational costs of the system. The last statement is part of the conclusions in the paper mentioned before.

3.4.2. Hydrogen Peroxide and UV Radiation (H_2O_2/UV)

In this system, hydroxyl radicals are formed through two mechanisms: hydrogen peroxide photolysis and hydrogen peroxide decomposition [61]. The formation of hydroxyl radicals from hydrogen peroxide photolysis is shown in the following reaction:

$$H_2O_2 \overset{h\gamma}{\rightarrow} 2\,HO\cdot \quad (23)$$

The wavelength for this reaction is 250 nm. This reaction depends on the pH and, as it increases, the number of hydroxyls also raises, and so does the number of hydroxyl radicals. The propagation reactions of hydrogen peroxide are the following:

$$H_2O_2 + HO\cdot \rightarrow HO_2\cdot + H_2O \quad (24)$$

$$H_2O_2 + HO_2\cdot \rightarrow HO\cdot + O_2 + H_2O \quad (25)$$

$$2\,HO_2\cdot \rightarrow H_2O_2 + O_2 \quad (26)$$

Finally, the chain reaction finishes:

$$2\,HO\cdot \rightarrow H_2O_2 \quad (27)$$

Hydrogen peroxide follows two paths: the path of radiation and the path of decomposition. However, the path of radiation is more efficient as more free radicals are formed, consuming fewer peroxides. It is vital to keep the amount of peroxide as low as possible since this is an expensive material. Some of the critical parameters of this system are the intensity of the UV lamp, the contact time within the reactor, the operating pressure and temperature, and the dose of hydrogen peroxide. The intensity of the lamp is necessary to form hydroxyl radicals and is, mainly, a residue strength function.

The main advantage of this AOP system is that it reaches a higher removal of contaminants than the UV or H_2O_2 systems individually. However, as previously mentioned, the presence of a certain degree of turbidity harms the efficiency of the process. The non-contaminant compounds present in the treated medium can also absorb UV, reducing the efficiency of the process.

The use of hydrogen peroxide and UV radiation to treat ethene spent caustic also has been reported by Yu et al. [57]. The experiments showed almost 70% of COD removal efficiency when the ratio of 0.8 by H_2O_2/COD was used. This article mentioned a direct relation of H_2O_2 dosage with removal efficiencies of COD, but it reported the ratio amount mentioned before as the optimum condition. While the hydroxyl radicals from the photolysis of H_2O_2 reacts with COD, a higher amount of H_2O_2

inhibits this system. Additionally, it mentioned that the combination of H_2O_2 and UV showed better results than the use of these individual technologies.

In summary, this technology has not been widely studied and reported to treat spent caustic, even when it is cheaper than the O_3/UV system when it was compared in textile effluent [62].

3.4.3. Ozone with Hydrogen Peroxide

The addition of H_2O_2 to the system improves the production rate of hydroxyl radicals, as shown in Section 3.4.1, according to Equation (22). This system is similar to the UV–hydrogen peroxide system. However, the residue characteristics, such as the turbidity, influence the ozone–hydrogen peroxide system less [62]. No information reported means that further research using this technology must be studied to treat spent caustic.

3.4.4. Ozone, Hydrogen Peroxide and UV Radiation (O_3/H_2O_2/UV)

This system uses two oxidants, O_3 and H_2O_2, in order to increase the formation of hydroxyl radicals. At lab-scale Yu et al. [57] reported the optimum loaded of H_2O_2 and O_3 as 0.8 and 0.6 g/L, respectively, to reach 76% COD removal in the treatment of spent caustic from the ethene plant. This research mentioned the improvement in the BOD/COD ratio from 0.22 to 0.52 when this technology is used, compared to the H_2O_2/UV system. Even though the removal of the contaminants is increased in the same way, the treatment costs also rise rapidly due to the use of two oxidants [62].

3.4.5. Ozonation and Ultrasound

Ultrasound is used to form hydroxyl radicals [63]. Ultrasound waves are transmitted and introduced into the wastewater in alternate cycles of compression and expansion. This process creates microbubbles when the amplitude of the expansion cycle is big enough to increase the molecular distance to a distance higher than the required to remain in the liquid phase. The following compression cycle makes the microbubbles collapse, producing extremely high temperatures and pressures, which causes the breakdown of molecular water producing hydroxyl radicals. Several factors influence the ultrasound systems, among which are the ultrasound frequency and the temperature and pressure of the dissolved gas. The intensity of the ultrasound is related to the formation rate of hydroxyl radicals. Higher ultrasound frequencies reduce the microbubbles' collapse time resulting in a very low possibility of recombination of hydroxyl radicals, which entails a higher generation rate of hydroxyl radicals.

Occasionally, ozone is used along with an ultrasound to promote the formation of hydroxyl radicals and improve the removal of contaminants [51].

The reaction produced between ozone and ultrasound in order to generate hydroxyl radicals is as follows:

$$H_2O \overset{\text{ultrasound}}{\rightarrow} H\cdot + OH\cdot \quad (28)$$

$$O_3 \overset{\text{ultrasound}}{\rightarrow} O_2 + O(^3P) \quad (29)$$

$$O(^3P) + H_2O \rightarrow 2\,OH\cdot \quad (30)$$

$$O_2 + H\cdot \rightarrow HO_2\cdot \quad (31)$$

$$O_3 + HO_2\cdot \rightarrow 2\,O_2 + OH\cdot \quad (32)$$

$$O_3 + HO_2\cdot \rightarrow 2\,O_2 + OH\cdot \quad (33)$$

$$OH\cdot + OH\cdot \rightarrow H_2O_2 \quad (34)$$

$$HO_2\cdot + OH\cdot \rightarrow H_2O + O_2 \quad (35)$$

The main advantage of this system is that the energy required to form the hydroxyl radicals is lower than the other AOPs, especially the ones that use UV, reducing the operational costs. The main disadvantage is that the amount of oxidant required, either ozone or hydrogen peroxide, is high, which raises the operational costs [63]. There is no information reported about this system, meaning that further research using this technology must be studied to treat spent caustic.

3.4.6. Fenton Reagent

The AOP system uses hydrogen peroxide as an oxidant agent and mixes it with a catalyst, which is ferrous ion [64]. The reaction of hydrogen peroxide and ferrous ion is called Fenton chemistry. Its practical application was carried out until the 60s [65], and it was used for the treatment of textile liquid residues, liquid effluents, pulp and paper liquid effluents, and acid waters of petroleum refineries [66].

The main advantages of this AOP system are its high efficiency and capacity to treat several contaminants. Some organics of low molecular weight, such as paraffin, chlorinated alkanes, and some short-chain carboxylic acids, are resistant to Fenton's oxidation.

The reactions involved in this system for the formation of hydroxyl radicals are the following:

$$Fe^{2+} + H_2O_2 \rightarrow Fe^{3+} + 3\,OH\cdot + OH^- \quad (36)$$

This is the initial reaction of the Fenton reaction. The formation of hydroxyl radicals is developed through a complex chain reaction. Once the ferrous ion is formed, it can react with the hydrogen peroxide according to:

$$Fe^{3+} + H_2O_2 \leftrightarrow Fe - OOH^{2+} + H^+ \quad (37)$$

$Fe\text{-}OOH^{2+}$ is an intermediate that is decomposed to form HO_2 radicals according to the following reaction:

$$Fe - OOH^{2+} \rightarrow HO_2\cdot + Fe^{2+} \quad (38)$$

HO_2 can oxidize contaminants, however, its oxidation power is much lower than other radicals. Other chain reactions that occur in this system are the following:

$$Fe^{2+} + HO_2\cdot \rightarrow Fe^{3+} + HO^{2-} + OH^- \quad (39)$$

$$Fe^{2+} + HO_2\cdot \rightarrow Fe^{2+} + O_2 + H^+ \quad (40)$$

$$OH\cdot + H_2O_2 \rightarrow H_2O + HO_2\cdot \quad (41)$$

Finally, the last step is carried out through the following reaction:

$$OH\cdot + Fe^{2+} \rightarrow OH^- + Fe^{3+} \quad (42)$$

The global reaction can thus be formulated as:

$$2Fe^{2+} + H_2O_2 + 2H^+ \rightarrow 2\,Fe^{3+} + 2\,H_2O \quad (43)$$

As it can be observed, the Fenton reaction requires an acid medium for it to occur. Once the hydroxyl radicals are formed, the oxidation of contaminants takes place. If the oxidation is complete, the contaminants can decompose in water, CO_2, and some innocuous inorganic salts [67].

The hydroxyl radicals can oxidize organic radicals (RH) through the removal of the organic radical protons that form organic radicals, for example, according to:

$$RH + OH\cdot \rightarrow H_2O + R\cdot \quad (44)$$

This reaction is a chain propagation reaction. Once the organic hydroxyls are formed, the subsequent oxidation can occur [68].

$$R\cdot + H_2O_2 \rightarrow ROH + OH\cdot \quad (45)$$

$$R\cdot + O_2 \rightarrow ROO\cdot \tag{46}$$

Organic radicals can follow other paths. They can be oxidized with ferrous ions, reduced with ferrous ions, or two radicals can be dimerized, as shown below:

Oxidation

$$R\cdot + Fe^{3+} \rightarrow R^{+} + Fe^{2+} \tag{47}$$

Reduction

$$R\cdot + Fe^{2+} \rightarrow R^{-} + Fe^{3+} \tag{48}$$

Dimerization

$$2\,R\cdot \rightarrow R - R \tag{49}$$

In addition to the oxidation of the contaminants through the Fenton reagent, this method can show coagulation effects. Adjusting the pH in the 5–7 range, the precipitation of dissolved ferrous particles can be achieved. The precipitated ferrous particles are combined to form flocs that can contribute to the removal of the dissolved solids. As a result, in this system, it is possible to obtain the treatment of residues by chemical oxidation and coagulation [69]. The chemical oxidation prevails when the ratio of hydrogen peroxide:ferrous salts is above 2. On the other hand, the coagulation effect prevails when this ratio is below 0.2. It is essential to select the appropriate ratio correctly in order to guarantee the destruction of the contaminants since the goal of this system is to destroy the contaminants and not just physically separate them.

Other essential parameters in the Fenton reaction that are to be assessed are pH, temperature, ferrous salts concentration, and reaction time [70,71]. The typical values of pH are within the 3–5 range. Regarding temperature, it is necessary to know that this reaction is exothermic.

Ferrous salts (ferrous sulfate) are responsible for the generation of hydroxyl radicals, which convert the ferrous ion to ferrous particles, known as coagulant agents. As the concentration of ferrous salts increases, the coagulation is incremented. The reaction time is crucial in order to guarantee a sufficient residence time that allows oxidation to be completed in the required degree. Once the Fenton reaction is completed, high strength contaminants are oxidized into organic acids of low molecular weight, which are easily treated in biological treatments.

The main advantage of this process is its capacity to treat residues with different flows and organic contents. Another advantage is the COD removal that can generally reach up to 95%, which is higher than other AOP processes. The system also can destroy some pollutants such as sulfides and mercaptans. In contrast, the process has low capital costs and is relatively easy to install.

3.5. Electrochemical Processes

Recently, some methods based on electrochemistry have been studied for the treatment of different residues, including the liquid residues produced in the oil industry [72,73]. In these methods, electric power is produced from a global reaction of oxidation-reduction (a half-reaction of oxidation that occurs in the anode and another of reduction, which occurs in the cathode), as shown in Figure 2. The three operational parameters that are always present in these methods, regardless of the modality applied, are the current density used, the sulfide load applied to the system, and the time used in the process.

Electrocoagulation is a variant of the electrochemical process that has been studied for the treatment of liquid effluents [74,75]. In this method, a flocculating agent is generated in the anode, at a certain pH, which allows removal of the sulfides, among other soluble materials, as in this case, since the hydroxyl ions produced in the cathode increase the pH in the medium and cause the precipitation of the sulfides [76]. Hariz et al. [17] obtained sulfides and organic matter removals over 80% in a simple electrocoagulation system, while operating with subsequent electrocoagulation, they reached removal efficiencies of over 95% for both contaminants, operating at pH 9 and 21.2 mA/cm^2 with initial concentrations of 34,517 mg/L and COD of 72,450 mg/L for 30 min.

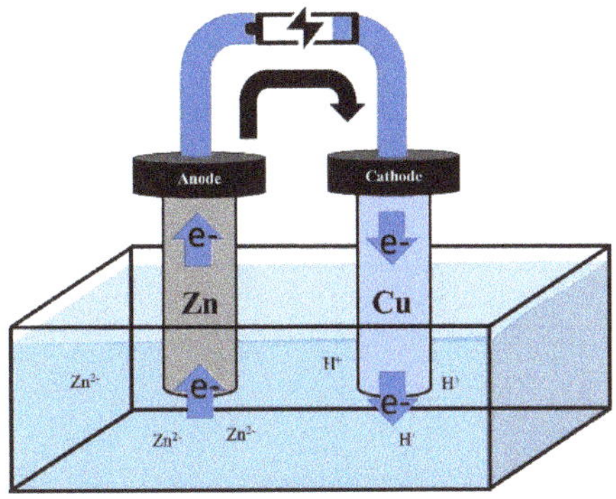

Figure 2. Chemical species of hydrogen sulfide in an aqueous medium according to pH.

Other electrochemical methods studied for the SC treatment are based on electrodialysis (ED), which consists of extracting ionized substances dissolved in an aqueous dissolution through selective membranes of ion exchange [77]. Based on this principle, the electro–electro-dialysis (EED) has been developed, which combines water electrodialysis and ions transport through selective membranes of ions exchange [78]. A more advanced electrodialysis process is the so-called bipolar membrane electro-dialysis (BMED), where a bipolar membrane is used to separate the water in H^+ and OH^-, producing caustic flows with acids [79]. This process adjusts the pH of a solution without adding acids or bases. In this case, the polarity of the electrodes is inverted periodically exchanging the outputs of the membrane accumulator of the freshwater produced and residual water. The ions are transferred in opposite directions through the membranes, which contributes to breaking the incrustations and washing and outside exiting of sludge and other deposits of the cells. In a research study, EED and BMED were applied for the regeneration of sodium hydroxide present in SSC [80]; and it was found that BMED was more efficient in terms of caustic soda recovery, consuming little more energy than in EED.

As it was described in Section 3.4.6, the system based on the Fenton reaction has been successfully applied for the removal of SSC. However, for this reaction to occur, it is necessary to add H_2O_2, in order to avoid this inconvenience. Hydrogen peroxide can be formed through an electrochemical process called electro-Fenton [81]. Davarnejad et al. [82] applied an electro-Fenton process to SSC. They assessed the effect of the operational parameters that influenced the removal of hydrogen sulfide: pH, current density, reaction time, among others, finding that, in order to have optimal operating conditions, at pH of 2.87 a reaction time of 65.52 min and a current density of 55.47 mA/cm^2 was required. The removal percentages were 81% and 100% for COD and sulfides, respectively.

As shown in Section 3.4, the use of UV rays in the AOP contributed significantly to the formation of free radicals that react and oxidize the sulfides and organic matter present in SCC. This is why, in the electro-Fenton process, applying UV increases the formation of free radicals. In this sense, Esfahani et al. [7] reached COD removals of 97%, operating at 4.5 pH at 60 °C for 80 min and a current density of 16.6 mA/cm^2.

Microbial fuel cell (MFC) is a process based on the fundamentals of electrochemistry, where the energy that contains the organic matter can be captured in the form of electricity from the work of microorganisms [83]. In this case, the movement of the electrons is produced through bacterial metabolism, where the electrons are transferred from a substrate to an electron acceptor at a higher redox potential, this means that while substrates are oxidized, the microorganisms produce protons and electrons, and their transfer to the cathode produces electricity [84,85]. Fazli et al. [86] operated an MFC with SSC at different retention times and suspended solids concentrations (SS), obtaining the higher efficiencies of COD and sulfides removal of 98% and 99%, respectively, for 9 days of retention time. The optimal concentration of SS for the MFC process was of 1500 mg/L, reducing COD and sulfides by 94.07% and 89.01%, respectively. The higher voltage for the same retention time was 82.1 mV.

The possibility to recover NaOH from an electrochemical process, in this case, electro-dialysis and electro-deionization, was proven by Keramati et al. [87].

3.6. Biological Processes

In general, the biological methods are environmentally more beneficial and more economical than the physic-chemical ones for most of the industrial processes, including residue treatment [87]. However, one of their limitations is that the microorganisms that operate in these processes can stand limited concentration levels of several substances, such as sulfides [88]. In order to overcome this limitation, several studies applied biological methods to SSC. De Graaf et al. [89] operated aerobic gas-lift reactors with retention times of 3.4–3.7 days with extremophile bacteria extracted from lake sediments with high levels of alkali to treat SC with 160–240 mM of sulfide contents. However, this concentration was reduced when mixing SSC with sodium carbonate solution at pH 9.5. Complete removal was reached up to 27 mM S/L·day through the conversion into sulfide by the halo-alkaline bacteria *Thioalkalivibrio*, also obtaining the removal of 93% of benzene. De Graff et al. [5] assessed a system in two aerobic stages, the first abiotic and the second with halo-alkaline bacteria inoculum, obtaining high removal rates of 33 mM/L·day operating the first reactor with 2 days of retention time and the second one with 3.5 days. The affluent to the first reactor was a synthetic solution simulating diluted SSC.

The microorganisms work correctly at absolute pH values depending on the specific species to be treated. For example, it has been determined that one of the few reported microorganisms known for the elimination of sulfides of SSC is *Thiobacillus denitrificans*, which operates correctly around neutral pH [90]. Considering the high values of pH of SSC, the dilution of these residues is required so they can be biologically treated. The use of the *Aspergillus* sp. mushroom has also been assessed for the SSC treatment, previously diluted, obtaining COD removals, total diluted solids, and sulfites, after 10 days, of 16%, 10%, and 55%, respectively [91].

3.7. Other Processes

Ionic liquids (ILs), also called green solvents, have been described by several authors as ionic salts and their melting point is less than 100 °C [92]. Due to these characteristics, however, they have been recently determined as a substitution alternative to organic solvents for the extraction of phenols and several organic compounds in aqueous systems [93]. Two quaternary ammonium-based ILs were applied to SSC, reaching in both cases efficiencies of 99% for the removal of phenols and mercaptans [16].

The treatment with oxidant agents such as sodium hypochlorite and spent sulfochromic mixtures have been assessed at a laboratory scale [94]. The reactions involved in these processes are the following:

$$HS^- + 4\,NaOCl \rightarrow 4\,NaCl + SO_4^{2-} + H^+ \tag{50}$$

$$29\,H^+ + 3HS^- + 4\,Cr_2O_7^{2-} \rightarrow 3\,SO_4^{2-} + 8\,Cr^{3+} + 16\,H_2O \tag{51}$$

With hypochlorite, no elemental sulfur was formed, reducing the COD by over 99%. The oxidation of organic matter explains why the consumption of sodium hypochlorite was higher (≈20%) than the stoichiometric amount required to oxidize all the HS^- ions.

The sulfochromic mixture also annihilated 99% of the COD discoloring the spent sodas solution. The free acidity was dramatically reduced from 8 to 0.5 mol/L (≈6% of the initial value).

In both cases, an effluent of those treatments without SH^- was obtained. It had to be neutralized with sulfuric acid at pH 7 before being ultimately arranged and, also, a residue with a high content of Na_2SO_4 remains.

The precipitation of insoluble sulfurs begins by obtaining precipitates of iron and manganese hydroxide precipitates, which can be obtained from the NaOH existent in the refinery according to the following reactions [94]:

$$FeCl_3 + 3\,NaOH \rightarrow Fe(OH)_3 \downarrow + 3\,NaCl \tag{52}$$

$$FeSO_4 + 2\,NaOH \rightarrow Fe(OH)_2 \downarrow +3\,Na_2SO_4 \tag{53}$$

$$MnCl_2 + 2\,NaOH \rightarrow Mn(OH)_2 \downarrow +2\,NaCl \tag{54}$$

In this last case, some drops of H_2O_2 (30%) are added, and the Mn(II) oxidizes to Mn(IV) according to:

$$Mn(OH)_2 + H_2O_2 \rightarrow MnO(OH)_2 \downarrow +H_2O \tag{55}$$

Thus, the reactions involved are:

$$2\,Fe(OH)_3 + 3\,NaSH \rightarrow 2\,FeS \downarrow +S \downarrow +3\,NaOH + 3\,H_2O \tag{56}$$

$$Fe(OH)_2 + NaSH \rightarrow FeS \downarrow +NaOH + H_2O \tag{57}$$

$$MnO(OH)_2 + NaSH \rightarrow Mn(OH)_2 + S \downarrow +NaOH \tag{58}$$

$$Mn(OH)_2 + NaSH \rightarrow MnS \downarrow +NaOH + H_2O \tag{59}$$

According to the stoichiometry of Equation (56), Fe(III) can react with 50% more SH^- ions than Fe(II) when compared with Equation (57). The experimental data shows that $Fe(OH)_3$ treats 45% more spent sulfidic sodas than $Fe(OH)_2$. However, Mn(IV) was more efficient than Fe(III). $MnO(OH)_2$ treated more spent sulfidic sodas than $Fe(OH)_3$, which is due to the higher oxidation state of manganese. Even though the sulfur components were removed, 10% of the COD remained in the spent sulfidic caustic soda treated. The solids produced, FeS or MnS, can be oxidized with hydrogen peroxide or hypochlorite at 25 °C in an alkaline medium allowing the recovery of manganese and iron hydroxides according to the following reactions:

$$4\,NaOH + 2\,FeS \downarrow +9\,H_2O_2 \rightarrow 8\,H_2O + 2\,Na_2SO_4 + 2\,Fe(OH)_3 \tag{60}$$

$$H_2O + 4\,NaOH + 2FeS \downarrow +9NaOCl \rightarrow 9NaCl + 2Na_2SO_4 + 2Fe(OH)_3 \tag{61}$$

$$2\,NaOH + 2\,MnS \downarrow +5\,H_2O_2 \rightarrow 5\,H_2O + 2\,Na_2SO_4 + 2\,MnO(OH)_2 \downarrow \tag{62}$$

$$2\,NaOH + 2\,MnS \downarrow +5\,NaOCl \rightarrow 5\,NaCl + 2\,Na_2SO_4 + 2\,MnO(OH)_2 \downarrow \tag{63}$$

For these reactions to occur correctly, the addition of oxidants must be slow and under stirring.

The treatments based on precipitation may be of interest, as they enable the reuse of treated sulfidic sodas, which reduces the formation of effluents and the consumption of water and chemical reagents. However, considerable amounts of chemical sludges (precipitated) are produced, which are, in general, hard to dispose of.

The use of exchange resins to treat spent sulfidic sodas is very limited, although, successful results have been obtained in the treatment of SSC [94]. The application of two sturdy anion exchange resins was studied, with a matrix of styrene-divinylbenzene. The behavior of both resins was very similar to the adsorption of 21.0–22.5 g S^{2-}/L. The robust anion exchange resins hold tightly the S^{2-} ions, which are not moved by the OH^-, even in high alkaline mediums. The pH adjustment to lower values is not a good practice since H_2S is released when the pH is adjusted under 9. Although the results were positive, the initial costs of investment must be seriously considered, as well as the maintenance costs due to the resin dirt. It would imply a process of regeneration of the resins with frequency, and, therefore, a relatively short duration time, incurring replacement costs for these materials.

Considering the diverse possibilities of use of treatment methods for spent sodas, in Figure 3 a summary is presented with the main possibilities of application of these methods according to the origin or characteristics of spent sodas.

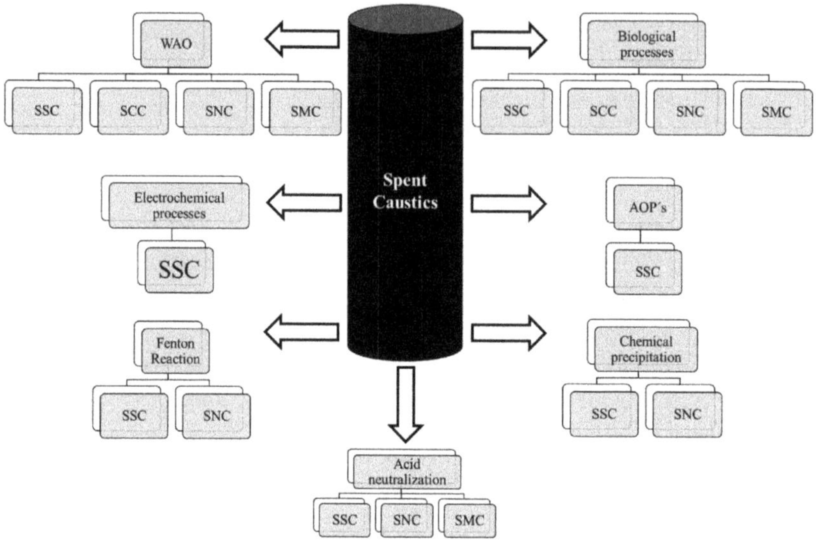

Figure 3. Application of diverse processes of spent sodas.

For the specific application of each of these methods, the subsequent biological treatment to which their effluent will be submitted is to be considered. It is essential to reconfirm that the direct application of these biological processes to the spent sodas cannot be carried out without the appropriate dilution of them.

4. Conclusions

The characteristics of spent sodas depend on the process where they are formed. High pollutant potential of these residues is caused by high pH and remarkable concentrations of sulfide compounds, mainly as a reduced form like sulfides and mercaptans. Additionally, organic compounds such as petroleum oils, phenols and benzenes, and significant amounts of salts are reported.

Wet air oxidation (WAO) is considered the safest, more resistant, and thoroughly proven technology applied at the laboratory, pilot, and plant level. The application of any of its alternatives at low, medium, and high temperatures, depends mainly on the characteristics of the spent sodas to be treated, i.e., sulfidic, naphthenic, cresylic, or a mixture of them. However, their high investment and operational costs have driven several types of research and specialists to seek other technological alternatives for the treatment of SC.

One of the technological alternatives is the Fenton process applied to refineries, as it is an option to be considered in the second level of applicability. Otherwise, acid neutralization could be applied in some cases due to the simplicity of the system. It brings advantages for low naphthenic acid content in spent sodas, for example they are separated in the process and not destroyed, as in the WAO process. A combination of this method with an advanced oxidation process or WAO at low temperatures should be assessed to replace the WAO that operates at high temperatures.

The application of processes to SC as they are produced is not possible due to its toxic characteristics. However, a combination of the correct degree of dilution and these processes is more environmentally friendly than the physical-chemical processes. Therefore, the alternative of a mixture with other waste should always be considered.

Other treatments of SC have been studied, such as electrochemical processes, precipitation, resins, and ionic liquids; they have several limitations on a real scale. Among these processes, the application of the so-called "Green Solvents" presents a new outlook. Their recovery after the treatment is still a problem to be solved in order to reduce the costs and residues. The electrochemical processes with membranes have also shown very positive aspects regarding their applicability for the treatment of SC. However, there are still technical scenarios to be studied in more detail and depth.

Author Contributions: E.P.-C., S.M. compiled the literature references. Writing—Original draft, S.M. Writing—Review and editing, E.P.-C., C.H., F.C., J.G. All authors have read and agreed to the published version of the manuscript.

Funding: This research received no external funding.

Acknowledgments: This work is dedicated to the memory of our well esteemed and wonderful colleague Dr. Silvio Montalvo Martínez who recently passed away. Thanks to Engineers of ENAP-Aconcagua Refinery, Chile, for their collaboration.

Conflicts of Interest: The authors declare no conflict of interest.

References

1. Abolghasem, K.; Kharaji, A.G.; Mehrabani-Zeinabad, A.; Faizi, V.; Kazemi, J.; Shariati, A. Synergy between two natural gas sweetening processes. *J. Unconv. Oil Gas Resour.* **2016**, *14*, 6–11.
2. Heidarinasab, A.; Hashemi, S.R. A study of biological treatment of spent sulfidic caustic. In Proceedings of the International Conference on Chemical, Ecology and Environmental Sciences (ICCEES'2011), Pattaya, Thailand, 17–18 December 2011.
3. Hashemi, S.R.; Heidarinasab, A. Spent Caustic Bioregeneration by using Thiobacillus denitrificans Bacteria. *World Acad. Sci. Eng. Technol.* **2012**, *67*, 417–419.
4. Hawari, A.; Ramadan, H.; Abu-Reesh, I.; Ouederni, M. A comparative study of the treatment of ethylene plant spent caustic by neutralization and classical and advanced oxidation. *J. Environ. Manag.* **2015**, *151*, 105–112. [CrossRef]
5. De Graaff, M.; Klok, J.B.M.; Bijmans, M.F.M.; Muyzer, G.; Janssen, A.J.H. Application of a 2-step process for the biological treatment of sulfidic spent caustics. *Water Res.* **2012**, *46*, 723–730. [CrossRef]
6. Conner, J.A.; Beitle, R.R.; Duncan, K.; Kolhatkar, R.; Sublette, K.L. Biotreatment of Refinery Spent-Sulfidic Caustic Using an Enrichment Culture Immobilized in a Novel Support Matrix. *Appl. Biochem. Biotechnol.* **2003**, *84*, 707–719.
7. Nasr Esfahani, K.; Farhadian, M.; Solaimany Nazar, A.R. Interaction effects of various reaction parameters on the treatment of sulfidic spent caustic through electro-photo-Fenton. *Int. J. Environ. Sci. Technol.* **2019**, *16*, 7165–7174. [CrossRef]
8. Karimi, A.; Fatehifar, E.; Alizadeh, R.; Ahadzadeh, I. Regeneration and treatment of sulfidic spent caustic using analytic hierarchy process. *Environ. Heal. Eng. Manag.* **2016**, *3*, 203–208. [CrossRef]
9. Lee, J.H.; Park, J.J.; Seo, K.S.; Choi, G.C.; Lee, T.H. Simultaneous autotrophic & heterotrophic denitrification by the injection of reformed spent sulfidic caustic (SSC) in a pilot-scale sewage treatment plant. *Korean J. Chem. Eng.* **2013**, *30*, 139–144.
10. Al Jabari, M. Spent Caustic Treatment Using Advanced Oxidation Processes. Master's Thesis, University of Sharjah, Sharjah, UAE, 2012.
11. Veerabhadraiah, G.; Mallika, N.; Jindal, S. Spent caustic management: Remediation review. *Hydrocarb. Process.* **2011**, *90*, 41–46.
12. Park, J.J.; Byun, I.G.; Park, S.R.; Lee, J.H.; Park, S.H.; Park, T.J.; Lee, T.H. Use of spent sulfidic caustic for autotrophic denitrification in the biological nitrogen removal processes: Lab-scale and pilot-scale experiments. *J. Ind. Eng. Chem.* **2009**, *15*, 316–322. [CrossRef]
13. Alnaizy, R. Economic analysis for wet oxidation processes for the treatment of mixed refinery spent caustic. *Environ. Prog.* **2008**, *27*, 295–301. [CrossRef]
14. Duesel, B.F.; Gibbons, J.P.; Rutsch, M.J. Treatment of Spent Caustic Refinery Effluents. U.S. Patent 7,214,290, 8 May 2007.
15. Sipma, J.; Svitelskaya, A.; Van Der Mark, B.; Hulshoff Pol, L.W.; Lettinga, G.; Buisman, C.J.N.; Janssen, A.J.H. Potentials of biological oxidation processes for the treatment of spent sulfidic caustics containing thiols. *Water Res.* **2004**, *38*, 4331–4340. [CrossRef] [PubMed]

16. Sabri, M.A.; Ibrahim, T.H.; Khamis, M.I.; Nancarrow, P.; Hassan, M.F. Spent caustic treatment using hydrophobic room temperatures ionic liquids. *J. Ind. Eng. Chem.* **2018**, *65*, 325–333. [CrossRef]
17. Ben Hariz, I.; Halleb, A.; Adhoum, N.; Monser, L. Treatment of petroleum refinery sulfidic spent caustic wastes by electrocoagulation. *Sep. Purif. Technol.* **2013**, *107*, 150–157. [CrossRef]
18. Kumfer, B.; Felch, C.; Maugans, C. Wet air oxidation treatment of spent caustic in petroleum refineries. In Proceedings of the National Petroleum Refiners Association Conference, Phoenix, AZ, USA, 21–23 March 2010; Volume 23, pp. 49–67.
19. Ahmad, W. Neutralization of Spent Caustic from LPG Plant at Preem AB Göteborg. Master's Thesis, Chalmers University of Technology, Gothenburg, Sweden, 2010.
20. Huaman, F.; Davila Villar, N. Disposal of Spent Caustic at the Repsol YPF Refinery in La Pampilla, Peru. In Proceedings of the Environmental Conference, Austin, TX, USA, 24–25 September 2007.
21. Ellis, C.E. Wet air oxidation of refinery spent caustic. *Environ. Prog.* **1998**, *17*, 28–30. [CrossRef]
22. Singh, S. Shikha Treatment and Recycling of Wastewater from Oil Refinery/Petroleum Industry. In *Advances in Biological Treatment of Industrial Waste Water and their Recycling for a Sustainable Future*; Springer: Singapore, 2019; pp. 303–332.
23. Tobiszewski, M.; Tsakovski, S.; Simeonov, V.; Namieśnik, J. Chlorinated solvents in a petrochemical wastewater treatment plant: An assessment of their removal using self-organising maps. *Chemosphere* **2012**, *87*, 962–968. [CrossRef]
24. Hasan, D.B.; Abdul Aziz, A.R.; Daud, W.M.A.W. Oxidative mineralisation of petroleum refinery effluent using Fenton-like process. *Chem. Eng. Res. Des.* **2012**, *90*, 298–307. [CrossRef]
25. Park, S.; Seon, J.; Byun, I.; Cho, S.; Park, T.; Lee, T. Comparison of nitrogen removal and microbial distribution in wastewater treatment process under different electron donor conditions. *Bioresour. Technol.* **2010**, *101*, 2988–2995. [CrossRef]
26. Al Zarooni, M.; Elshorbagy, W. Characterization and assessment of Al Ruwais refinery wastewater. *J. Hazard. Mater.* **2006**, *136*, 398–405. [CrossRef]
27. Copa, W.M.; Momont, J.A. Wet air oxidation of energetics and chemical agent surrogates. *J. Energy Mater.* **1995**, *13*, 235–258. [CrossRef]
28. DeAngelo, D.J.; Wilhelmi, A.R. Wet air oxidation of spent caustic liquors. *Chem. Eng. Prog* **1983**, *79*, 3.
29. Copa, W. Wet air oxidation of spent caustics. *Natl. Environ. J.* **1994**, *4*, 16–19.
30. Matthews, R. Performance update: Low pressure wet air oxidation unit at grangemouth, Scotland. *Environ. Prog.* **1997**, *16*, 9–12. [CrossRef]
31. Ellis, C.; Maugans, C.B. WAO treats spent caustic liquor in Asia and Brazil. *Water Wastewater Int.* **2004**, *19*, 35–36.
32. Huaman, F.D.; Villar, N.; Felch, C.; Maugans, C.; Olsen, S. A highly refined effort. *Water Wastewater Int.* **2009**, *24*, 28–32.
33. Chang, C.J.; Lin, J.-C.; Chen, C.-K. Effects of temperature and Cu2+ catalyst on liquid-phase oxidation of industrial wastewaters. *J. Chem. Technol. Biotechnol.* **1993**, *57*, 355–361. [CrossRef]
34. Jagushte, M.V.; Mahajani, V.V. Insight into spent caustic treatment: On wet oxidation of thiosulfate to sulfate. *J. Chem. Technol. Biotechnol.* **1999**, *74*, 437–444. [CrossRef]
35. Debellefontaine, H.; Foussard, J.N. Wet air oxidation for the treatment of industrial wastes. Chemical aspects, reactor design and industrial applications in Europe. *Waste Manag.* **2000**, *20*, 15–25. [CrossRef]
36. Seyedin, S.; Hassanzadeganroudsari, M. Evaluation of the Different Methods of Spent Caustic Treatment. *Int. J. Adv. Res. Sci. Eng. Technol.* **2018**, *5*, 5275–5283.
37. De Haan, S.; Howdeshell, M.; Maugans, C. Update: Spent caustic treatment. *Hydrocarb. Process.* **2010**, *9*, 61.
38. Kim, K.H.; Ihm, S.K. Heterogeneous catalytic wet air oxidation of refractory organic pollutants in industrial wastewaters: A review. *J. Hazard. Mater.* **2011**, *186*, 16–34. [CrossRef]
39. Zhu, W.; Bin, Y.; Li, Z.; Jiang, Z.; Yin, T. Application of catalytic wet air oxidation for the treatment of H-acid manufacturing process wastewater. *Water Res.* **2002**, *36*, 1947–1954. [CrossRef]
40. Janssen, A.J.; Lettinga, G.; Bontsema, J.; Van Straten, G.; Kuenen, J.G.; De Zwart, J.M.M. Biological Treatment of Spent Caustic. U.S. Patent 6,045,695, 4 April 2000.
41. Oliviero, L.; Barbier, J.; Duprez, D.; Guerrero-Ruiz, A.; Bachiller-Baeza, B.; Rodríguez-Ramos, I. Catalytic wet air oxidation of phenol and acrylic acid over Ru/C and Ru-CeO 2 /C catalysts. *Appl. Catal. B Environ.* **2000**, *25*, 267–275. [CrossRef]

42. Lei, L.; Hu, X.; Chu, H.P.; Chen, G.; Yue, P.L. Catalytic wet air oxidation of dyeing and printing wastewater. *Water Sci. Technol.* **1997**, *35*, 311–319. [CrossRef]
43. Núñez, F.; Del Angel, G.; Tzompantzi, F.; Navarrete, J. Catalytic Wet-Air Oxidation of p-Cresol on $Ag/Al_2O_3-ZrO_2$ Catalysts. *Ind. Eng. Chem. Res.* **2011**, *50*, 2495–2500. [CrossRef]
44. Zermeño-Montante, I.; Nieto-Delgado, C.; Sagredo-Puente, R.D.; Cárdenas-Galindo, M.G.; Handy, B.E. Catalytic Wet Air Oxidation of Sodium Sulfide Solutions. Effect of the Metal-Support and Acidity of the Catalysts. *Top. Catal.* **2011**, *54*, 579–586. [CrossRef]
45. Barge, A.S.; Vaidya, P.D. Kinetics of wet air oxidation of sodium sulfide over heterogeneous iron catalyst. *Int. J. Chem. Kinet.* **2019**, *52*, 92–98. [CrossRef]
46. Barge, A.S.; Vaidya, P.D. Wet air oxidation of cresylic spent caustic—A model compound study over graphene oxide (GO) and ruthenium/GO catalysts. *J. Environ. Manag.* **2018**, *212*, 479–489. [CrossRef]
47. Sheu, S.H.; Weng, H.S. Treatment of olefin plant spent caustic by combination of neutralization and fenton reaction. *Water Res.* **2001**, *35*, 2017–2021. [CrossRef]
48. Wang, S.; Liu, X.; Zhang, M. Reduction of Ammineruthenium(III) by Sulfide Enables In Vivo Electrochemical Monitoring of Free Endogenous Hydrogen Sulfide. *Anal. Chem.* **2017**, *89*, 5382–5388. [CrossRef]
49. Nuñez, P.; Hansen, H.K.; Rodriguez, N.; Guzman, J.; Gutierrez, C. Electrochemical Generation of Fenton's Reagent to Treat Spent Caustic Wastewater. *Sep. Sci. Technol.* **2009**, *44*, 2223–2233. [CrossRef]
50. Oprea, F.; Fendu, E.-M.; Nicolae, M.; Pantea, O.; Dunka, M. Experimental studies concerning treatment of the spent caustic solutions from refineries MEROX and EXOMER units. *Rev. Chim.* **2010**, *61*, 608–661.
51. Poyatos, J.M.; Muñio, M.M.; Almecija, M.C.; Torres, J.C.; Hontoria, E.; Osorio, F. Advanced oxidation processes for wastewater treatment: State of the art. *Water Air Soil Pollut.* **2010**, *205*, 187–204. [CrossRef]
52. Oller, I.; Malato, S.; Sánchez-Pérez, J.A. Combination of Advanced Oxidation Processes and biological treatments for wastewater decontamination—A review. *Sci. Total Environ.* **2011**, *409*, 4141–4166. [CrossRef]
53. Shon, H.K.; Vigneswaran, S.; Snyder, S.A. Effluent organic matter (EfOM) in wastewater: Constituents, effects, and treatment. *Crit. Rev. Environ. Sci. Technol.* **2006**, *36*, 327–374. [CrossRef]
54. Andreozzi, R.; Caprio, V.; Insola, A.; Marotta, R. Advanced oxidation processes (AOP) for water purification and recovery. *Catal. Today* **1999**, *53*, 51–59. [CrossRef]
55. Mabury, S. Hydroxyl Radical in Natural Waters. Ph.D. Thesis, University of California, Davis, CA, USA, 1993.
56. Gurol, M.D.; Vatistas, R. Oxidation of phenolic compounds by ozone and ozone + u.v. radiation: A comparative study. *Water Res.* **1987**, *21*, 895–900. [CrossRef]
57. Yu, Z.-Z.; Sun, D.-Z.; Li, C.-H.; Shi, P.-F.; Duan, X.-D.; Sun, G.-R.; Liu, J.-X. UV-catalytic treatment of spent caustic from ethene plant with hydrogen peroxide and ozone oxidation. *J. Environ. Sci.* **2004**, *16*, 272–275.
58. Alaton, I.A.; Balcioglu, I.A.; Bahnemann, D.W. Advanced oxidation of a reactive dyebath effluent: Comparison of O_3, H_2O_2/UV-C and TiO_2/UV-A processes. *Water Res.* **2002**, *36*, 1143–1154. [CrossRef]
59. Gottschalk, C.; Libra, J.A.; Saupe, A. *Ozonation of Water and Waste Water: A Practical Guide to Understanding Ozone and Its Applications*; John Wiley & Sons: Hoboken, NJ, USA, 2008; ISBN 9783527613342.
60. Hernández, F.; Geissler, G. Photooxidative Treatment of Sulfurous Water for Its Potabilization. *Photochem. Photobiol.* **2005**, *81*, 636.
61. Glaze, W.H.; Kang, J.W.; Chapin, D.H. The chemistry of water treatment processes involving ozone, hydrogen peroxide and ultraviolet radiation. *Ozone Sci. Eng.* **1987**, *9*, 335–352. [CrossRef]
62. Yonar, T.; Yonar, G.K.; Kestioglu, K.; Azbar, N. Decolorisation of textile effluent using homogeneous photochemical oxidation processes. *Color. Technol.* **2005**, *121*, 258–264. [CrossRef]
63. Shemer, H.; Narkis, N. Trihalomethanes aqueous solutions sono-oxidation. *Water Res.* **2005**, *39*, 2704–2710. [CrossRef]
64. Fan, X.; Hao, H.; Shen, X.; Chen, F.; Zhang, J. Removal and degradation pathway study of sulfasalazine with Fenton-like reaction. *J. Hazard. Mater.* **2011**, *190*, 493–500. [CrossRef]
65. Neyens, E.; Baeyens, J. A review of classic Fenton's peroxidation as an advanced oxidation technique. *J. Hazard. Mater.* **2003**, *98*, 33–50. [CrossRef]
66. Bianco, B.; De Michelis, I.; Vegliò, F. Fenton treatment of complex industrial wastewater: Optimization of process conditions by surface response method. *J. Hazard. Mater.* **2011**, *186*, 1733–1738. [CrossRef]
67. Isarain-Chávez, E.; Rodríguez, R.M.; Garrido, J.A.; Arias, C.; Centellas, F.; Cabot, P.L.; Brillas, E. Degradation of the beta-blocker propranolol by electrochemical advanced oxidation processes based on Fenton's reaction chemistry using a boron-doped diamond anode. *Electrochim. Acta* **2010**, *56*, 215–221. [CrossRef]

68. Lipczynska-Kochany, E.; Sprah, G.; Harms, S. Influence of some groundwater and surface waters constituents on the degradation of 4-chlorophenol by the Fenton reaction. *Chemosphere* **1995**, *30*, 9–20. [CrossRef]
69. Vilve, M.; Hirvonen, A.; Sillanpää, M. Effects of reaction conditions on nuclear laundry water treatment in Fenton process. *J. Hazard. Mater.* **2009**, *164*, 1468–1473. [CrossRef]
70. Kulik, N.; Trapido, M.; Goi, A.; Veressinina, Y.; Munter, R. Combined chemical treatment of pharmaceutical effluents from medical ointment production. *Chemosphere* **2008**, *70*, 1525–1531. [CrossRef]
71. Kang, Y.W.; Hwang, K.-Y. Effects of reaction conditions on the oxidation efficiency in the Fenton process. *Water Res.* **2000**, *34*, 2786–2790. [CrossRef]
72. El-Naas, M.H.; Al-Zuhair, S.; Al-Lobaney, A.; Makhlouf, S. Assessment of electrocoagulation for the treatment of petroleum refinery wastewater. *J. Environ. Manag.* **2009**, *91*, 180–185. [CrossRef]
73. Tezcan Un, U.; Koparal, A.S.; Bakir Ogutveren, U. Electrocoagulation of vegetable oil refinery wastewater using aluminum electrodes. *J. Environ. Manag.* **2009**, *90*, 428–433. [CrossRef]
74. Vepsalainen, M.; Selin, J.; Rantala, P.; Pulliainen, M.; Sarkka, H.; Kuhmonen, K.; Bhatnagar, A.; Sillanpaa, M. Precipitation of dissolved sulphide in pulp and paper mill wastewater by electrocoagulation. *Environ. Technol.* **2011**, *32*, 1393–1400. [CrossRef]
75. Asselin, M.; Drogui, P.; Brar, S.K.; Benmoussa, H.; Blais, J.-F. Organics removal in oily bilgewater by electrocoagulation process. *J. Hazard. Mater.* **2008**, *151*, 446–455. [CrossRef]
76. Moreno-Casillas, H.A.; Cocke, D.L.; Gomes, J.A.G.; Morkovsky, P.; Parga, J.R.; Peterson, E. Electrocoagulation mechanism for COD removal. *Sep. Purif. Technol.* **2007**, *56*, 204–211. [CrossRef]
77. Kabay, N.; Demircioglu, M.; Ersöz, E.; Kurucaovali, I. Removal of calcium and magnesium hardness by electrodialysis. *Desalination* **2002**, *149*, 343–349. [CrossRef]
78. Kumar, M.; Tripathi, B.P.; Saxena, A.; Shahi, V.K. Electrochemical membrane reactor: Synthesis of quaternary ammonium hydroxide from its halide by in situ ion substitution. *Electrochim. Acta* **2009**, *54*, 1630–1637. [CrossRef]
79. Feng, H.; Huang, C.; Xu, T. Production of Tetramethyl Ammonium Hydroxide Using Bipolar Membrane Electrodialysis. *Ind. Eng. Chem. Res.* **2008**, *47*, 7552–7557. [CrossRef]
80. Wei, Y.; Wang, Y.; Zhang, X.; Xu, T. Comparative study on regenerating sodium hydroxide from the spent caustic by bipolar membrane electrodialysis (BMED) and electro-electrodialysis (EED). *Sep. Purif. Technol.* **2013**, *118*, 1–5. [CrossRef]
81. Kubo, D.; Kawase, Y. Hydroxyl radical generation in electro-Fenton process with in situ electro-chemical production of Fenton reagents by gas-diffusion-electrode cathode and sacrificial iron anode. *J. Clean. Prod.* **2018**, *203*, 685–695. [CrossRef]
82. Davarnejad, R.; Bakhshandeh, M. Olefin plant spent caustic wastewater treatment using electro-Fenton technique. *Egypt. J. Pet.* **2018**, *27*, 573–581. [CrossRef]
83. Guerrero, L.; Montalvo, S.; Huiliñir, C.; Campos, J.L.; Barahona, A.; Borja, R. Advances in the biological removal of sulphides from aqueous phase in anaerobic processes: A review. *Environ. Rev.* **2015**, *24*, 84–100. [CrossRef]
84. Sawasdee, V.; Pisutpaisal, N. Simultaneous pollution treatment and electricity generation of tannery wastewater in air-cathode single chamber MFC. *Int. J. Hydrogen Energy* **2016**, *41*, 15632–15637. [CrossRef]
85. Logan, B.E.; Rabaey, K. Conversion of wastes into bioelectricity and chemicals by using microbial electrochemical technologies. *Science* **2012**, *337*, 686–690. [CrossRef]
86. Fazli, N.; Mutamim, N.S.A.; Jafri, N.M.A.; Ramli, N.A.M. Microbial Fuel Cell (MFC) in treating spent caustic wastewater: Varies in Hydraulic Retention Time (HRT) and Mixed Liquor Suspended Solid (MLSS). *J. Environ. Chem. Eng.* **2018**, *6*, 4339–4346. [CrossRef]
87. Keramati, N.; Moheb, A.; Ehsani, M.R. Effect of operating parameters on NaOH recovery from waste stream of Merox tower using membrane systems: Electrodialysis and electrodeionization processes. *Desalination* **2010**, *259*, 97–102. [CrossRef]
88. Delgado Vela, J.; Dick, G.J.; Love, N.G. Sulfide inhibition of nitrite oxidation in activated sludge depends on microbial community composition. *Water Res.* **2018**, *138*, 241–249. [CrossRef]
89. De Graaff, M.; Bijmans, M.F.M.; Abbas, B.; Euverink, G.-J.W.; Muyzer, G.; Janssen, A.J.H. Biological treatment of refinery spent caustics under halo-alkaline conditions. *Bioresour. Technol.* **2011**, *102*, 7257–7264. [CrossRef]
90. Rajganesh, B.; Sublette, K.L.; Camp, C.; Richardson, M.R. Biotreatment of Refinery Spent Sulfidic Caustics. *Biotechnol. Prog.* **1995**, *11*, 228–230. [CrossRef]

91. Gholipour, S.; Mehrkesh, P.; Azin, E.; Nouri, H.; Rouhollahi, A.A.; Moghimi, H. Biological treatment of toxic refinery spent sulfidic caustic at low dilution by sulfur-oxidizing fungi. *J. Environ. Chem. Eng.* **2018**, *6*, 2762–2767. [CrossRef]
92. Welton, T. Room-Temperature Ionic Liquids. Solvents for Synthesis and Catalysis. *Chem. Rev.* **1999**, *99*, 2071–2084. [CrossRef] [PubMed]
93. Abulhassani, J.; Manzoori, J.L.; Amjadi, M. Hollow fiber based-liquid phase microextraction using ionic liquid solvent for preconcentration of lead and nickel from environmental and biological samples prior to determination by electrothermal atomic absorption spectrometry. *J. Hazard. Mater.* **2010**, *176*, 481–486. [CrossRef] [PubMed]
94. Paulino, J.F.; Afonso, J.C. New strategies for treatment and reuse of spent sulfidic caustic stream from petroleum industry. *Química Nov.* **2012**, *35*, 1447–1452. [CrossRef]

processes

MDPI

Review

Applications of Electrolyzed Water as a Sanitizer in the Food and Animal-By Products Industry

Juan C. Ramírez Orejel [1] and José A. Cano-Buendía [2,*]

1 Facultad de Medicina Veterinaria y Zootecnia, Department of Animal Nutrition and Biochemistry, Universidad Nacional Autónoma de Mexico (UNAM), Ciudad Universitaria, Mexico D.F. 04510, Mexico; jrorejel@unam.mx
2 Facultad de Medicina Veterinaria y Zootecnia, Department of Microbiology and Immunology, Universidad Nacional Autónoma de Mexico (UNAM), Ciudad Universitaria, Mexico D.F. 04510, Mexico
* Correspondence: jcano@unam.mx; Tel.: +52-(55)-5622-5900

Received: 31 March 2020; Accepted: 29 April 2020; Published: 2 May 2020

Abstract: Food demand is increasing every year and, usually animal-derived products are generated far from consumer-places. New technologies are being developed to preserve quality characteristics during processing and transportation. One of them is electrolyzed water (EW) that helps to avoid or decrease the development of foodborne pathogens, or losses by related bacteria. Initially, EW was used in ready-to-eat foods such as spinach, lettuce, strawberries, among others; however, its application in other products is under study. Every product has unique characteristics that require an optimized application of EW. Different sanitizers have been developed; unfortunately, they could have undesirable effects like deterioration of quality or alterations in sensory properties. Therefore, EW is gaining popularity in the food industry due to its characteristics: easy application and storage, no corrosion of work surfaces, absence of mucosal membrane irritation in workers handling food, and it is considered environmentally friendly. This review highlights the advantages of using EW in animal products like chicken, pork, beef, eggs and fish to preserve their safety and quality.

Keywords: electrolyzed water; foodborne pathogens; sanitization

1. Introduction

The human population is continuously growing [1] and, consequently, food demand is also increasing. Animal-based protein plays an important role in the human diet. However, animal products are very susceptible to contamination by foodborne pathogens like *Escherichia coli* [2], *Salmonella* [3,4], *Listeria monocytogenes* [5,6], *Campylobacter jejuni* [7], and so forth which are related with processing plants, slaughterhouses and outbreaks in different countries.

Food can become contaminated during processing, transportation or storage, causing losses of 75% in developing countries [8]. The abundance of biological macromolecules (carbohydrates, proteins, lipids, nucleic acids) offers a perfect environment for the development of different pathogens that can spoil and lead to different diseases, if contaminated food is consumed.

Safety is an important concern for consumers and the food industry, hence different technologies have been developed to preserve quality. The food market is demanding products with less processing and fewer alterations in the organoleptic properties. One alternative is the use of electrolyzed water, considered a non-thermal chemical technology [9].

2. Electrolyzed Water

Electrolyzed water (EW) is a sanitizer that contains mostly hypochlorous acid (HOCl), which is responsible for the bactericidal effect [10]. It is gaining popularity because of its physical and chemical properties. EW manufacturing requires NaCl and water; it can be applied in different fields. Initially, it was used to disinfect medical supplies (e.g., dialyzers) [11]; afterwards, different applications were reported like disinfection of ready-to-eat foods (fruits and vegetables), where it helps to control food contamination and microbial spoilage, as well as improving safety and shelf life, without affecting organoleptic properties [12]. Different antimicrobial effect against viruses [13,14], bacteria, and toxins [15–17] through short periods of exposure (5 to 20 s) have been reported. EW has many advantages; the most important is that it is considered environmentally friendly; after reaction with bacteria and organic material, it reverts to water and salt and, because EW kills bacteria physically, accordingly, they do not generate resistance [18].

EW is generated by an electrolysis process of saline solution (NaCl or KCl) contained within an electrolytic chamber with positively and negatively charged electrodes, with or without a membrane [19]. During electrolysis, NaCl dissociates into Na^+ and Cl^- ions and water into OH^- and H^+. Negatively charged ions (Cl^- and OH^-) are attracted to the anode, where these reagents are oxidized and generate hypochlorous acid, hypochlorite ion, hydrochloric acid, oxygen gas, and chlorine gas; positive ions (Na^+ and H^+) are attracted to the cathode and reduced, producing hydrogen and sodium hydroxide (Figure 1).

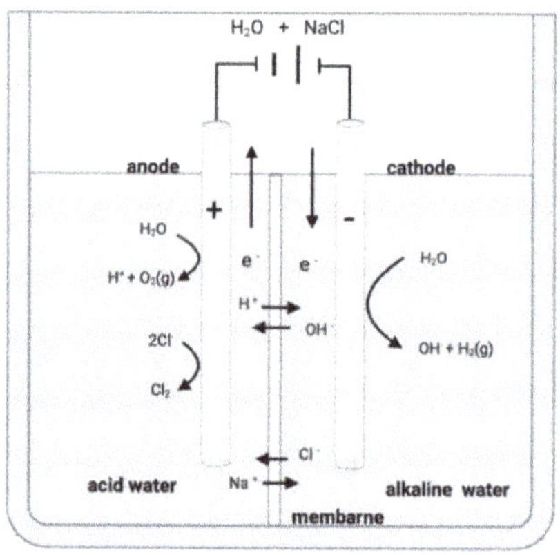

Figure 1. Schematic diagram of electrolyzed water (EW) generation.

There are many factors that affect the generation of all the reagent species [20]. The main species that provide bactericidal characteristics are hypochlorous acid (HOCl) and hypochlorite ions ($^-$OCl); it has been reported that concentrations of both species are related with pH [10] and oxidation-reduction potential (ORP), which is related to its ability to be oxidized or reduced, generating its bactericidal effect. Loss of bacterial membrane integrity increases permeability and generates bacterial lysis as well as DNA destruction [21].

Electrolysis produces three types of EW [20] (Table 1). Acidic EW (AEW), pH (2.3 to 2.7), and ORP (>1000 mV) is produced at the anode side. Basic EW (BEW), pH (10 to 11.5), and ORP (800 to 900 mV) is generated at the cathode side. Neutral EW (NEW) is produced using different protocols [9]. It could be generated when the electrolytic cell does not have a separative membrane [20] or by mixing the catholyte with a diluted NaCl solution [22,23]. It has been reported that NEW is more stable and keeps its antibacterial activity after storage in comparison to AEW and BEW [24,25].

Table 1. Properties of different types of electrolyzed water.

Type of Electrolyzed Water	pH	ORP [a] (mV)
Acidic electrolyzed water	2–3	>1100
Basic electrolyzed water	10–13	−800 to −900
Neutral electrolyzed water	6.5–7.5	700 to 800

[a] Oxidation reduction potential.

The bactericidal effect of EW is a combination of different activities; however, all of them focus on the loss of bacterial membrane integrity. Some reports show this effect against different foodborne pathogens [26,27]. AEW has strong antibacterial activity, which is due to high ORP and low pH. HOCl can penetrate bacterial membranes [28] and oxidizes proteins involved in important metabolic pathways [29]. It can damage bacterial genetic material [20]; however, antibacterial effect by free chlorine decreases with increasing pH [10]. It has been reported that HOCl is produced by phagocytic cells through the oxidative burst pathway, also producing a hydroxyl radical that acts on different pathogens [10,30]. BEW has a pH higher than 11, and its bactericidal effect has been reported to be caused by a strong ORP which reduces free bacterial radicals [18]. ORP causes modifications in metabolic flux and ATP production; it inhibits glucose oxidation; disrupts protein synthesis; and inhibits oxygen uptake and oxidative phosphorylation, which is coupled with the leakage of some macromolecules [31] and damage to bacterial membranes [32]. NEW has been described as the less corrosive EW, and it can be stored longer than AEW [33]. In general, all active chlorine forms (Cl_2, HOCl and $^{-}$OCl) damage the outer bacterial membrane, allowing HOCl to penetrate bacteria, and oxidize proteins and enzymes involved in metabolic processes (e.g., phosphate acetyltransferase-acetate kinase, ribose-5-phosphate, and others) [34,35] (Figure 2).

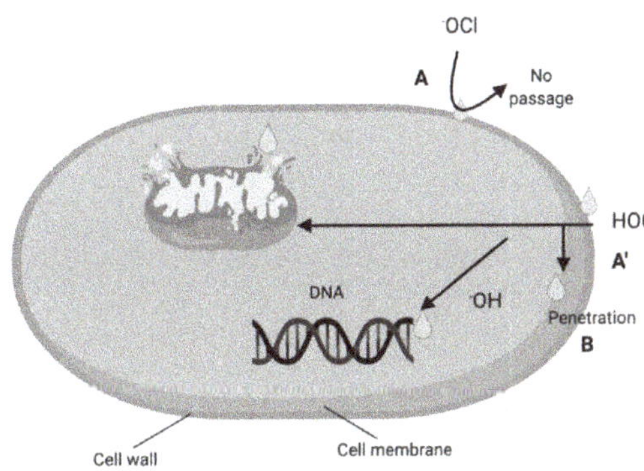

Figure 2. EW bactericidal activity. The hypochlorite ion ($^{-}$OCl) cannot penetrate bacterial membrane A. Hypochlorous acid (HOCl) can diffuse through bacterial membrane and disrupt outer membrane A′, internal membrane B, and bacterial proteins C. Alteration of bacterial DNA has been reported (Adapted from Rahman [20]).

The use of EW has shown many advantages against other disinfectants; however, some reported limitations are short lifespan; AEW is corrosive [36] to metal [37] and leaves a salt residue on products affecting texture and taste. NEW has not shown these limitations [36]. Len et al. [30] have reported that EW is sensitive to light and should be stored in a closed container.

The use of EW is gaining popularity. In this review, we will put an emphasis on EW use on different animal meats and sub-products to maintain food safety and avoid spoilage.

3. Pork

Pork is a very popular food around the world and its consumption represents about 40% of the global amount, compared to other animal meats [38]. One of the main goals in porciculture is preservation of freshness. To maintain this characteristic, many enhancers are used to improve palatability [39] and preserve quality.

In the following evaluations, most have been made using AEW (Table 2). Fabrizio [40] evaluated the treatment of pork belly artificially inoculated with 8 mL of manure with *Listeria monocytogenes, Salmonella typhimurium* and *Campylobacter coli*. One modification was the evaluation of aged AEW to identify any difference with fresh AEW. EW effect was compared with the bactericidal effect of lactic acid, chlorinated water, and plain water. There was no significant difference between treatments, the only difference was between the use of treatment and no treatment with respect to total viable counts (TVC): *E. coli, Salmonella typhimurium, Listeria monocytogenes* and total coliform survival numbers after zero, two, and seven days post treatment. In the case of *Campylobacter coli*, AEW showed similar patterns as described previously at days zero and two, no significant difference was detected at day 7. This work showed that 15 s spray treatment can eliminate bacterial contamination; however, for other pathogens, it is necessary to increase contact time over 10 min. In a follow up of this study by the same researchers, they showed the effect of AEW and BEW on frankfurters and ham, given their popularity as a food type. Dipping versus spraying treatments were compared using *Listeria monocytogenes, Salmonella typhymurium* and *Campylobacter coli* as contaminants [41]. AEW decreased the *L. monocytogenes* population at days zero and seven; however, there was no significant difference between treatments when they were evaluated against aerobic mesophilic bacteria. AEW did not change bacterial counts on frankfurters after seven days of treatment. Treated ham showed a slight decrease in the *L. monocytogenes* survival population after treatments (~0.5 $\log_{10}$ CFU/g) at days zero, three and seven; AEW did not affect the color of ham. Once more, this group confirmed that 15 s exposure time is enough to eliminate *Campylobacter sp* but it is not enough time to eliminate other pathogens attached to the surface of pork.

Mansur et al. [42] evaluated the use of AEW and slightly AEW (SAEW) (pH 6.29) on 10 g of fresh pork loins. As a novelty, they evaluated the combination of SAEW with 0.5% fumaric acid (FA). Pork samples were contaminated with 100 μL of a bacterial cocktail containing 8 Log CFU/mL of *E. coli O157:H7, Listeria monocytogenes, Staphylococcus aureus* and *Salmonella Typhimurium*. Meat was stored at 4 °C or 10 °C. Samples were dipped in evaluated solutions for 3 min. There was no significant difference between AEW and SAEW treatments. This result suggested that the bactericidal effect of EW is due to the concentration of available chlorine rather than pH and ORP. The combination of FA with SAEW yielded higher reductions (≥2.5 log CFU/g) against all evaluated pathogens compared to all treatments. At the same time, this combination increased shelf life in pork by five to six days.

Other research group [43] evaluated the effect of sprayed AEW on pork loins at different pressures and time of treatments. AEW decreased lactic acid-producing bacteria at day one after treatment. No differences were reported on the rest of the sampling days. Additionally, AEW decreased mesophilic bacterial counts at day one but not at days 15 and 29. When BEW was evaluated in combination with AEW or SAEW, the combination worked much better against mesophilic and psychrotrophic bacteria. EW did not affect pH, ORP or the red color of pork, and EW did not accelerate lipid oxidation. However, AEW and SAEW oxidize pork protein shortly after application.

Table 2. Evaluations of electrolyzed water in pork.

Material	Type of Electrolyzed Water	Concentration of EW (ppm)	Microorganisms	Inoculum Concentration [b]	Type of Treatment	Duration of Treatment	Reference
Pork belly	AEW, AEW + lactic acid	50	*L. monocytogenes* *Salmonella* Typhimurium *Campylobacter coli*	7 log CFU/mL	Spray	15 s	[40]
Frankfurters Ham	AEW and BEW	50	*L. monocytogenes*	5 log CFU/mL	Spray, dip	15 s	[41]
Fresh pork loin	AEW + fumaric acid	30	*E. coli* *L. monocytogenes* *Staph. Aureus* *Salmonella sp*	8 log CFU/mL	Dip	5 min	[42]
Carcass	low concentration EW (NEW)	10	*E. coli O157:H7*	9 log CFU/mL	Dip	5 min	[38]
Pork loin	AEW and slight AEW	74 51	*L. monocytogenes* Mesophilic and psycrotrophs	4.12 log10 CFU g^{-1}	Spray	20–40 s	[43]
Pork loin	BEW	ND [a]	ND	ND	Injection	15 min	[44]
Pork loin	NEW	16.6	Aerobic bacteria Psychrotrophs Yeast and moulds	0.55 to 0.57 log CFU/cm^{-2} 0.49 to 0.54 log CFU/cm^{-2} 0.52 to 0.60 log CFU/cm^{-2}	Spray	120 s	[45]

[a] Non determined; [b] For mesophilic, psycchrotrophs, yeast and mould, inoculum concentration values were obtained from the no-treatment group.

EW has been tested as an enhancer [44] to improve water holding capacity. However, AEW did not improve pork tenderness or sensorial characteristics.

In 2013, a new EW was evaluated [38]. It was called low concentration EW (pH 6.8), but it can be considered a form of NEW. This solution was evaluated with 3% calcium lactate (CaL) on contaminated carcasses. Treatments were performed by immersion at room temperature for 5 min. The greatest bactericidal reduction (2.2 log CFU/g) was achieved with a combination of the EW with CaL. Shelf life was increased from six to 12 days with treatment. The combination of EW with CaL did not alter pH during pork storage at 4 °C. The combination retarded microbial growth during storage and reduced surface microbial counts. Another research group [45] evaluated NEW at different concentrations on pork (*Longissimus thoracis*). They evaluated different concentrations, the highest of which (0.1%) inactivated the highest number of bacteria on pork surfaces at days zero and seven. Previously, we described how Fabrizio [40] reported that 15 s exposure time reduced Campylobacter counts. However, this evaluation showed that treatment affected total haem pigment (THP) content and the variability of myoglobin forms in pork. Treatment decreased THP values by 8.3% and 14.2% at days zero and seven of storage, respectively. These results mean that the bright red fresh color was lost (discoloration). This effect could be due to the available chlorine concentration because available chlorine oxidized myoglobin into metmyoglobin. Treatment also affected *L* value from CIELab space. They reported an effect of lowest *L* value when NaCl concentration decreased and an opposite effect was observed on *a* value. However, these color changes were not significant for consumers after seven days of treatment. Treated pork had the best sensory scores (color, odor, and appearance).

4. Fish

Fish is an important food worldwide because of its nutritional components [46]; this characteristic increases its demand in wealthy countries. This could be the reason why this product has the higher number of studies with EW (Table 3), especially with AEW. All the cited studies reported two types of treatments: immersion in solution and immersion in ice. This could be due to the way fish is produced/captured. In this document, we divided treatment method by fish type.

Table 3. Evaluations of electrolyzed water in fish.

Material	Type of Electrolyzed Water	Concentration of EW (ppm)	Microorganisms	Inoculum Concentration[b]	Type of Treatment	Duration of Treatment	Reference
Salmon	**AEW and**	**76.9**	*E. coli O157:H7*	**8.7 log CFU/mL**	**Immersion**	**64 min**	[47]
	BEW	ND	*L. monocytogenes*				
Salmon, mahi mahi (*Coryphaena hippurus*)	AEW	50	*L. monocytogenes* *Morganella morganii*	4.47 log CFU/g 4.02 log CFU/g	Immersion	5 min	[48]
Salmon	AEW	65	*L. monocytogenes*	6 log CFU/mL	Immersion	5 min	[49]
			Aerobic bacteria	4.2 to 5.9 log CFU/g			
			Coliforms	2.8 to 4 log CFU/g			
			Yeast and moulds	1.3 to 2.9 log CFU/g			
Salmon	AEW, NEW	60	*L. monocytogenes*	7.70 log CFU/g	Immersion	10 min	[50]
Smoked salmon	AEW	60	*L. monocytogenes*	8.48 log CFU/mL	Immersion	10 min	[51]
Salmon, tuna fish skin	AEW	100	*Enterobacter aerogenes* *Enterobacter cloacae* *Klebsiella pneumoniae* *Morganella morganii* *Proteus hauser*	8 to 9 log CFU/mL	Soaking in ice	120 min to 24 h	[52]
Tuna	AEW + CO gas	10, 50 and 100	Aerobic bacteria	3.14 log CFU/g	Immersion	5 min	[53]
Tuna	AEW	41	Aerobic bacteria	< 3 log CFU/mL	Immersion	15 min	[54]
Catfish	AEW	300	*L. monocytogenes* *Salmonella spp*	5 log CFU/g	Wash	3 min	[55]
Catfish	BEW + polyphosphate	ND[a]	ND	ND	Immersion	2 h	[56]
Tilapia	AEW	120	*E. coli* *Vibrio parahaemolyticus*	8 log CFU/mL	Immersion	10 min	[57]
Tilapia	NEW + PROSAN	150	*Listeria innocua* *E. coli K12* *Pseudomona putida*	6 to 7 log CFU/g	Soaking in ice	72 h	[58]
Carp	BAE	0.87	Aerobic bacteria	6 log CFU/mL	Immersion	15 min	[59]
	AEW	40.8					
Pacific saury (*Cololabis saira*)	weak AEW	34.2 to 47.2	Aerobic bacteria Psychrotrophic bacteria	3 log CFU/g	Soaking in ice	30 days	[60]
Trout	AEW	38	Aerobic bacteria	9 log CFU/mL	Immersion	5 to 10 min	[61]
American shad (*Alosa sapidissima*)	AEW + chitosan	70 to 80	Aerobic bacteria	3.71 to 3.94 log CFU/g	Immersion	15 min	[62]
Bombay duck (*Harpadon nehereus*)	slightly AEW +ebony-bamboo leaves complex extracts	27.37	Aerobic bacteria	1.5 log CFU/g	Immersion	5 min	[63]
Shrimp	AEW	66	*V. parahaemolyticus*	9 log CFU/mL	Immersion	2.5 min	[64]
Shrimp	AEW	44	Aerobic bacteria	6.04 log CFU/g	Soaking in ice	7 days	[65]
Oyster	AEW	30	*V. parahaemolyticus* *Vibrio vulnificus*	8.94 log CFU/mL	Immersion	4 to 6 h	[66]
Clams and mussels	AEW	20	*E. coli* O104:H4	9 log CFU/mL	Immersion	1 to 2 h	[67]
	BEW	10	*L. monocytogenes* *V. parahaemolyticus* *Aeromonas hydrophila*				

[a] Non determined; [b] For aerobic, coliforms, yeast and mould, inoculum concentration values were obtained from the no-treatment group.

4.1. Salmon

This fish type has high economical value. Ozer evaluated the use of AEW or a bitreatment of BEW + AEW by immersion at different times and temperatures against *E. coli* and *L. monocytogenes*. AEW decrease bacterial counts after treatments at 35 °C and 64 min; furthermore, the use of BEW did not increase the anti-microbial effect [47]. Another study evaluated the use of AEW against various strains of *L. monocytogenes* and *Morganella morganii* on artificially contaminated salmon and mahi-mahi fish (*Coryphaena hippurus*); here, in contrast, EW treatment did not reduce bacterial counts on fish [48].

Miks-Krajnik [49] performed a similar study where EW treatment was combined with ultrasound (US), UV, or US+UV. Treatments were evaluated against *L. monocytogenes* and natural microbiota. Their results showed that US+UV and US+UV+AEW caused bacterial reduction in salmon.

In another study, they use of AEW or AEW+NEW was evaluated on raw Atlantic salmon fillets contaminated with *L. monocytogenes*. Results showed that NEW has better bactericidal activity than AEW. NEW caused reductions of 5.6 log CFU/g after a 10 min treatment at 65 °C. An important characteristic was that NEW has less negative impact on salmon protein than AEW; this could be due to the pH of each solution. This group detected that temperature and time of contact are important factors that affect bactericidal activity of EW [50]. Cold smoked salmon is very popular; therefore AEW was evaluated with this food; it was inoculated with *L. monocytogenes* at different temperatures and contact times. A different research group [51] concluded that treatment at 40 °C for 10 min, before curing, can reduce bacterial numbers by 3.0 log/g without affecting sensorial properties. If treatment was performed at different temperatures, bactericidal effect can decrease to 1.5 to 2.0 log/g.

A different approach was used when AEW was evaluated during ice treatment against histamine-producing bacteria found on Atlantic salmon and yellowfin tuna fish skins [52]. Fish skin was soaked in 50 ppm AEW and *Enterobacter cloacae*, *Kebsiela pneumoniae* and *Proteus hauseri* did not survive after treatment; however, *Enterobacter aerogenes* and *Morganella morganii* did survive. In a second experiment, researchers increased concentration and exposure time. They evaluated EW in ice solution at 50 and 100 ppm during different exposure times against *E. aerogenes* and *M. morganii*. *M. morganii* counts were reduced after 6 h of treatment by 0.91 and 1.4 log CFU/cm^2 using 50 or 100 ppm treatment solutions, respectively. *E. aerogenes* showed better reduction counts (1.27 and 1.62 log CFU/cm^2) using 50 and 100 ppm treatments after 24 h, respectively. Results were similar when experiments were performed with yellowfin tuna skin.

4.2. Tuna Fish

Tuna is used to prepare sashimi, and to preserve the quality of this fish, tuna was soaked using different concentrations of AEW in combination with CO_2 treatment. AEW at 50 ppm with CO_2 showed the best antibacterial effect because aerobic bacterial counts were lower compared to the other treatment groups and total volatile basic nitrogen (TVB-N) values were the lowest during storage. This combination did not affect tuna color [53]. In a different study [54], AEW was combined with an edible solution containing 0.5% eugenol and 0.5% linalool (essential oils). TVB-N values were lower in treated tuna after 20 days of storage and treatment extended shelf life of semi fried tuna fillets.

4.3. Catfish

Near neutral EW (pH 6.4) was evaluated on catfish fillets at 300 ppm. A solution was used to wash fillets, but treatment did not reduce *L. monocytogenes* counts. However, near neutral EW treatment 3 min treatment showed a bactericidal effect against *Salmonella sp* and this reduction was maintained through 13 days after treatment [55]. In another study [56], BEW (pH 11.6) was combined with polyphosphate which has properties such as moisture retention, oxidation prevention, and cryoprotectant, and it extends shelf life. BEW alone or in combination were used to treat catfish fillets before the freezing process. Both treatments helped catfish muscle to retain water after 90 days of storage; this effect was not detected when no treatment or tap water was used. BEW treatment helped reduce water loss during cooking after freezer storage. BEW + triphosphate did not affect color and caused lower lipid oxidation (TBARS) production in comparison to no treatment or tap water treatment. The combination of BEW with polyphosphate improved weight gain, moisture retention, and oxidation resistance after freezer storage.

4.4. Tilapia

Tilapia fillets are gaining popularity, and its production is cheaper when compared to other types of fish. To study the bactericidal effect on tilapia, fillets were artificially contaminated with *Escherichia coli* and *Vibrio parahaemolyticus*. Fish were soaked for 10 min with AEW. Treatment caused reduction values of 0.7 and 1.5 log CFU/cm2 for *E. coli* and *V. parahaemolyticus*, respectively. This study established that agitation helps EW to react with bacterial cells more efficiently [57]. Feliciano evaluated the combination of NEW and an organic acidic sanitizer PRO-SAN [58], both as ice flakes on filleted tilapia fish. Fillets were inoculated with *Listeria innocua*, *E. coli* K-12, and *Pseudomona putida*. The initial finding was that tap water ice melted faster than NEW- and PRO-SAN-treated ice. When fish were kept in ice, bacterial numbers were smaller for *E. coli* and *P. putida*; however, no significant difference was detected. In the case of *L. innocua* there was no difference found. Water collected from treated melted ice showed lower bacterial counts than non-treated ice. Authors concluded that EW has the potential to reduce bacterial load when it is used as ice; however, results showed no significant reduction.

4.5. Other Types of Fish

AEW (pH 2.22) and BEW (pH 11.6) were evaluated on carp fillets [59]. Fillets were dipped in EW and 16 pure cultures of aerobic bacteria were used to evaluate treatment. The use of both solutions caused important bacterial reductions; however, under AEW treatment, bacterial counts were below the detection limit (10^2 CFU/mL).

Weak AEW (pH 5.1) was evaluated as ice on pacific saury fish (*Cololabis saira*) [60]. When ice (pH 4.9) was formed, 30% of the active chlorine was lost, thus explaining the change in pH. Aerobic and psychrotropic bacterial counts were lower than tap water ice treatment. EW as ice showed a bactericidal effect on pacific saury fish during refrigerated storage; TVB-N content was lower and lipid oxidation was suppressed on ice-treated fish.

AEW was evaluated on raw trout skin [61]. Trout was inoculated with *Salmonella* Typhimurium, *E. coli* O157:H7 and *L. monocytogenes*; the skin was soaked at room temperature and bacterial survival count was performed. Bacteria were susceptible to treatment; *E. coli* was found to be the most susceptible (1 to 1.5 log CFU/g). The authors described a time dependent inhibitory pattern when AEW was used against *Salmonella sp*. The authors also concluded that it is important to reduce organic matter present on fish before treatment; they suggest pre-treating fish with clean water to remove cell debris prior to treatment with EW.

A sensorial evaluation was performed with American shad (*Alosa sapidissima*) treated with AEW and chitosan by immersion [62]. Treatment inhibited bacterial growth (total viable counts), protein decomposition, and lipid oxidation. At the same time, AEW treatment extended shelf life by nine to 10 days under refrigerated storage. During sensory analysis, treated fish received high scores and non-treated fish received unacceptable scores on days 10 and 20. In a different study [63], bombay duck (*Harpadon nehereus*) was treated with slightly AEW (pH 5.5) and ebony-bamboo leaves complex extracts (EBLCE). Fish were dipped for 5 min at 25 °C with shaking. The non-treated group showed a total viable count of 1.5 log10 CFU/g higher than the treated group. The increase in TVB-N in the treated group was slower than the non-treated group and lipid oxidation was inhibited in the treated group. In this study, AEW with EBLCE extended shelf life 16 days, while shelf life without treatment was only 4 days. This study demonstrated the use of EW on a fish that, generally, does not have a long shelf life.

4.6. Shrimp

Vibrio parahaemolyticus has a negative impact in shrimp production. AEW was evaluated in inoculated shrimp to test for susceptibility to different strains of *V. parahaemolyticus*. Shrimp that were immersed in AEW for 2.5 min, completely suppressed bacterial proliferation at 4 °C [64].

AEW can be used as ice. Wang [65] included shrimp muscle fiber analysis, and the experimental design included TVC analysis and protein degradation by an SDS-PAGE fingerprint. AEW as ice was renewed every 12 h. Results included reduction of TVC from days 0 to 3, showing potential to delay bacterial growth. Bacterial diversity analysis showed less diversity with AEW than without treatment. There was no protein degradation, and melanosis was observed to a lower degree with EW than without treatment. These results show a promising use for AEW in the shrimp industry.

4.7. Bivalve Mollusk

Oysters are consumed raw in many countries and can be contaminated or spoiled by different bacteria. Ren [66] evaluated the use of AEW on infected oysters with *V. parahaemolyticus* and *V. vulnificus*. In vitro analysis showed that EW decreased >6.6 log CFU/mL in 15 s. Inoculated oysters were immersed in AEW, and the best treatment exposure time was determined to be 4 to 6 h. Treatments longer than 12 h were detrimental to oysters. AEW at 30 ppm caused a decrease in bacterial numbers by 1.58 and 0.83 most probable numbers (MPN) / g of *V. parahaemolyticus* and *V. vulnificus* respectively.

AEW (pH 3.55) and a strong AEW (pH 3.1) were used on *E. coli*, *L monocytogenes*, *V. parahaemolyticus*, *C. jejuni* and *Aeromona hydrophila* contaminated clams and mussels [67]. Clams and mussels were kept in EW for 1 or 2 h. Results showed that there was significant difference between treatment time; however, strong AEW showed better results at 10 to 20 ppm during 2 h treatment in live clams and mussels without affecting quality. This showed that the bactericidal effect could be due to the available chlorine concentration rather than exposure time.

5. Chicken

Easy production, price, low fat content, high nutritional value, and easy access make chicken a popular meat around the word. These characteristics force the poultry industry to find new methodologies to preserve chicken without affecting organoleptic qualities and production cost. Different applications on poultry products are depicted in Table 4. Chicken breast has been treated with slightly (pH 6.2–6.5) or strong (pH 2.54) AEW by immersion for 10 min at 22 °C [68]. Chicken samples were stored at 5 °C mimicking retail display. Treatment decreased TVC by 1.5 log CFU/g and at day 10 never reached the limit of 7.0 log CFU/g. There was no statistical difference between strong and slight AEW treatments; however, *L. monocytogenes* was more susceptible than *Salmonella* Typhimurium. pH and TBARS were kept at low levels in the treated group and shelf life was extended from 1 to 4 days without affecting sensory quality. Shimamura [69] evaluated BEW and strong AEW dipping treatment on fresh chicken breast contaminated with *Salmonella enteritidis*, *E. coli* and *Staphylococcus aureus*. Treatment was applied at 4 °C and 25 °C for 3 min. Treatment with both solutions inhibited transcription of staphylococcal enterotoxin A and significantly reduced bacterial counts. There were no differences between the treated group and non-treated group for pH, lipid oxidation, color, and amino acid content. A different approach on prechilled chicken breast cylinders was the use of slightly AEW (pH 6.0) with ultra sound treatment [70] for 10 min at 10 °C. No significant differences between disinfection solutions and tap water treatment were detected for psychrotropic, lactic acid, enterobacteria, and mesophilic bacteria. This could be because of the size of the produced bubbles during cavitation. All treatments presented low levels of lipid and protein oxidation with no modifications in muscle fiber structure.

Table 4. Evaluations of electrolyzed water in poultry.

Material	Type of Electrolyzed Water	Concentration of EW (ppm)	Microorganisms	Inoculum Concentration [b]	Type of Treatment	Duration of Treatment	Reference
Chicken breast	Slightly AEW	10	*L monocytogenes*	9 log CFU/mL	Immersion	10 min	[68]
	strong AEW	50	*S.* Typhimurium				
Chicken breast	AEW 4 °C	30	*Salmonella* Enteritidis	9 log CFU/mL	Immersion	3 min	[69]
	AEW 25 °C	14	*E. coli*				
			Staph. aureus				
Chicken breast	Slightly AEW + ultrasound	5	Mesophilic bacteria	3.8 log CFU/g	Immersion	30 min	[70]
			Psychrotrophic bacteria	3.47 log CFU/g			
			Lactic acid bacteria	3.22 log CFU/g			
			Enterobacteria	2.1 log CFU/g			
			Staph. aureus	2.25 log CFU/g			
Chicken carcass	AEW, BEW	50	*S.* Typhimurium	5 log CFU/mL	Immersion	45 min	[71]
			E. coli		Spray wash	15 s	
			Total coliforms				
Chicken carcass	AEW, slightly AEW	58	Aerobic bacteria	4 log CFU/cm^2	Spray wash	15 s	[72,73]
		30	Total coliforms				
Chicken carcass	NEW + lactic acid	50	*C. jejuni*	9 log CFU/g	Immersion Spray wash	3 min	[74]
Chicken wings	AEW	50	*C. jejuni*	7 to 8 log CFU/mL	Immersion	10 or 30 min	[75]
Egg	AEW	8	*S.* Typhimurium	Different values	Spray	4 × 15 s	[76]
			Staph. aureus				
			L monocytogenes				
			E. coli				
Egg	slightly AEW	26	*Salmonella* Enteritidis	8 log CFU/mL	Immersion	3 min	[77]
			E. coli				
Egg	AEW, BAW	70 to 80	*Salmonella* Enteritidis	6 log CFU/mL	Immersion	1 to 5 min	[78]
			E. coli K12				
Egg	AEW	ND	Enterobacteriaceae		Spray	ND [a]	[79]
			Aerobic bacteria	3.5 log CFU/cm^2			
Egg	NEW	46	*L monocytogenes*	6 log CFU/mL	Spray	30 s	[27]
Egg	NEW	60	*Salmonella enterica*	6 log CFU/mL	Spray	30 s	[26]
			E. coli				

[a] Non determined; [b] For aerobic mesophilic, psychrotrophic bacteria, Staph. aureus, Enterobacteriaceae and total coliforms, inoculum concentration values were obtained from the no-treatment group.

For chicken carcass disinfection, Fabrizio tested different solutions like AEW (pH 2.6), BEW (pH 11.6), ozonated water, acetic acid, and trisodium phosphate solution against *Salmonella* Typhimurium, *E. coli* and total coliforms [71]. Carcasses were submerged in tested solutions for 45 min at 4 °C or spray washed for 15 s at 85 psi at 25 °C or a combination of both treatments (multiple intervention) were used. The immersion chilling AEW treatment was more effective than chlorinated water (control treatment) after seven days of refrigeration storage. Spray treatment showed no statistical difference with the control treatment. Authors explained that this performance could be the result of a 15 s interaction versus a 45 min immersion treatment. However, multiple intervention treatment showed better bactericidal effect against *Salmonella* at days 0 and 7. In a similar study, chicken carcasses from white and yellow feathered flocks were treated by spraying different solutions like chlorine dioxide, 2% lactic acid, sodium hypochlorite, AEW or slightly AEW for 15 s [72]. TVC and coliform counts were monitored. Both types of EW showed TVC reductions by 0.63 log CFU/cm^2; similar results were obtained from samples from different parts of evaluated carcasses. Lactic acid and AEW treatments showed lower counts compared to slightly AEW during the storage period and pH dropped due to their acidic origin/nature. Furthermore, these treatments maintained lower TVB-N levels during storage. Slightly AEW showed no pH alteration and AEW and slightly AEW did not show pro-oxidant potential. As a conclusion of this study, AEW helped maintain the quality of chicken carcasses. Authors recommended decontaminating chicken after chilling. In a continuation of this study, they evaluated the use of AEW in a newly designed spray cabinet [73], and the treatment reduced microbial numbers by 1.0 log CFU/cm^2.

NEW and lactic acid (pH 2.0) were evaluated by Rasschaert [74] on carcasses. Chickens were submerged or sprayed after scalding. Unfortunately, this treatment showed no significant effect on carcasses contaminated with Campylobacter. Authors explained that bacteria may be in crevices and feather follicles. These regions are difficult to reach by EW.

AEW was evaluated in chicken wings against *C. jejuni* during the washing process [75]. Treatment was applied by immersion for 10 to 30 min at different temperatures. EW reduced *C. jejuni* counts by 3.0 log CFU/g after 30 min of immersion. Bacterial counts after 30 min of treatment were lower than with 10 min of treatment; however, no significant difference was detected. No viable cells were detected in wash solution and bacteria were detected in wash control solution.

6. Egg

Egg is an important source of protein; however, its contamination with *Salmonella sp* is related with the natural production process (laying) and there are related factors such as the equipment used during the handling of eggs. Therefore, the use of a sanitizer in this industry has great relevance to eliminate foodborne pathogens that can cause cross-contamination and produce further infection through their dispersion in the environment of the hatchery.

AEW was evaluated on the surface of eggs containing *Listeria monocytogenes, Escherichia coli, Staphylococcus aureus* and *Salmonella* Typhimurium. EW was applied using an electrostatic atomization technique. Treatments were evaluated on complete eggs and four repetitions were performed. AEW completely eliminated *Salmonella* Typhimurium on ranges from 6.7 to 53.3%. In average, *Salmonella* reduction counts were 4.0 log10. *Staphylococcus aureus* was completely eliminated on ranges from 80 to 73.3% and reduction counts were around 3.0 log10. *L. monocytogenes* was totally eliminated in 93.3 to 53.3%. Authors explained that the antibacterial mechanism of action of AEW is based on its composition and can be considered an electrical method in conjunction with electrostatic atomization, involving conductance and impedance [76].

In another study, the bactericidal activity of slightly AEW (sAEW) (pH 6.53), AEW (pH 2.81), and sodium hypochlorite (pH 10.12) solutions were evaluated against *Salmonella* Enteritidis and *Escherichia coli* on artificially inoculated eggshells. The best result was in the detection of smaller amounts of survival populations obtained after treatments with sAEW. Internal egg quality attributes like Haugh units, yolk index, weight loss, yolk pH, and albumin were evaluated.

The best results were obtained with sAEW, indicating that it could be a sanitizing method for eggs to not only decrease microbial load, but also preserve egg quality and thus its shelf life. This is attributed to the less corrosive effects exerted by sAEW on the eggshell, compared to other treatments, without affecting the egg´s internal composition. However, in this work, there was no cuticle analysis; researchers compare initial weight versus the end of storage. Weight loss was used as an indicator for cuticle damage. For this reason, they suggest the usage of sAEW in combination with other techniques such as UV light, coating, and freezer storage [77].

Bialka et. al. [78] evaluated the effect of applying BEW (pH 11.4) first and AEW (pH 2.7) after, during the egg washing process. Bactericidal effect was compared using a commercial detergent/sanitizer. Different temperatures and treatment times were evaluated. From the in vitro evaluation, the best treatment was applied at 45 °C for 3 min. The best bactericidal results were obtained using both EW against *Salmonella* Enteritidis and *E. coli* K12, when the treatment was compared against single solution treatment or commercial detergent. Afterwards, they performed a pilot commercial scale evaluation using a commercial egg washer and they only evaluated the bactericidal effect against *E. coli*. However, results showed that the combination of BEW and AEW, as well as the use of a commercial detergent could affect cuticle integrity.

Another research group evaluated AEW on hatching eggs [79]. They evaluated a sprayed treatment and quantified the total aerobic bacteria on eggshells. AEW treatment decreased bacterial counts on eggshells by 1.0 log CFU/cm^2. This group claimed no effect on cuticle and broiler mortality decreased in the treated group; cuticle integrity was evaluated by egg weight.

The bactericidal effect of NEW (pH 6.86) on contaminated eggshells with *Listeria monocytogenes* [27], *Salmonella enterica* and *Escherichia coli* [26] has been evaluated. Treatments were applied by spraying for 15 s. *L. monocytogenes* counts were reduced by 2.0 Log_{10}/egg meanwhile *E. coli* and *S. enterica* counts were reduced by 6.3 and 1.4 Log_{10}/egg respectively. Additionally, cuticle analysis was performed by electron microscopy where the structure of NEW-treated eggshells looked like non-treated and NaCl-treated eggshells.

7. Cattle Products

7.1. Beef

Microbial contamination is an important public concern in the food industry because it can shorten shelf life and increase the risk of food safety in fresh meat and its derived products.

Temperature is the most important environmental factor that affects bacterial growth in beef; this factor constantly changes during processing, storage, and distribution of meat products. Therefore, the most commonly used disinfectants in this area are those based on chlorine, such as sodium hypochlorite, due to its antimicrobial efficacy, convenience, and low price. However, some previous studies have warned users about the limited efficacy of chlorine for reducing microorganisms in meat and surfaces that come into contact with meat [80–82]. Furthermore, there is a potential health risk in chlorine consumption for consumers, e.g., cancer [82]. *Listeria monocytogenes* is an important organism because it has been identified in many ready-to-eat meat products. In order to eliminate this pathogen, the food industry should apply a post-lethality treatment to the product to inhibit or reduce bacterial growth before packaging [41,81,82]. Different studies of EW in meat and milk are listed in Table 5.

Table 5. Evaluations of electrolyzed water on cattle products.

Material	Type of Electrolyzed Water	Concentration of EW (ppm)	Microorganisms	Inoculum Concentration [b]	Type of Treatment	Duration of Treatment	Reference
Beef meat	Slightly AEW	38	*E. coli* 0157:H7 *S.* Typhimurium *L. monocytogenes*	9 log CFU/mL	Immersion	10 min	[61]
Fresh meat	NEW	27 to 39, 50	*L. monocytogenes* *E. coli* O157:H7 *Salmonella sp*	8 log CFU/mL	Spray	30 s	[81]
Rib meat	AEW, slightly AEW	50 5	*E. coli* O157:H7	9 log CFU/mL	Immersion	3 min	[83]
Meat	AEW, slightly AEW	ND [a]	Aerobic bacteria	4.78 log CFU/g	Immersion	ND	[84]
			Fungi and yeast	3.71 log CFU/g			
Beef meat	Slightly AEW + tea polyphenols	40	Aerobic bacteria	3.06 log CFU/g	Immersion	5 min	[85]
Beef fillets	BEW	100	Aerobic bacteria Total coliforms Yeast Lactic acid bacteria	3.82 log CFU/cm^2 1.94 log CFU/cm^2 2.21 log CFU/cm^2 2.64 log CFU/cm^2	Spray	90 s	[86]
Beef head	AEW	60	*E. coli* O157:H7	6 log CFU/cm^2	Spray	12 s	[87]
Bovine carcass	BEW, AEW	400	Aerobic bacteria	5 log CFU/400 cm^2	Spray	ND[a]	[88]
			E. coli O157:H7 Coliforms	0.60 log CFU/400 cm^2 0.83 log CFU/400 cm^2			
Milk	AEW	ND	Aerobic bacteria	2.48 log CFU/mL	Mix	15 min	[89]

[a] Non determined; [b] For aerobic mesophilic, total coliforms, lactic acid bacteria, fungi and yeast, inoculum concentration values were obtained from the no-treatment group.

One study reported the evaluation of AEW and slight AEW on contaminated meat with *Escherichia coli* O157:H7 using 5.0 log UFC/g. Both treatments reduced 3.36 log UFC/g and 3.28 log UFC/g, respectively. Authors worked on a mathematical model to study the effect of storage temperature; however, further studies were needed to accomplish this goal [83]. In a different study, AEW was used on fresh meat, processed meat (frankfurters), and meat-contact surfaces having *L. monocytogenes*, *E. coli* O157:H7, and *Salmonella sp.* No bactericidal effect was observed in meat and frankfurters due to the presence of organic matter buffering the action of hypochlorous acid against bacteria. The bactericidal analysis on surfaces of equipment handling ready-to-eat meat products compared the effect on clean and dirty stainless-steel cutting blades. AEW was applied using three different concentrations: 5, 25, and 250 ppm. On clean blades, *L. monocytogenes* had bacterial reductions of 1.4, 3.6, and 5.7 log (CFU/mL) when 5, 25, and 250 ppm AEW were used, respectively. Unlike these results, for dirty blades, *L. monocytogenes* reductions were 0, 0.64, and 3.3 log (CFU/mL) using the same concentrations previously mentioned. These results showed again that the presence of organic matter significantly reduces the effectiveness of EW. Despite these results, the authors believe that the best use for AEW in fresh meat and ready-to-eat meat is by applying EW directly to the product wrapped in waterproof containers, where there can be a minimum of organic material in contact with EW [81].

In another study, the use of AEW was compared against conventional defrosting methods (air, tap water, and microwaves) to assess microbiological safety and quality attributes of thawed meat [84]. TVC, fungi, and yeast counts were significant reduced by 0.83 log UFC/g and 1.16 log UFC/g, respectively, compared to the control group. AEW showed effectiveness as a thawing medium for controlling microbial contamination during thawing. Regarding the physicochemical characteristics of thawed meat, AEW has a negligible impact on moisture loss, surface meat color was paler than control, but internal color did not show differences. Lipid and protein oxidation were retarded as well as protein aggregation and degradation.

A similar study was performed using slight AEW on beef stored by refrigeration. Microbicidal efficacy and shelf life were evaluated. The results showed that the TVC of the treated group decreased to 2.28 log CFU/g and control group viable counts were 3.06 log CFU/g. After three days, the microbial population increased in all samples but, in different proportions, since the TVC of the sAEW treated group was 2.89 log CFU/g, being the lowest value obtained in the results. Bacterial counts in the sAEW group were acceptable by day 14 after treatment, indicating that the use of sAEW in beef could preserve meat. sAEW showed an ability to keep meat in good condition for a longer period of time. Its bactericidal effect could be because it slows the increase of pH and generation of TVB-N. Sensory scores for odor, appearance, texture, and acceptability were better when EW was used [85]. In a different study, sAEW was used on trout, chicken, and beef. Its application caused *E. coli* O157:H7, *Salmonella* Typhimurium, and *L. monocytogenes* reductions compared to the non-treated group or the sterile distilled water treated group [61].

Near NEW was evaluated against *Salmonella* Typhimurium DT104 and *E. coli* O157:H7 on fresh hides. Treatments involve use of near NEW (pH 6.5 at room temperature), BEW (pH 11.6), hot BEW (43 °C), BEW spray followed by NEW spray (both at room temperature), 0.02% peroxyacetic acid (room temperature), 5% lactic acid (pH 2.04 at room temperature), deionized water (W), and no treatment. All these strategies were evaluated because near NEW has bactericidal activity but, it is not corrosive and is stable. Results showed that *S.* Typhimurium DT104 and *E. coli* O157:H7 were reduced by 1.09 y 0.65 log UFC/ cm^2, respectively [61].

Another study evaluated the influence of BEW on the microbiota present in beef fillets. Meat was treated with EW before being vacuum packed and stored at 4 °C. Results showed that there was no impact on the initial microbiological situation after treatment or during storage; this could be due to the fact that the analysis did not consider the initial microbiota composition, which can react with EW. Some of the identified microorganisms before treatment were *Pseudomonas* sp., *Brochothrix* sp., *Psychrobacter* sp., *Lactobacillus* sp., and *Acinetobacter* sp. Which are commonly reported as meat contaminants from processing environments. After treatment, during the first day, *Psychrobacter* sp. And *Acinetobacter Iwoffii* were able to survive. However, *Pseudomonas fragi* was active and predominated.

Though, after day 5, lactic acid bacteria, *Lactobacillus sakei*, *Leuconostoc gasicomitatum*, and *Lactococcus piscium* were the most abundant population, probably due to limited oxygen conditions [86].

EW can be applied sequentially to optimize its antimicrobial activity. This methodology was carried out during the treatment of fresh hides, where BEW and AEW were used sequentially. This strategy reduced aerobic bacteria and Enterobacteriaceae counts to 3.5 and 4.3 log CFU/100 cm^2, respectively, while *E. coli* O157:H7 was reduced from 82 to 35%. The effect of time and temperature of EW treatments was also evaluated, showing that they contributed to some extent to the bactericidal effect.

Some AEW evaluations have shown poor results. When AEW was used in the treatment of bovine heads (hide removed), *E. coli* O157:H7 was reduced by less than 0.5 log CFU/cm^2 [87]; in another report, EW was used to decontaminate bovine carcasses, but unfortunately, EW did not reduce aerobic bacterial counts significantly [88].

7.2. Milk

There is only one report about the use of AEW in milk. Kalit et al. [89] used a commercial EW on milk against aerobic mesophilic bacteria. In this study, EW was diluted, and treatments were applied for 15 min. Results showed that the highest concentration of AEW caused the highest bactericidal effect. Bacterial counts were reduced from 2.48 log CFU/mL to 0.33 log CFU/mL. The requirement of high concentration of AEW is due to available chlorine reacts with proteins and vitamins of milk. However, it is useful in organic production/processing of milk or in areas where water resources are limited because EW can be used to disinfect and rinse equipment.

8. Conclusions

The importance of the use of electrolyzed water lies in its easy production and the versatility of its presentations (i.e., concentrations) that it can have, depending on the production methods. EW can be produced on-site, decreasing storage, and transportation cost. An important attribute of EW is that its bactericidal effect is maintained at different temperatures, allowing its application under cold or warm environments. The most important characteristic is that it can be applied in different products from the food industry. In this document, different applications were reviewed. The most common evaluated procedures were immersion and spray. Immersion is a common practice in poultry and fish industry; however, this practice requires the use of large volumes of water; if EW solutions are used for immersion treatments, it is important to know the residual effect of a used EW; nevertheless, there are no reports about the efficacy of residual effect after EW been used more than once with animal by products. The fish industry uses spray treatments frequently, though there are no reports of the use of sprayed EW on fish, which allow the use of smaller amounts of water compared with dip treatments. The use of EW as ice is an interesting approach for fish and shrimp, however, it needs to be evaluated for every type of food because EW is affected by intrinsic components. In different studies, it has been reported that fish were treated at different temperatures and/or with agitation; fish is a delicate product and it is important to evaluate if this treatment is feasible to apply in this industry. The importance of the beef, pork, and poultry industry has impulse studies to evaluate spray treatments. These industries are exploring new uses as enhancement solutions or evaluating the use with different food types like milk. Although there have already been many studies on the use of EW, more scientific work is still needed to elucidate certain details, such as the possible food sensory changes caused by its use. One possible use is by sequential application (for example, BEW followed by AEW), or in combination with other bactericidal substances to obtain better disinfection and conservation results without affecting quality. However, it is still highly relevant to venture into the use of other cleaning methods, through specific studies, that in combination with EW could achieve optimal performance. The use of electrolyzed water is a promising strategy to preserve different raw, ready-to-eat meat, chicken, fish and others without affecting sensory characteristics. EW uses can be applied in different types of food and against different pathogens. It could be interesting to see its effect on virus contaminated food like milk and expand its uses. A variety of products can be candidates for the application of EW to increase shelf life and decrease the incidence of foodborne diseases.

Author Contributions: Conceptualization, J.C.R.O. and J.A.C.-B.; writing—original draft preparation, J.C.R.O. and J.A.C.-B.; writing—review and editing, J.A.C.-B. All authors have read and agreed to the published version of the manuscript.

Funding: This manuscript received no external funding.

Acknowledgments: We are very thankful to Jesus Guzman for providing language help and discussion of the manuscript.

Conflicts of Interest: The authors declare no conflict of interest.

References

1. Department of Economic and Social Affairs, UN. *World Population Prospects 2019. Volume II: Demographic Profiles*; United Nations: New York, NY, USA, 2019; Volume 2, ISBN 9789211483284.
2. Mylius, M.; Dreesman, J.; Pulz, M.; Pallasch, G.; Beyrer, K.; Claußen, K.; Allerberger, F.; Fruth, A.; Lang, C.; Prager, R.; et al. Shiga toxin-producing Escherichia coli O103:H2 outbreak in Germany after school trip to Austria due to raw cow milk, 2017—The important role of international collaboration for outbreak investigations. *Int. J. Med. Microbiol.* **2018**, *308*, 539–544. [CrossRef] [PubMed]
3. Morton, V.K.; Kearney, A.; Coleman, S.; Viswanathan, M.; Chau, K.; Orr, A.; Hexemer, A. Outbreaks of Salmonella illness associated with frozen raw breaded chicken products in Canada, 2015–2019. *Epidemiol. Infect.* **2019**, *147*, 2019–2021. [CrossRef] [PubMed]
4. Friesema, I.; de Jong, A.; Hofhuis, A.; Heck, M.; van den Kerkhof, H.; de Jonge, R.; Hameryck, D.; Nagel, K.; van Vilsteren, G.; van Beek, P.; et al. Large outbreak of Salmonella Tompson related to smoked salmon in the Netherlands, August to December 2012. *Eurosurveillance* **2014**, *19*, 1–8. [CrossRef] [PubMed]
5. Tchatchouang, C.K.; Fri, J.; De Santi, M.; Ateba, C.N. Listeriosis Outbreak in South Africa: A Comparative Analysis with Previously Reported Cases Worldwide. *Microorganisms* **2020**, *8*, 135. [CrossRef] [PubMed]
6. Cabal, A.; Allerberger, F.; Huhulescu, S.; Kornschober, C.; Springer, B.; Schlagenhaufen, C.; Wassermann-Neuhold, M.; Fötschl, H.; Pless, P.; Krause, R.; et al. Listeriosis outbreak likely due to contaminated liver pâté consumed in a tavern, Austria, December 2018. *Euro Surveill.* **2019**, *24*, 1–6. [CrossRef]
7. Sinulingga, T.S.; Aziz, S.A.; Bitrus, A.A.; Zunita, Z.; Abu, J. Occurrence of Campylobacter species from broiler chickens and chicken meat in Malaysia. *Trop. Anim. Health Prod.* **2019**, *52*, 151–157. [CrossRef]
8. Lung, H.M.; Cheng, Y.C.; Chang, Y.H.; Huang, H.W.; Yang, B.B.; Wang, C.Y. Microbial decontamination of food by electron beam irradiation. *Trends Food Sci. Technol.* **2015**, *44*, 66–78. [CrossRef]
9. Deng, L.Z.; Mujumdar, A.S.; Pan, Z.; Vidyarthi, S.K.; Xu, J.; Zielinska, M.; Xiao, H.W. Emerging chemical and physical disinfection technologies of fruits and vegetables: A comprehensive review. *Crit. Rev. Food Sci. Nutr.* **2019**, 1–28. [CrossRef]
10. Len, S.V.; Hung, Y.C.; Erickson, M.; Kim, C. Ultraviolet spectrophotometric characterization and bactericidal properties of electrolyzed oxidizing water as influenced by amperage and pH. *J. Food Prot.* **2000**, *63*, 1534–1537. [CrossRef]
11. Tanaka, H.; Hirakata, Y.; Kaku, M.; Yoshida, R.; Takemura, H.; Mizukane, R.; Ishida, K.; Tomono, K.; Koga, H.; Kohno, S.; et al. Antimicrobial activity of superoxidized water. *J. Hosp. Infect.* **1996**, *34*, 43–49. [CrossRef]
12. Khan, I.; Tango, C.N.; Miskeen, S.; Lee, B.H.; Oh, D.H. Hurdle technology: A novel approach for enhanced food quality and safety—A review. *Food Control* **2017**, *73*, 1426–1444. [CrossRef]
13. Morita, C.; Sano, K.; Morimatsu, S.; Kiura, H.; Goto, T.; Kohno, T.; Hong, W.; Miyoshi, H.; Iwasawa, A.; Nakamura, Y.; et al. Disinfection potential of electrolyzed solutions containing sodium chloride at low concentrations. *J. Virol. Methods* **2000**, *85*, 163–174. [CrossRef]
14. Tagawa, M.; Yamaguchi, T.; Osamu, Y.; Matsutani, S.; Maeda, T.; Saisho, H. Inactivation of a hepadnavirus by electrolysed acid water. *J. Antimicrob. Chemother.* **2000**, *46*, 363–368. [CrossRef]

15. Gómez-Espinosa, D.; Cervantes-Aguilar, F.J.; Río-García, J.C.; Villarreal-Barajas, T.; Vázquez-Durán, A.; Méndez-Albores, A. Ameliorative effects of neutral electrolyzed water on growth performance, biochemical constituents, and histopathological changes in Turkey poults during aflatoxicosis. *Toxins* **2017**, *9*, 104. [CrossRef] [PubMed]
16. Fan, S.; Zhang, F.; Liu, S.; Yu, C.; Guan, D.; Pan, C. Removal of aflatoxin B1 in edible plant oils by oscillating treatment with alkaline electrolysed water. *Food Chem.* **2013**, *141*, 3118–3123. [CrossRef] [PubMed]
17. Suzuki, T.; Itakura, J.; Watanabe, M.; Ohta, M.; Sato, Y.; Yamaya, Y. Inactivation of staphylococcal enterotoxin-A with an electrolyzed anodic solution. *J. Agric. Food Chem.* **2002**, *50*, 230–234. [CrossRef]
18. Al-Haq, M.I.; Sugiyama, J.; Isobe, S. Applications of Electrolyzed Water in Agriculture & Food Industries. *Food Sci. Technol. Res.* **2005**, *1*, 135–150.
19. Ngnitcho, P.F.K.; Khan, I.; Tango, C.N.; Hussain, M.S.; Oh, D.H. Inactivation of bacterial pathogens on lettuce, sprouts, and spinach using hurdle technology. *Innov. Food Sci. Emerg. Technol.* **2017**, *43*, 68–76. [CrossRef]
20. Rahman, S.; Khan, I.; Oh, D.H. Electrolyzed Water as a Novel Sanitizer in the Food Industry: Current Trends and Future Perspectives. *Compr. Rev. Food Sci. Food Saf.* **2016**, *15*, 471–490. [CrossRef]
21. Hati, S.; Mandal, S.; Minz, P.S.; Vij, S.; Khetra, Y.; Singh, B.P.; Yadav, D. Electrolyzed Oxidized Water (EOW): Non-Thermal Approach for Decontamination of Food Borne Microorganisms in Food Industry. *Food Nutr. Sci.* **2012**, *3*, 760–768. [CrossRef]
22. Yang, H.; Feirtag, J.; Diez-Gonzalez, F. Sanitizing effectiveness of commercial "active water" technologies on Escherichia coli O157:H7, Salmonella enterica and Listeria monocytogenes. *Food Control* **2013**, *33*, 232–238. [CrossRef]
23. Zhao, L.; Zhang, Y.; Yang, H. Efficacy of low concentration neutralised electrolysed water and ultrasound combination for inactivating Escherichia coli ATCC 25922, Pichia pastoris GS115 and Aureobasidium pullulans 2012 on stainless steel coupons. *Food Control* **2017**, *73*, 889–899. [CrossRef]
24. Cui, X.; Shang, Y.; Shi, Z.; Xin, H.; Cao, W. Physicochemical properties and bactericidal efficiency of neutral and acidic electrolyzed water under different storage conditions. *J. Food Eng.* **2009**, *91*, 582–586. [CrossRef]
25. Nagamatsu, Y.; Chen, K.K.; Tajima, K.; Kakigawa, H.; Kozono, Y. Durability of Bactericidal Activity in Electrolyzed Neutral Water by Storage. *Dent. Mater. J.* **2002**, *21*, 93–104. [CrossRef] [PubMed]
26. Medina-Gudiño, J.; Rivera-Garcia, A.; Santos-Ferro, L.; Ramirez-Orejel, J.C.; Agredano-Moreno, L.T.; Jimenez-Garcia, L.F.; Paez-Esquiliano, D.; Martinez-Vidal, S.; Andrade-Esquivel, E.; Cano-Buendia, J.A. Analysis of Neutral Electrolyzed Water anti-bacterial activity on contaminated eggshells with Salmonella enterica or Escherichia coli. *Int. J. Food Microbiol.* **2020**, *320*, 108538. [CrossRef] [PubMed]
27. Rivera-Garcia, A.; Santos-Ferro, L.; Ramirez-Orejel, J.C.; Agredano-Moreno, L.T.; Jimenez-Garcia, L.F.; Paez-Esquiliano, D.; Andrade-Esquivel, E.; Cano-Buendia, J.A. The effect of neutral electrolyzed water as a disinfectant of eggshells artificially contaminated with Listeria monocytogenes. *Food Sci. Nutr.* **2019**, *7*, 2252–2260. [CrossRef] [PubMed]
28. Albrich, J.M.; Gilbaugh, J.H.; Callahan, K.B.; Hurst, J.K. Effects of the putative neutrophil-generated toxin, hypochlorous acid, on membrane permeability and transport systems of Escherichia coli. *J. Clin. Invest.* **1986**, *78*, 177–184. [CrossRef]
29. Hurst, J.K.; Barrette, W.C.; Michel, B.R.; Rosen, H. Hypochlorous acid and myeloperoxidase-catalyzed oxidation of iron-sulfur clusters in bacterial respiratory dehydrogenases. *Eur. J. Biochem.* **1991**, *202*, 1275–1282. [CrossRef]
30. Len, S.V.; Hung, Y.C.; Chung, D.; Anderson, J.L.; Erickson, M.C.; Morita, K. Effects of storage conditions and pH on chlorine loss in electrolyzed oxidizing (EO) water. *J. Agric. Food Chem.* **2002**, *50*, 209–212. [CrossRef]
31. Huang, Y.R.; Hung, Y.C.; Hsu, S.Y.; Huang, Y.W.; Hwang, D.F. Application of electrolyzed water in the food industry. *Food Control* **2008**, *19*, 329–345. [CrossRef]
32. Liao, L.B.; Chen, W.M.; Xiao, X.M. The generation and inactivation mechanism of oxidation-reduction potential of electrolyzed oxidizing water. *J. Food Eng.* **2007**, *78*, 1326–1332. [CrossRef]
33. Rahman, S.M.E.; Jin, Y.G.; Oh, D.H. Combined Effects of Alkaline Electrolyzed Water and Citric Acid with Mild Heat to Control Microorganisms on Cabbage. *J. Food Sci.* **2010**, *75*, 111–115. [CrossRef] [PubMed]
34. Liu, Q.; Wu, J.; Lim, Z.Y.; Aggarwal, A.; Yang, H.; Wang, S. Evaluation of the metabolic response of Escherichia coli to electrolysed water by 1H NMR spectroscopy. *LWT Food Sci. Technol.* **2017**, *79*, 428–436. [CrossRef]
35. Liu, Q.; Wu, J.; Lim, Z.Y.; Lai, S.; Lee, N.; Yang, H. Metabolite profiling of Listeria innocua for unravelling the inactivation mechanism of electrolysed water by nuclear magnetic resonance spectroscopy. *Int. J. Food Microbiol.* **2018**, *271*, 24–32. [CrossRef] [PubMed]

36. Zhang, J.; Yang, H.; Chan, J.Z.Y. Development of Portable Flow-Through Electrochemical Sanitizing Unit to Generate Near Neutral Electrolyzed Water. *J. Food Sci.* **2018**, *83*, 780–790. [CrossRef]
37. Abadias, M.; Usall, J.; Oliveira, M.; Alegre, I.; Viñas, I. Efficacy of neutral electrolyzed water (NEW) for reducing microbial contamination on minimally-processed vegetables. *Int. J. Food Microbiol.* **2008**, *123*, 151–158. [CrossRef] [PubMed]
38. Rahman, S.M.E.; Wang, J.; Oh, D.H. Synergistic effect of low concentration electrolyzed water and calcium lactate to ensure microbial safety, shelf life and sensory quality of fresh pork. *Food Control* **2013**, *30*, 176–183. [CrossRef]
39. Tyszkiewicz, I.; Kłossowska, B.M.; Wieczorek, U.; Jakubiec-Puka, A. Mechanical Tenderisation of Pork Meat: Protein and Water Release due to Tissue Damage. *J. Sci. Food Agric.* **1997**, *73*, 179–185. [CrossRef]
40. Fabrizio, K.A.; Cutter, C.N. Comparison of electrolyzed oxidizing water with other antimicrobial interventions to reduce pathogens on fresh pork. *Meat Sci.* **2004**, *68*, 463–468. [CrossRef]
41. Fabrizio, K.A.; Cutter, C.N. Application of electrolyzed oxidizing water to reduce Listeria monocytogenes on ready-to-eat meats. *Meat Sci.* **2005**, *71*, 327–333. [CrossRef]
42. Mansur, A.R.; Oh, D.H. Combined effects of thermosonication and slightly acidic electrolyzed water on the microbial quality and shelf life extension of fresh-cut kale during refrigeration storage. *Food Microbiol.* **2015**, *51*, 154–162. [CrossRef] [PubMed]
43. Athayde, D.R.; Flores, D.R.M.; da Silva, J.S.; Genro, A.L.G.; Silva, M.S.; Klein, B.; Mello, R.; Campagnol, P.C.B.; Wagner, R.; de Menezes, C.R.; et al. Application of electrolyzed water for improving pork meat quality. *Food Res. Int.* **2017**, *100*, 757–763. [CrossRef]
44. Rigdon, M.; Hung, Y.C.; Stelzleni, A.M. Evaluation of alkaline electrolyzed water to replace traditional phosphate enhancement solutions: Effects on water holding capacity, tenderness, and sensory characteristics. *Meat Sci.* **2017**, *123*, 211–218. [CrossRef] [PubMed]
45. Brychcy, E.; Malik, M.; Drozdzewski, P.; Ulbin-Figlewicz, N.; Jarmoluk, A. Low-concentrated acidic electrolysed water treatment of pork: Inactivation of surface microbiota and changes in product quality. *Int. J. Food Sci. Technol.* **2015**, *50*, 2340–2350. [CrossRef]
46. Belton, B.; Bush, S.R.; Little, D.C. Not just for the wealthy: Rethinking farmed fi sh consumption in the Global South. *Glob. Food Sec.* **2018**, *16*, 85–92. [CrossRef]
47. Ozer, N.P.; Demirci, A. Electrolyzed oxidizing water treatment for decontamination of raw salmon inoculated with Escherichia coli O157:H7 and Listeria monocytogenes Scott A and response surface modeling. *J. Food Eng.* **2006**, *72*, 234–241. [CrossRef]
48. McCarthy, S.; Burkhardt, W. Efficacy of electrolyzed oxidizing water against Listeria monocytogenes and Morganella morganii on conveyor belt and raw fish surfaces. *Food Control* **2012**, *24*, 214–219. [CrossRef]
49. Miks-Krajnik, M.; Feng, L.X.J.; Bang, W.S.; Yuk, H.-G. Inactivation of Listeria monocytogenes and natural microbiota on raw salmon fillets using acidic electrolyzed water, ultraviolet light or/and ultrasounds. *Food Control* **2017**, *74*, 54–60. [CrossRef]
50. Ovissipour, M.; Shiroodi, S.G.; Rasco, B.; Tang, J.; Sablani, S.S. Electrolyzed water and mild-thermal processing of Atlantic salmon (Salmo salar): Reduction of Listeria monocytogenes and changes in protein structure. *Int. J. Food Microbiol.* **2018**, *276*, 10–19. [CrossRef]
51. Ghorban Shiroodi, S.; Ovissipour, M.; Ross, C.F.; Rasco, B.A. Efficacy of electrolyzed oxidizing water as a pretreatment method for reducing Listeria monocytogenes contamination in cold-smoked Atlantic salmon (Salmo salar). *Food Control* **2016**, *60*, 401–407. [CrossRef]
52. Phuvasate, S.; Su, Y. Effects of electrolyzed oxidizing water and ice treatments on reducing histamine-producing bacteria on fish skin and food contact surface. *Food Control* **2010**, *21*, 286–291. [CrossRef]
53. Huang, Y.R.; Shiau, C.Y.; Hung, Y.C.; Hwang, D.F. Change of hygienic quality and freshness in tuna treated with electrolyzed water and carbon monoxide gas during refrigerated and frozen storage. *J. Food Sci.* **2006**, *71*, M127–M133. [CrossRef]
54. Abou-Taleb, M.; Kawai, Y. Shelf life of semifried tuna slices coated with essential oil compounds after treatment with anodic electrolyzed NaCl solution. *J. Food Prot.* **2008**, *71*, 770–774. [CrossRef] [PubMed]
55. Rajkowski, K.T.; Sommers, C.H. Effect of anolyte on background microflora, salmonella, and listeria monocytogenes on catfish fillets. *J. Food Prot.* **2012**, *75*, 765–770. [CrossRef] [PubMed]

56. Lin, H.-M.; Hung, Y.-C.; Deng, S.-G. Effect of partial replacement of polyphosphate with alkaline electrolyzed water (AEW) on the quality of catfish fillets. *Food Control* **2020**, *112*, 1–8. [CrossRef]
57. Huang, Y.R.; Hsieh, H.S.; Lin, S.Y.; Lin, S.J.; Hung, Y.C.; Hwang, D.F. Application of electrolyzed oxidizing water on the reduction of bacterial contamination for seafood. *Food Control* **2006**, *17*, 987–993. [CrossRef]
58. Feliciano, L.; Lee, J.; Lopes, J.A.; Pascall, M.A. Efficacy of Sanitized Ice in Reducing Bacterial Load on Fish Fillet and in the Water Collected from the Melted Ice. *J. Food Sci.* **2010**, *75*, 231–238. [CrossRef]
59. Mahmoud, B.S.M.; Yamazaki, K.; Miyashita, K.; Il-Shik, S.; Dong-Suk, C.; Suzuki, T. Decontamination effect of electrolysed NaCl solutions on carp. *Lett. Appl. Microbiol.* **2004**, *39*, 169–173. [CrossRef]
60. Kim, W.T.; Lim, Y.S.; Shin, I.S.; Park, H.; Chung, D.; Suzuki, T. Use of electrolyzed water ice for preserving freshness of pacific saury (Cololabis saira). *J. Food Prot.* **2006**, *69*, 2199–2204. [CrossRef]
61. Al-holy, M.A.; Rasco, B.A. The bactericidal activity of acidic electrolyzed oxidizing water against Escherichia coli O157:H7, Salmonella Typhimurium, and Listeria monocytogenes on raw fish, chicken and beef surfaces. *Food Control* **2015**, *54*, 317–321. [CrossRef]
62. Xu, G.; Tang, X.; Tang, S.; You, H.; Shi, H.; Gu, R. Combined effect of electrolyzed oxidizing water and chitosan on the microbiological, physicochemical, and sensory attributes of American shad (Alosa sapidissima) during refrigerated storage. *Food Control* **2014**, *46*, 397–402. [CrossRef]
63. Chen, J.; Xu, B.; Deng, S.; Huang, Y. Effect of Combined Pretreatment with Slightly Acidic Electrolyzed Water and Botanic Biopreservative on Quality and Shelf Life of Bombay Duck (Harpadon nehereus). *J. Food Qual.* **2016**, *39*, 116–125. [CrossRef]
64. Wang Jing, J.; Sun, W.S.; Jin, M.T.; Liu, H.Q.; Zhang, W.; Sun, X.H.; Pan, Y.J.; Zhao, Y. Fate of Vibrio parahaemolyticus on shrimp after acidic electrolyzed water treatment. *Int. J. Food Microbiol.* **2014**, *179*, 50–56. [CrossRef] [PubMed]
65. Wang, M.; Wang, J.J.; Sun, X.H.; Pan, Y.J.; Zhao, Y. Preliminary mechanism of acidic electrolyzed water ice on improving the quality and safety of shrimp. *Food Chem.* **2015**, *176*, 333–341. [CrossRef]
66. Ren, T.; Su, Y.C. Effects of electrolyzed oxidizing water treatment on reducing Vibrio parahaemolyticus and Vibrio vulnificus in raw oysters. *J. Food Prot.* **2006**, *69*, 1829–1834. [CrossRef]
67. Al-Qadiri, H.M.; Al-Holy, M.A.; Shiroodi, S.G.; Ovissipour, M.; Govindan, B.N.; Al-Alami, N.; Sablani, S.S.; Rasco, B. Effect of acidic electrolyzed water-induced bacterial inhibition and injury in live clam (Venerupis philippinarum) and mussel (Mytilus edulis). *Int. J. Food Microbiol.* **2016**, *231*, 48–53. [CrossRef]
68. Rahman, S.M.E.; Park, J.; Song, K.B.; Al-Harbi, N.A.; Oh, D.H. Effects of slightly acidic low concentration electrolyzed water on microbiological, physicochemical, and sensory quality of fresh chicken breast meat. *J. Food Sci.* **2012**, *77*, M35–M41. [CrossRef]
69. Shimamura, Y.; Shinke, M.; Hiraishi, M.; Tsuchiya, Y.; Masuda, S. The application of alkaline and acidic electrolyzed water in the sterilization of chicken breasts and beef liver. *Food Sci. Nutr.* **2016**, *4*, 431–440. [CrossRef]
70. Cichoski, A.J.; Flores, D.R.M.; De Menezes, C.R.; Jacob-Lopes, E.; Zepka, L.Q.; Wagner, R.; Barin, J.S.; de Moraes Flores, É.M.; da Cruz Fernandes, M.; Campagnol, P.C.B. Ultrasound and slightly acid electrolyzed water application: An efficient combination to reduce the bacterial counts of chicken breast during pre-chilling. *Int. J. Food Microbiol.* **2019**, *301*, 27–33. [CrossRef]
71. Fabrizio, K.A.; Sharma, R.R.; Demirci, A.; Cutter, C.N. Comparison of electrolyzed oxidizing water with various antimicrobial interventions to reduce Salmonella species on poultry. *Poult. Sci.* **2002**, *81*, 1598–1605. [CrossRef] [PubMed]
72. Duan, D.; Wang, H.; Xue, S.; Li, M.; Xu, X. Application of disinfectant sprays after chilling to reduce the initial microbial load and extend the shelf-life of chilled chicken carcasses. *Food Control* **2017**, *75*, 70–77. [CrossRef]
73. Wang, H.; Qi, J.; Duan, D.; Dong, Y.; Xu, X.; Zhou, G. Combination of a novel designed spray cabinet and electrolyzed water to reduce microorganisms on chicken carcasses. *Food Control* **2018**, *86*, 200–206. [CrossRef]
74. Rasschaert, G.; Piessens, V.; Scheldeman, P.; Leleu, S.; Stals, A.; Herman, L.; Heyndrickx, M.; Messens, W. Efficacy of electrolyzed oxidizing water and lactic acid on the reduction of Campylobacter on naturally contaminated broiler carcasses during processing. *Poult. Sci.* **2013**, *92*, 1077–1084. [CrossRef] [PubMed]
75. Park, H.; Hung, Y.-C.; Brackett, R.E. Antimicrobial effect of electrolyzed water for inactivating Campylobacter jejuni during poultry washing. *Int. J. Food Microbiol.* **2002**, *72*, 77–83. [CrossRef]
76. Russell, S.M. The effect of electrolyzed oxidative water applied using electrostatic spraying on pathogenic and indicator bacteria on the surface of eggs. *Poult. Sci.* **2003**, *82*, 158–162. [CrossRef]

77. Zang, Y.T.; Bing, S.; Li, Y.J.; Shu, D.Q.; Huang, A.M.; Wu, H.X.; Lan, L.T.; Wu, H.D. Efficacy of slightly acidic electrolyzed water on the microbial safety and shelf life of shelled eggs. *Poult. Sci.* **2019**, *98*, 5932–5939. [CrossRef]
78. Bialka, K.; Demirci, A.; Knabel, S.; Patterson, P.; Puri, V. Efficacy of electrolyzed oxidizing water for the microbial safety and quality of eggs. *Poult. Sci.* **2004**, *83*, 2071–2078. [CrossRef]
79. Fasenko, G.M.; O'Dea Christopher, E.E.; McMullen, L.M. Spraying hatching eggs with electrolyzed oxidizing water reduces eggshell microbial load without compromising broiler production parameters. *Poult. Sci.* **2009**, *88*, 1121–1127. [CrossRef]
80. Killinger, K.M.; Kannan, A.; Bary, A.I.; Cogger, C.G. Validation of a 2 percent lactic acid antimicrobial rinse for mobile poultry slaughter operations. *J. Food Prot.* **2010**, *73*, 2079–2083. [CrossRef]
81. Veasey, S.; Muriana, P. Evaluation of Electrolytically-Generated Hypochlorous Acid ('Electrolyzed Water') for Sanitation of Meat and Meat-Contact Surfaces. *Foods* **2016**, *5*, 42. [CrossRef]
82. Wang, H.; Duan, D.; Wu, Z.; Xue, S.; Xu, X.; Zhou, G. Primary concerns regarding the application of electrolyzed water in the meat industry. *Food Control* **2019**, *95*, 50–56. [CrossRef]
83. Ding, T.; Rahman, S.M.E.; Purev, U.; Oh, D.H. Modelling of Escherichia coli O157:H7 growth at various storage temperatures on beef treated with electrolyzed oxidizing water. *J. Food Eng.* **2010**, *97*, 497–503. [CrossRef]
84. Liao, X.; Xiang, Q.; Cullen, P.J.; Su, Y.; Chen, S.; Ye, X.; Liu, D.; Ding, T. Plasma-activated water (PAW) and slightly acidic electrolyzed water (SAEW) as beef thawing media for enhancing microbiological safety. *LWT* **2020**, *117*, 108649. [CrossRef]
85. Sheng, X.; Shu, D.; Tang, X.; Zang, Y. Effects of slightly acidic electrolyzed water on the microbial quality and shelf life extension of beef during refrigeration. *Food Sci. Nutr.* **2018**, *6*, 1975–1981. [CrossRef] [PubMed]
86. Botta, C.; Ferrocino, I.; Cavallero, M.C.; Riva, S.; Giordano, M.; Cocolin, L. Potentially active spoilage bacteria community during the storage of vacuum packaged beefsteaks treated with aqueous ozone and electrolyzed water. *Int. J. Food Microbiol.* **2018**, *266*, 337–345. [CrossRef]
87. Kalchayanand, N.; Arthur, T.M.; Bosilevac, J.M.; Brichta-Harhay, D.M.; Guerini, M.N.; Wheeler, T.L.; Koohmaraie, M. Evaluation of Various Antimicrobial Interventions for the Reduction of Escherichia coli O157:H7 on Bovine Heads during Processing. *J. Food Prot.* **2008**, *71*, 621–624. [CrossRef]
88. Signorini, M.; Costa, M.; Teitelbaum, D.; Restovich, V.; Brasesco, H.; García, D.; Superno, V.; Petroli, S.; Bruzzone, M.; Arduini, V.; et al. Evaluation of decontamination efficacy of commonly used antimicrobial interventions for beef carcasses against Shiga toxin-producing Escherichia coli. *Meat Sci.* **2018**, *142*, 44–51. [CrossRef] [PubMed]
89. Kalit, S.; Kos, T.; Kalit, M.T.; Kos, I. The Efficacy of Electrolysed Oxidizing Water as a Disinfectant in the Dairy Industry. *J. Hyg. Eng. Des.* **2015**, *12*, 28–32.

processes

MDPI

Article

Effect of Heating Oxidation on the Surface/Interface Properties and Floatability of Anthracite Coal

Guoqiang Rong [1,2], Mengdi Xu [1,2,*], Dongyue Wang [3], Xiahui Gui [1,2,*] and Yaowen Xing [1,2,*]

1 Chinese National Engineering Research Center of Coal Preparation and Purification, China University of Mining and Technology, Xuzhou 221116, China; cumtrgq@126.com
2 School of Chemical Engineering and Technology, China University of Mining and Technology, Xuzhou 221116, China
3 School of Mineral Processing and Bioengineering, Central South University, Changsha 41000, China; dywang@csu.edu.cn
* Correspondence: cumtxmd@outlook.com (M.X.); guixiahui1985@163.com (X.G.); cumtxyw@cumt.edu.cn (Y.X.)

Received: 20 April 2019; Accepted: 22 May 2019; Published: 6 June 2019

Abstract: Oxidation processes of coal surfaces are both fundamental and interesting from academic and engineering points of view. In this work, we comprehensively analyzed the mechanism of heating oxidation at 200 °C on the surface/interface characters and the floatability of anthracite coal. The variations of surface/interface characters were studied using SEM (scanning electron microscopy), FTIR (Fourier transform infrared spectroscopy), and XPS (X-ray photoelectron spectroscopy). The floatability was further identified using Induction Time and Bubble-Particle Wrap Angle. It was found that, after heating oxidation at 200 °C, both surface ravines and oxygen-containing groups were increased. The degradation of hydroxyl on anthracite could be neglected during the heating, while the oxidation of hydrocarbon chains dominated the balance of hydrophobicity and hydrophilicity on coal surface. The induction time significantly increased from 200 ms to 1200 ms and 2000 ms after 10 h and 20 h of heating oxidation at 200 °C, respectively. Additionally, raw coal exhibited the fastest kinetics of bubble-particle attachment and the largest wrap angle, directly proving that the floatability decreased after oxidation.

Keywords: heating oxidation; surface/interface properties; floatability; induction time; bubble-particle wrap angle

1. Introduction

In China, the coal is currently the important energy source. Flotation is one of the most common methods for fine coal de-ashing that cannot be achieved by gravity separation [1–4]. It is based on different wetting properties between organic matters and gangue particles. Coal is inherently hydrophobic. Therefore, it can be easily captured by air bubbles in the water phase. Conversely, gangue particles such as silica and clays are inherently hydrophilic. As a result, they cannot get attached onto a bubble surface, sinking to the bottom of the flotation cell to become the tailings [3].

Oxidation processes of coal surfaces are both fundamental and interesting from academic and engineering point of view [5]. During the storage of coal slime, coal can be oxidized easily due to weather processes. On the other hand, coal spontaneous combustion can also make the coal particles oxidized. A great deal of heat is released during spontaneous combustion, leading to a serious oxidation of coal particles. Coal floatability and recovery decreased significantly after oxidation and cannot even be economically recovered via a traditional flotation method [6–11]. It is well known that oily collectors have been always used for coal flotation. However, the adsorption efficiency of oil collectors on oxidized coal surfaces was decreased significantly. To float oxidized coals, a number of works on polar collector, pretreatment, and new flotation procedures have been reported [6,12–16].

To understand the formation process of oxidized coal, the effect of high-temperature oxidation by heating (>500 °C) on the surface/interface characters and wettability of high-ranked coal and anthracite coal has been widely investigated [11,17]. It was found that the hydrophobicity always decreases after oxidation by heating since the content of hydrophobic functional groups decreases, while the number of hydrophilic groups increases [11,17]. One opposite finding is that, for low-ranked coal, the contact angle increased after oxidation by heating after decreasing in hydrophilic functional groups [15,18,19]. It is well known that two opposing phenomena occur that influence the hydrophobicity at the coal surface during oxidation processes with heat treatment. One is the degradation of oxygen-containing groups such as hydroxyl and carboxyl, thereby increasing the hydrophobicity. The other is the oxidation of hydrocarbon chains on the coal surface, producing new carboxyl and phenol groups. This in turn makes coal more hydrophilic [18,20]. Çınar [18] found that the improved floatability of low-ranked coal after oxidation by heating was mainly attributed to the removal of OH groups. It is speculated that heating for too long could oxidize low-ranked coal, hence decreasing its floatability. An open question is whether the hydrophobicity of high-ranked coal such as anthracite could also be enhanced by oxidation by heating. In previous works, it is noted that the heating temperature for treating anthracite was very high (>500 °C). Heating oxidation at 200 °C for anthracite coal has not been reported.

The objective of this study is to comprehensively understand the effect of oxidation on the surface/interface characters and floatability of anthracite by heating (200 °C). The changes of surface/interface characters of coal particles were confirmed using SEM, FTIR and XPS. The floatability of coal particles was further identified using induction time and the bubble-particle wrap angle.

2. Materials and Methods

2.1. Materials and Heating Oxidation Process

The anthracite sample was obtained from Xuehu Plant, China. The raw samples were crushed into small particles by a hammer mill, whose particle size was smaller than 0.5 mm. The ash content of anthracite was 21.30%. A muffle furnace was used for heating the coal sample. The interval of heating was 10 h and 20 h at 200 °C, whose process was the heating oxidation, respectively.

2.2. SEM

An SEM system was used to study the changes on surface morphology and elemental composition of coal before and after heating oxidation at 200 °C. The type of SEM system was FEI Quanta TM 250 (Hillsboro, OR, USA) which was equipped with Energy Dispersive X-ray Spectrometer (EDS). Gold sputtering treatment was carried out before the SEM test.

2.3. FTIR

FTIR (Nicolet is5, Thermo Scientific, USA) tests were performed in a wave number range of 4000 cm^{-1} to 400 cm^{-1} before and after heating oxidation at 200 °C. A coal sample of 2 mg mixed with KBr of 300 mg was ground into −2 μm. The pressure that was applied to prepare a thin circular plate was 30 MPa.

2.4. XPS

XPS system with 900 μm light spot size and Al Ka radiation (hv = 1486.6 eV) were used to examine the surface chemistry changes before and after heating oxidation at 200 °C. The type of XPS system was ESCALAB 250Xi, Thermo Scientific, Waltham, MA, America. XPS Peakfit software was used for data analysis [21]. The binding energies were calibrated with respect to the C1s hydrocarbon (-CH_2-CH_2-bonds) peak set to 284.6 eV.

2.5. Induction Time Test

A homemade induction time measuring system [21,22] was employed for determining the induction time between bubble and anthracite coal particles. This system contains a controller, high-speed camera, micro-syringe, micro-displacement sensor, actuator, illumination device, measuring cell, etc., as shown in Figure 1. The coal, whose particle size was 0.5–0.25 mm, was sieved for preparing a flat particle bed. A bubble produced at the end of capillary tube was driven to contact with the particle bed and then kept in contact for a prescribed time. The induction time was defined as the minimum contact time at which the attachment happened.

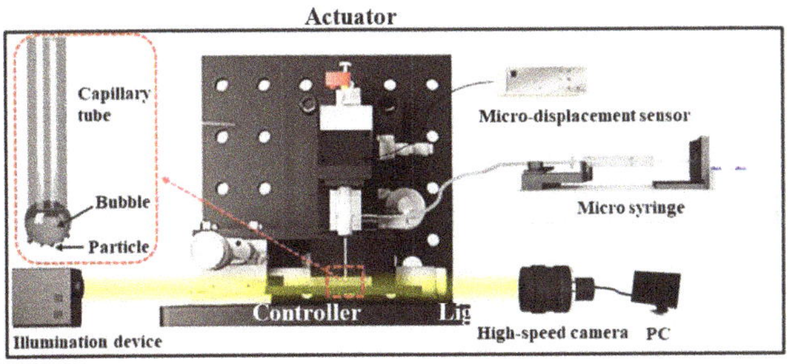

Figure 1. The induction time measuring system ([22], Reproduced with permission from Xia Y., Fuel; published by Elsevier, 2019).

2.6. Bubble-Particle Wrap Angle Experiment

The bubble-particle wrap angle was used to reveal coal floatability before and after heating oxidation at 200 °C. A schematic of the procedure to measure the Bubble-particle wrap angle (θ) is shown in Figure 2. In order to simulate the flotation process and indicate the situation of particles who attached onto bubble surfaces, the bubble-particle wrap angle experiment was introduced by Chu et al. firstly [23]. It provides the average coverage angle of the coal particles on bubble. The bubble-particle wrap angle (BPWA) was directly connected to the floatability of the coal particles. Attachment kinetics showed that better floatability occurred following the larger BPWA. 2 g of coal (particle size was 0.25–0.125 mm) was put into a 5 × 5 × 10 cubic cm container, together with 200 mL of ultrapure water. A rotor speed of 250 rpm was set for the magnetic stirrer during the experiment. The stirring time with rotatory period was set as 20, 40, 80 and 160 s, and bubble images were recorded accordingly. MB Ruler software was used to analysis and measure the wrap angle

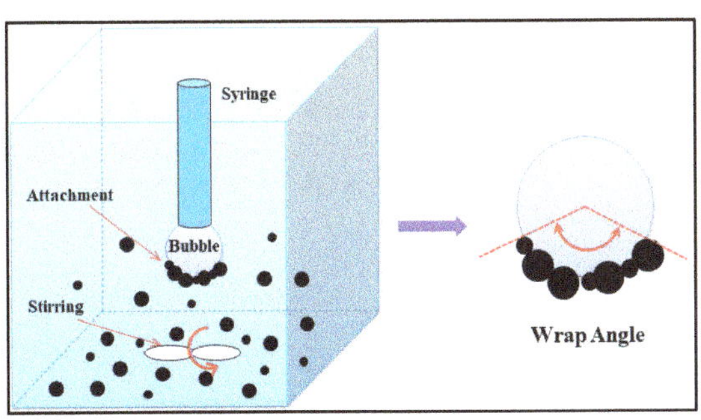

Figure 2. Schematic of bubble-particle wrap angle experiment ([24], Reproduced with permission from Xia Y., Fuel; published by Elsevier, 2019).

3. Results and Discussion

3.1. Effect of Heating Oxidation at 200 °C on Surface Morphology and Functional Groups of Anthracite Coal

The surface morphologies of anthracite coal before and after heating oxidation at 200 °C are shown in Figure 3. The EDS plane sweep and the chemical compositions before and after heating oxidation at 200 °C from the EDS plane sweep are depicted in Figures 4 and 5. LH-10 and LH-20 represented the coal sample after heating oxidation for 10 h and 20 h, respectively. As shown in Figure 3, the raw coal surface was quite smooth, while the surface after heating was characterized by lots of ravines and asperities. According to the Wenzel model, the hydrophilicity can be enhanced by surface roughness on the hydrophilic surface. The contact angle of the raw coal was approximately 80–90°, showing a relative high hydrophobicity, while the increased surface roughness was detrimental to the floatability. From EDS results, the content of organic carbon decreased from 74.58% to 51.32% and 46.01% after 10 h and 20 h of heating, respectively. In contrast, the oxygen content increased from 17.65% to 35.55% and 38.07%. This proves that the coal surface was significantly oxidized during the heating process. The organic matters were attacked by oxygen, generating oxygen-containing groups such as hydroxyl, carbonyl, and carboxyl. These oxygen-containing groups further decomposed into gases such as carbon dioxide, leaving inorganic minerals on the surface of anthracite coal. As a result, a lot of ravines were formed.

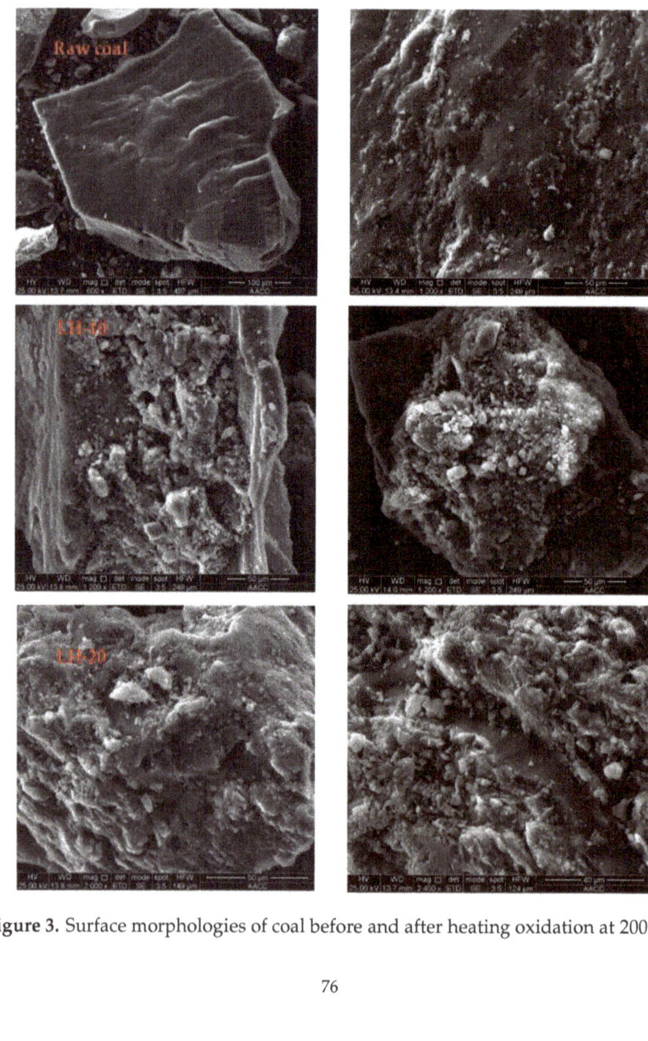

Figure 3. Surface morphologies of coal before and after heating oxidation at 200 °C.

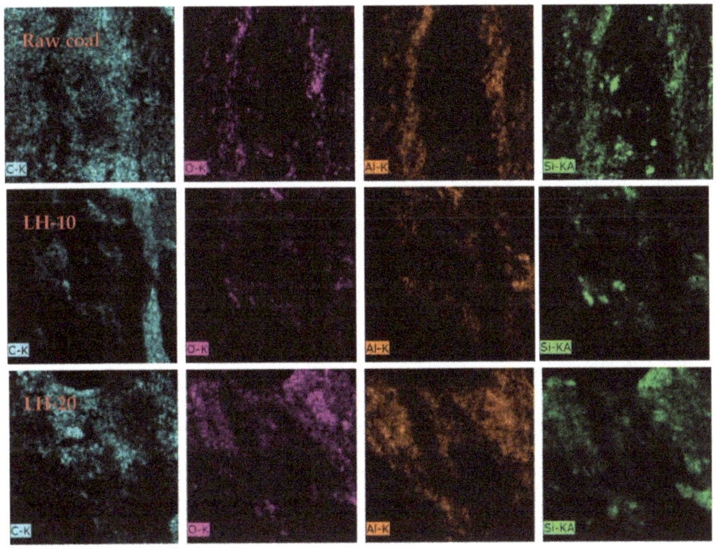

Figure 4. EDS plane sweep of the coal sample before and after heating oxidation at 200 °C.

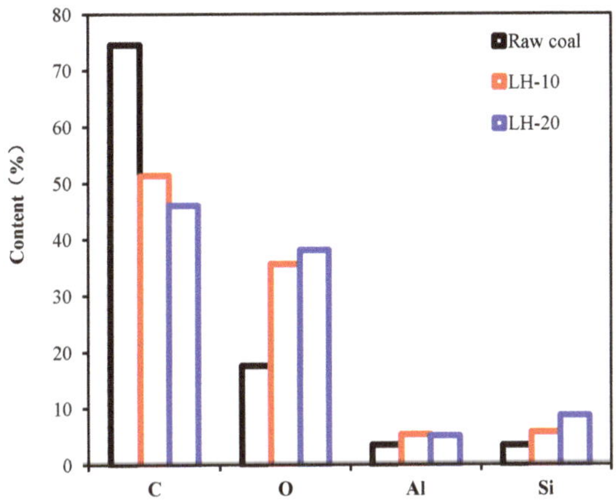

Figure 5. Chemical compositions before and after heating oxidation at 200 °C from EDS plane sweep.

FTIR of the coal before and after heating oxidation at 200 °C are shown in Figure 6. The peak at 3440 cm^{-1} can be attributed to the –OH group, whereas the peak at 1700 cm^{-1} can be attributed to the C=O group. The peaks at 3440 cm^{-1} and 1700 cm^{-1} both enhanced after heating. This means that the number of hydrophilic oxygen-containing groups increased, which is consistent with EDS results. Other research also indicated that the oxidization will lead to the increase in the hydrophilic oxygen-containing groups by using FTIR [25,26]. Unlike low-ranked coal, the number of –OH and C=O groups on anthracite surface did not decrease after heating, mainly because a small amount of these groups pre-existed on raw anthracite surface. Therefore, the degradation of hydroxyl could be neglected while the oxidation of hydrocarbon chains dominated the balance of hydrophobicity and hydrophilicity on an anthracite coal surface.

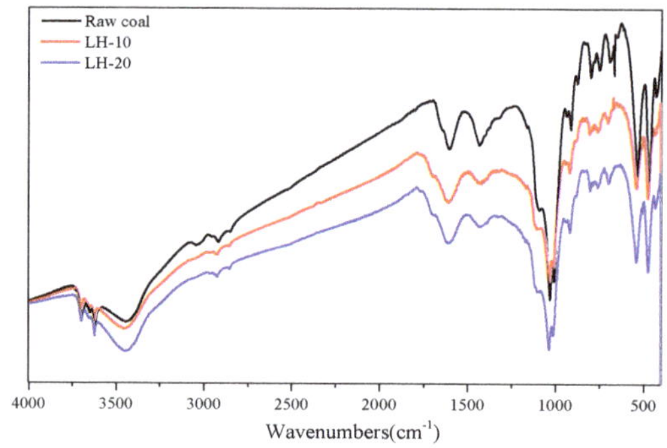

Figure 6. FTIR of the coal before and after heating oxidation at 200 °C.

XPS spectrums of the coal before and after heating oxidation at 200 °C are shown in Figure 7. The semi-quantitative results of the surface chemical composition from XPS wide spectrums show that the content of organic carbon decreased from 45.68% to 34.11% and to 28% after heating intervals of 10 h and 20 h, respectively. In contrast, the oxygen content increased from 36.54% to 44.27% and to 48.69%. Peak fitting of the XPS C1s peak provided quantitative results about the functional groups on the coal surface, as shown in Figure 8. The binding energies of O=C-O and C=O, C-O, C-H, and C-C were 289.1 eV, 286.6 eV, 285.60 eV, and 284.60 eV, respectively [23]. Functional groups of the surface of coal before and after heating oxidation at 200 °C from peak fitting of the XPS C1s peak are shown in Figure 9. The hydrophobic functional group C-C/C-H decreased after heating, while the total content of hydrophilic oxygen-containing groups (C-O, C=O, and COOH) increased. Remarkably, the content of COOH increased from 3.49% to 6.65% and to 6.67% after heating intervals of 10 h and 20 h, respectively, which is consistent with FTIR results. Xia et al. [11] also found that the heating process reduces the content of hydrophobic functional groups (C-H and C-C) but increases the content of hydrophilic functional groups (C-O, C=O and COOH). This change in the chemical groups led to the decrease in the hydrophobicity, which is closely related to the coal flotation. Based on the above analysis, the floatability of anthracite coal after heating oxidation should decrease due to the increased surface ravines and oxygen-containing groups.

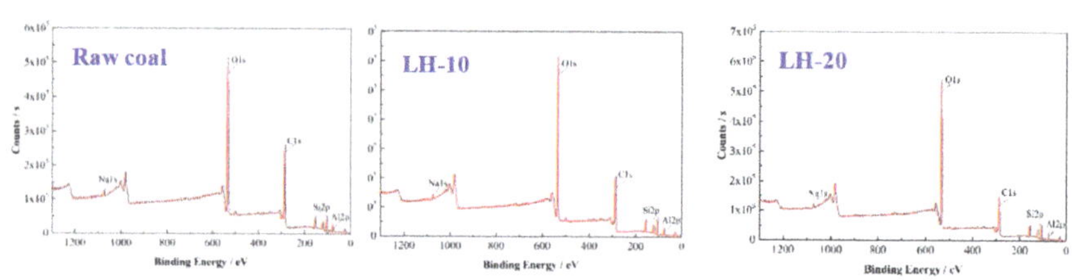

Figure 7. XPS wide energy spectrums of coal before and after heating oxidation at 200 °C.

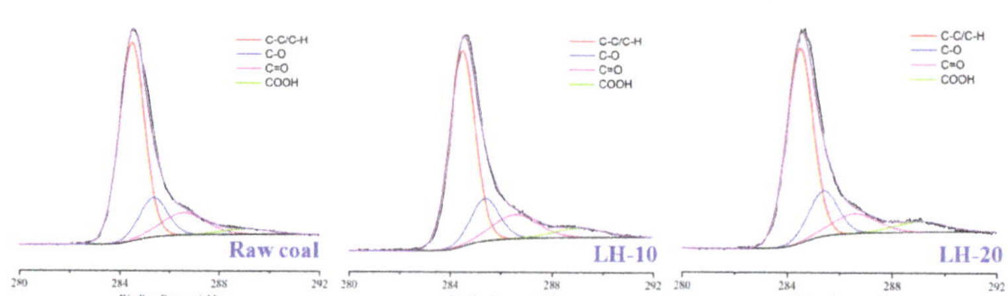

Figure 8. Peak fitting of the XPS C1s peak.

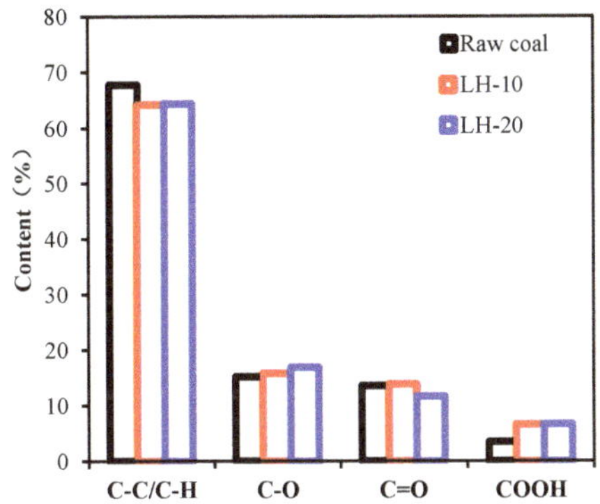

Figure 9. Functional groups of coal surface before and after heating oxidation at 200 °C from peak fitting of the XPS C1s peak.

3.2. *Effect of Heating Oxidation at 200 °C on the Floatability of Anthracite Coal*

The effect of oxidation on the induction time of anthracite coal is shown in Figure 10. The induction time significantly increased from 200 ms to 1200 ms and to 2000 ms, illustrating that coal floatability decreased after heating. On the one hand, hydrogen bonding could be easily formed between water molecules and the oxygen-containing groups on coal surfaces. On the other hand, the surface ravines were filled by water, preventing the bubble-particle attachment. These two factors led to film thinning and made a rupture between coal particle and bubble more difficult. More time is needed to complete the attachment process.

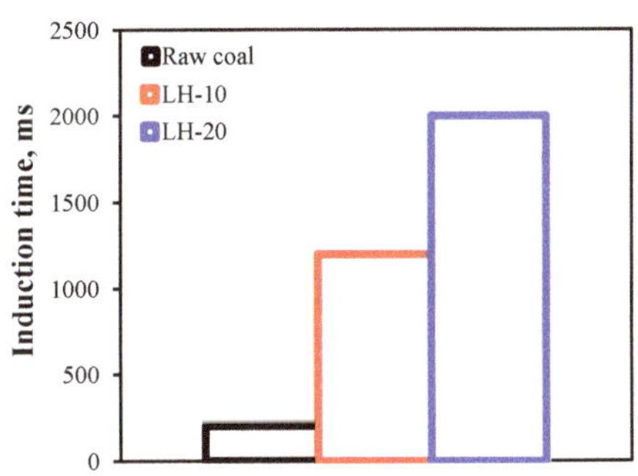

Figure 10. Induction time before and after heating oxidation at 200 °C.

The bubble-particle wrap angle before and after heating oxidation at 200 °C is shown in Figure 11. Given that the hydrodynamics and bubble-particle diameter were kept constant in all tests, the attachment angle kinetics was only determined by the competition between bubble-particle attachment and detachment efficiency, which is more suitable to describe the floatability compared with the induction time. The Bubble-particle warp angle had been used by Xia et al. [22,27] to confirm the floatability of low-ranked coal in the presence of different collectors. Raw coal clearly exhibited the fastest kinetics of the bubble-particle wrap angle compared with LH-10 and LH-20. The maximum attachment angle of raw coal was 61.0° while the maximum angle of LH-10 and LH-20 decreased to 40.2° and 30.3°, respectively. The increased oxygen-containing groups make coal hydrophilic, thereby leading to a decreased attachment and an increased detachment efficiency. These results directly illustrated that the floatability decreased after heating. In the study of Çınar [18], the floatability had been investigated by low-temperature heat treatment. After heating it was found that the floatability, hydrophobicity, and separation efficiency of hydrophilic coal changed dramatically. Chang [12] reported that coal flotation performance decreased with an increase in the degree of surface oxidation, which is consistent with our results.

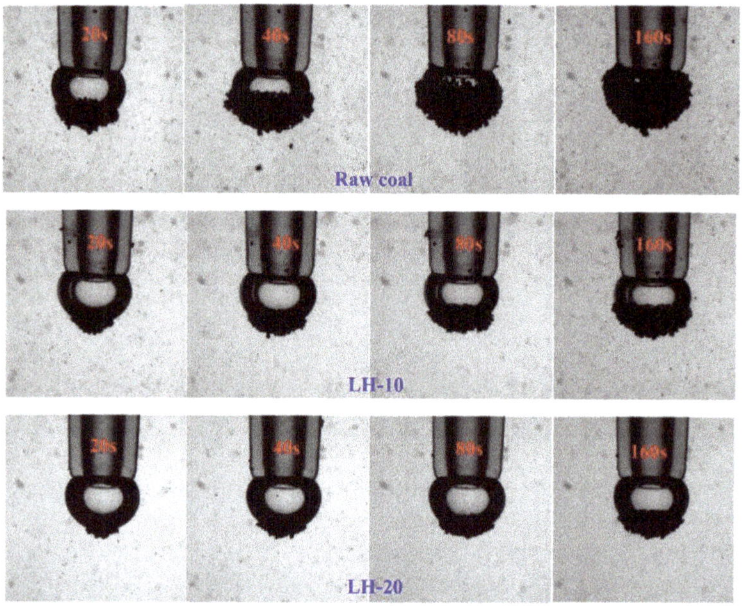

Figure 11. Bubble-particle wrap angle kinetics before and after heating oxidation at 200 °C.

4. Conclusions

The effect of heating oxidation at 200 °C on the surface/interface characters and floatability of anthracite coal were comprehensively studied in this paper. After heating oxidation, both surface ravines and oxygen-containing groups were observed to increase. The degradation of hydroxyl on anthracite could be neglected during the heating, while the oxidation of hydrocarbon chains dominated the balance of hydrophobicity and hydrophilicity on coal surface. The induction time significantly increased from 200 ms to 1200 ms and 2000 ms after oxidation intervals of 10 h and 20 h by heating at 200 °C, respectively. Furthermore, raw coal exhibited the fastest kinetics of bubble-particle attachment and the largest attachment angle, directly illustrating that the floatability decreased after heating oxidation at 200 °C. Overall, it was observed that the effect of heating oxidation at 200 °C on the floatability of anthracite coal was similar to that of high-temperature oxidation.

Author Contributions: Y.X. and G.R. conceived and designed the experiments; M.X. and D.W., performed the experiments; X.G. and G.R. analyzed the data; Y.X. and G.R. wrote the paper.

Funding: National Key R&D Program of China (No.: 2018YFC0604702), Jiangsu Province Science Fund for Distinguished Young Scholars (No.: BK20180032), China Postdoctoral Science Foundation funded project (2018M642369), Young Elite Scientists Sponsorship Program by CAST (2018QNRC001), and the National Nature Science Foundation of China (Grant No.: 51774286, 51574236).

Acknowledgments: This research was supported by National Key R&D Program of China (No.: 2018YFC0604702), Jiangsu Province Science Fund for Distinguished Young Scholars (No.: BK20180032), China Postdoctoral Science Foundation funded project (2018M642369), Young Elite Scientists Sponsorship Program by CAST (2018QNRC001), and the National Nature Science Foundation of China (Grant No.: 51774286, 51574236), for which the authors express their appreciation.

Conflicts of Interest: The authors declare no conflict of interest.

References

1. Xing, Y.; Gui, X.; Cao, Y.; Wang, D.; Zhang, H. Clean low-rank-coal purification technique combining cyclonic-static microbubble flotation column with collector emulsification. *J. Clean. Prod.* **2017**, *153*, 657–672. [CrossRef]
2. Ozkan, S.G. Further investigations on simultaneous ultrasonic coal flotation. *Minerals* **2017**, *7*, 177. [CrossRef]
3. Xing, Y.; Gui, X.; Pan, L.; Pinchasik, B.E.; Cao, Y.; Liu, J.; Kappl, M.; Butt, H.J. Recent experimental advances for understanding bubble–particle attachment in flotation. *Adv. Colloid Interface Sci.* **2017**, *246*, 105–132. [CrossRef]
4. Zhu, G.; Zhang, B.; Zhao, P.; Duan, C.; Zhao, Y.; Zhang, Z.; Yan, G.; Zhu, X.; Ding, W.; Rao, Z. Upgrading low-quality oil shale using high-density gas-solid fluidized bed. *Fuel* **2019**, *252*, 666–674. [CrossRef]
5. Li, E.; Lu, Y.; Cheng, F.; Wang, X.; Miller, J.D. Effect of oxidation on the wetting of coal surfaces by water: Experimental and molecular dynamics simulation studies. *Physicochem. Probl. Miner. Process.* **2018**, *54*, 1039–1051.
6. Armed, H.A.; Drzymala, J. Effect of flotation procedure and composition of reagents on yield of a difficult-to-float coal. *Physicochem. Probl. Miner. Process.* **2004**, *38*, 53–63.
7. Bolat, E.; Saglam, S.; Piskin, S. The effect of oxidation on the flotation properties of a Turkish bituminous coal. *Fuel Process. Technol.* **1998**, *55*, 101–105. [CrossRef]
8. Fuerstenau, D.W.; Diao, J. Characterization of coal oxidation and coal wetting behavior by film flotation. *Coal Prep.* **1992**, *10*, 1–17. [CrossRef]
9. Fuerstenau, D.W.; Yang, G.; Laskowski, J.S. Oxidation phenomena in coal flotation part I. Correlation between oxygen functional group concentration, immersion wettability and salt flotation response. *Coal Prep.* **1987**, *4*, 161–182. [CrossRef]
10. Sarikaya, M. Flotation test as a method for studying coal weathering. *Int. J. Miner. Process.* **1995**, *43*, 31–35. [CrossRef]
11. Xia, W.; Xie, G. Changes in the hydrophobicity of anthracite coals before and after high temperature heating process. *Powder Technol.* **2014**, *264*, 31–35. [CrossRef]
12. Chang, Z.; Chen, X.; Peng, Y. Understanding and improving the flotation of coals with different degrees of surface oxidation. *Powder Technol.* **2017**, *321*, 190–196. [CrossRef]
13. Gui, X.; Xing, Y.; Wang, T.; Cao, Y.; Miao, Z.; Xu, M. Intensification mechanism of oxidized coal flotation by using oxygen-containing collector α-furanacrylic acid. *Powder Technol.* **2017**, *305*, 109–116. [CrossRef]
14. Jia, R.; Harris, G.H.; Fuerstenau, D.W. An improved class of universal collectors for the flotation of oxidized and/or low-rank coal. *Int. J. Miner. Process.* **2000**, *58*, 99–118. [CrossRef]
15. Ozdemir, O.; Ersoy, O.F.; Guven, O.; Turgut, H.; Cinar, M.; Çelik, M.S. Improved flotation of heat treated lignite with saline solutions containing mono and multivalent ions. *Physicochem. Probl. Miner. Process.* **2018**, *54*, 1070–1082.
16. Sokolovic, J.M.; Stanojlovic, R.D.; Markovic, Z.S. Activation of oxidized surface of anthracite waste coal by attrition. *Physicochem. Probl. Miner. Process.* **2012**, *48*, 5–18.
17. Xia, W. The effect of heating on the wettability of lignite. *Energy Sources Part A* **2016**, *38*, 3521–3526. [CrossRef]
18. Çinar, M. Floatability and desulfurization of a low-rank (Turkish) coal by low-temperature heat treatment. *Fuel Process. Technol.* **2009**, *90*, 1300–1304. [CrossRef]
19. Ye, Y.; Jin, R.; Miller, J.D. Thermal treatment of low-rank coal and its relationship to flotation response. *Int. J. Coal Prep. Util.* **1988**, *6*, 1–16. [CrossRef]
20. Çelik, M.S.; Seyhan, K. Effect of heat treatment on the flotation of Turkish lignites. *Int. J. Coal Prep. Util.* **1995**, *16*, 65–79. [CrossRef]
21. Xu, M.; Xing, Y.; Cao, Y.; Gui, X. Effect of dodecane and oleic acid on the attachment between oxidized coal and bubbles. *Minerals* **2018**, *8*, 29. [CrossRef]
22. Xia, Y.; Zhang, R.; Xing, Y.; Gui, X. Improving the adsorption of oily collector on the surface of low-rank coal during flotation using a cationic surfactant: An experimental and molecular dynamics simulation study. *Fuel* **2019**, *235*, 687–695. [CrossRef]
23. Chu, P.; Mirnwzami, M.; Finch, J.A. Quantifying particle pick up at a pendant bubble: A study of non-hydrophobic particle–bubble interaction. *Miner. Eng.* **2014**, *55*, 162–164. [CrossRef]

24. Xia, Y.; Wang, L.; Zhang, R.; Yang, Z.; Xing, Y.; Gui, X.; Cao, Y.; Sun, W. Enhancement of flotation response of fine low-rank coal using positively charged microbubbles. *Fuel* **2019**, *245*, 505–513. [CrossRef]
25. Wang, B.; Peng, Y.; Vink, S. Diagnosis of the surface chemistry effects on fine coal flotation using saline water. *Energy Fuels* **2013**, *27*, 4869–4874. [CrossRef]
26. Wang, Y.; Cao, Y.; Li, G.; Liao, Y.; Xing, Y.; Gui, X. Combined effect of chemical composition and spreading velocity of collector on flotation performance of oxidized coal. *Powder Technol.* **2018**, *325*, 1–10. [CrossRef]
27. Xia, Y.; Yang, Z.; Zhang, R.; Xing, Y.; Gui, X. Enhancement of the surface hydrophobicity of low-rank coal by adsorbing DTAB: An experimental and molecular dynamics simulation study. *Fuel* **2019**, *239*, 145–152. [CrossRef]

processes

MDPI

Communication

Reusing Cow Manure for the Production of Activated Carbon Using Potassium Hydroxide (KOH) Activation Process and Its Liquid-Phase Adsorption Performance

Wen-Tien Tsai [1,*], Po-Cheng Huang [2] and Yu-Quan Lin [2]

1 Graduate Institute of Bioresources, National Pingtung University of Science and Technology, Pingtung 912, Taiwan
2 Department of Environmental Science and Engineering, National Pingtung University of Science and Technology, Pingtung 912, Taiwan; mike299123@gmail.com (P.-C.H.); wsx55222525@gmail.com (Y.-Q.L.)
* Correspondence: wttsai@mail.npust.edu.tw

Received: 18 September 2019; Accepted: 8 October 2019; Published: 14 October 2019

Abstract: In this work, cow manure (CM) was reused as a potential precursor in the production of activated carbon (AC) using a potassium hydroxide activation process at different temperatures (i.e., 500, 600 and 700 °C). The optimal activated carbon from cow manure (CM-AC) with high specific surface area (*ca.* 950 m^2/g) was further investigated for its adsorption performance in the removal of a model compound (i.e., methylene blue) from aqueous solution with various initial concentrations and adsorbent dosages at 25 °C. It was found that the resulting AC could be an effective adsorbent for removal of cationic dye from aqueous solution in comparison with a commercial coal-based AC. Based on the observations of the energy dispersive X-ray spectroscopy and Fourier transform infrared spectroscopy (FTIR), the CM-AC adsorbent has a stronger interaction with the cationic compound due to its more oxygen-containing complex on the surface. Furthermore, the adsorption kinetic parameters fitted using the pseudo-second order model with high correlations were in accordance with their pore properties.

Keywords: cow manure; chemical activation process; activated carbon; pore property; cationic pollutant; adsorption performance

1. Introduction

Water pollution caused by organic pollutants (e.g., dye, pesticide) is a serious issue because it can degrade water quality and render it toxic to public health and the environment. In general, there are two advanced treatment processes for removal of contaminants from the water bodies [1]. One belongs to destructive processes such as ozone oxidation and the Fenton method [2]. Other processes are recuperative. Among the recuperative process, adsorption may be the most used method due to its features of easy operation, high efficiency and simple design [3]. Therefore, carbon adsorption by granular activated carbon (GAC) or powdered activated carbon (PAC) has been considered as one of the best available technologies in the water and wastewater treatment processes for removal of organic pollutants [4,5]. However, the adsorption process is not a low-cost method because commercial activated carbon (AC) products have higher prices and the regeneration or disposal when they are exhausted will generate additional issues. In recent years, the use of low-cost precursors in the preparation of AC for the removal of various types of pollutants from water and wastewater has been reviewed [6–11]. Obviously, utilizing organic wastes or residues as AC precursors poses several advantages such as climate change mitigation (reduction of greenhouse gas emissions), biomass energy production (reusing exhausted AC as auxiliary fuel), and waste management (improvement of environmental quality).

Cattle manure or dung is a renewable resource because it mainly comes from the undigested residue of cellulose-based feeds being excreted by livestock animal species such as cow (dairy cattle) and bull. Generally, cattle manure was converted into fertilizer or soil amendment by composting because composted manure contains many dissolved organic matter (DOM) fractions [12]. Without proper treatment or management, it can cause some environmental problems, including air quality deterioration (e.g., odor), public hazards (e.g., infectious pathogen and asphyxia poisoning), greenhouse gas emissions (e.g., carbon dioxide and methane), and water pollution (e.g., eutrophication) [13]. Due to its high carbon content, this biomass resource can be thermally converted into various forms of energy sources (i.e., heat or electricity) [13–20] and biochar used as an adsorbent [21–26], but this approach may encounter some drawbacks, including low energy yield and air pollutants (e.g., sulfur oxides, nitrogen oxides, carbon monoxide and particulate matters) being emitted [27]. In order to upgrade its additional value of reuse, the production of AC from cow manure and its adsorption applications have been reported in recent years [28–36].

AC may be the frequently used carbon material in different state-of-the-art applications due to its physical and electrochemical properties, including high surface area, wide pore distribution and large specific capacitance [10,37]. In general, there are two methods for producing AC: physical and chemical activation processes [38]. Physical activation involves carbonization and activation in a single step or two separate steps at higher temperature under the atmosphere of gasification gas (i.e., steam or carbon dioxide). The resulting AC possesses poor pore properties as compared to AC prepared from chemical activation. By contrast, AC production using chemical activation is performed in a single step at a lower temperature using precursor impregnated with chemical reagents (i.e., potassium hydroxide, phosphoric acid, or zinc chloride). However, some impurities like potassium (K), zinc (Zn) or phosphorus (P), depending on the chemical reagents used, can be found in AC products, thus raising the production costs due to the additional processes for removing these residual activation reagents. As reviewed above, little research has performed the reuse of cow manure or cattle manure as a feedstock for producing AC. In the reports by Machida's group [28–36], they first studied the production of AC from cattle manure compost (CMC) using chemical activation with zinc chloride ($ZnCl_2$). Furthermore, the CMC-derived activated carbon was tested to evaluate the liquid-phase adsorption performances of various adsorption targets, including phenol, methylene blue, water vapor, metal ions (i.e., Cu, Pb), and 2-methylisoborneol. In another recent study by Li et al. [36], AC was also produced from cow manure using chemical activators like potassium carbonate (K_2CO_3) and $ZnCl_2$, showing that the values of specific surface area range from 114 to 893 m^2/g. The authors also treated the wastewater from a cow farm with the resulting ACs.

Although many studies have reported the utilization of cow manure as a precursor for producing ACs [28–36], they were mostly produced using chemical activation with $ZnCl_2$ and potassium carbonate (K_2CO_3). On the other hand, the adsorption performances of the resulting AC products were not compared with commercial AC. Thus, the main purposes of this paper were to produce highly porous AC from cow manure using a potassium hydroxide (KOH) activation process at different temperatures. Subsequently, the resulting AC with the optimal pore properties was performed to evaluate its effectiveness in the removal of cationic dye (i.e., methylene blue) from the aqueous solution. Finally, the adsorption performances for the removal of methylene blue from aqueous solution were compared between the CM-based AC and commercial coal-based AC.

2. Materials and Methods

2.1. Materials

The precursor for producing AC was obtained from the Livestock Research Institute (Tainan, Taiwan). The cow (Holstein) manure (CM) sample was dried using a hot air circulating oven for about one week. The methods for characterizing its main thermochemical properties were similar to the previous studies [26,39], showing that the dried CM is composed mainly of volatile matter (about 80 wt%)

with 42 wt% carbon. The potassium hydroxide (KOH; Merck Co., Darmstadt, Germany) was used to impregnate CM to produce AC using chemical activation. As methylene blue was commonly selected as a probe compound for determining the adsorption capacity of AC quickly [37], this cationic dye, which was purchased from Merck Co., was used as an organic pollutant target in the work. In order to compare the adsorption performance of the resulting AC with commercial AC, a product (Filtrasorb-300; Calgon Carbon Co., Moon Township, PA, USA) for liquid-phase applications was used to remove methylene blue (MB) at the same adsorption conditions.

2.2. Impregnation and Activation Experiments

The preparation of AC from CM was carried out using the KOH activation process referred to in a previous study [40]. In the present study, the CM impregnated with KOH (KOH-CM) was described as follows: 10 g of CM was mixed with 5 g of KOH in 150 cm^3 of de-ionized water and stirred using a magnetic bar at about 75 °C for 30 min, and then decanted into the supernatant prior to being dried at about 105 °C for 24–48 h. Subsequently, the KOH-CM (about 3 g) was activated by passing nitrogen gas (500 cm^3/min) at a heating rate of about 10 °C/min up to the specified temperature (i.e., 500, 600, and 700 °C, respectively). Similarly to the procedures in the impregnation with KOH, the AC product was poured into 3 N HCl solution, then heated at about 75 °C for 30 min and finally rinsed with deionized water (150 cm^3) three times to remove the residual KOH and inherent ash minerals effectively. Prior to the determination of pore properties, the resulting AC products (denoted as CM-AC-500, CM-AC-600 and CM-AC-700) were dried at about 105 °C for 24 h.

2.3. Characterization of Resulting and Commercial ACs

The pore properties of the resulting AC (CM-AC) products and the commercial AC (Filtrasorb-300) were determined by its N_2 adsorption-desorption isotherms at the temperature of liquid nitrogen (i.e., −196 °C) on the Micromeritics ASAP 2020 accelerated surface area and porosimetry system. Based on the analytical methods applied, the types of pore characterization included specific surface area, pore volume and pore size. For example, the specific surface area of the resulting AC can be obtained using the Brunauer–Emmett–Teller (BET) method in the relative pressure (P/P_0) range of 0.05–0.30. The total pore volume is indicative of the liquid nitrogen amount adsorbed at P/P_0 of about 0.99 [41]. According to the values of the BET surface area and total pore volume, the mean pore diameter or width can be further estimated by assuming the cylindrical geometry [42].

In order to elucidate the elemental compositions of AC on the surface, energy dispersive X-ray spectroscopy (Swift ED3000, Oxford Instruments, Abingdon, UK) was conducted to observe their contents of carbon, oxygen and other elements. The data could be associated with the interaction between the AC adsorbent and target adsorbate in the liquid-phase adsorption system. Using the potassium bromide (KBr) pellet technique, the Fourier infrared spectrometer (FT/IR-4600, JASCO Co., Tokyo, Japan) was further used to measure the functional groups on the surface of the resulting AC (CM-AC) and the commercial AC (Filtrasorb-300) in the range 4000–400 cm^{-1}.

2.4. Adsorption Performance Experiments

Following the adsorption experiments referred to in a previous study [43], the batch adsorption kinetics of the CM-AC-700 were carried out in a 3-L mixing tank with four baffles. In the work, the data on the adsorption uptake of methylene blue (MB) by the MC-AC-700 from aqueous solution (2 liters) were obtained at the fixed solution temperature of 25 °C and agitation speed of 200 rpm under the determining adsorption process parameters, including initial MB concentrations (i.e., C_0 = 5, 10, 15 and 20 mg/L) and AC dosages (i.e., 0.1, 0.3 and 0.5 g, based on the solution volumes of 2 L, which were abbreviated by 0.1, 0.3 and 0.5 g/2 L). During the adsorption experiments, an aliquot solution (about 10 cm^3) was taken out at specified intervals (i.e., 5, 10, 20, 30, 40, 50 and 60 min). After filtrating with mixed cellulose esters membrane filter with a size of 25 mm, the analysis of dye concentration (i.e., C_t) in the remaining filtrate solution was immediately measured with a UV/Visible spectrophotometer (U-2900, Hitachi Co., Tokyo, Japan) at the maximum absorption wavelength (i.e., 661 nm). It should be noted that about half of the solution sample after the filtration was first discarded to dye the filter completely. The adsorbed amount of MB (q_t, mg/g) was obtained by the AC mass, solution volume (i.e., 2 L), and the difference between C_0 and C_t. On the other hand, this work also compared the adsorption performance of a commercial AC (i.e., Filtrasorb-300) with the CM-AC-700. The adsorption conditions were fixed at an initial MB concentration of 10 mg/L, AC dosage of 0.3 g/2 L, agitation speed of 200 rpm, and temperature of 25 °C.

3. Results and Discussion

3.1. Pore Properties of Resulting and Commercial ACs

Table 1 listed the pore properties of the resulting ACs (CM-AC) and commercial AC (Filtrasorb 300), including surface area, pore volume and average pore diameter. Obviously, an increase in the activation temperature created the higher pore properties due to the development of more micropores. In addition, CM-AC-700 possessed favorable pore properties by comparison with Filtrasorb-300. For instance, the BET surface area of CM-AC-700 (i.e., 946 m^2/g) is slightly higher than that of Filtrasorb-300 (i.e., 660 m^2/g). Based on the data of surface area, regardless of the calculation methods, the ratio of that of Filtrasorb-300 to that of CM-AC-700 is close to 0.70. Using the data of total pore volume, the ratio of that of Filtrasorb-300 to that of CM-AC-700 was 0.68, close to 0.70. This result gave an indication of the pore size difference between CM-AC-700 and Filtrasorb-300. As mentioned above, we confidently inferred that the contribution to total pore volume of Filtrasorb-300 should be mostly from micropores. These results were in accordance with their average pore diameters (3.04 vs. 2.32 nm, listed in Table 1). As shown in Table 1, the BET surface area values of the resulting AC products (CM-AC-600 and CM-AC-700) are higher than that (445 m^2/g) of biomass-based AC prepared from plant (*Jatropha curcas* L.) root with similar KOH activation conditions [44]. For large adsorbate molecules having similar dimensions to MB, CM-AC-700 could be preferable to remove them from the aqueous system due to the steric sizes of adsorbate molecules [45].

Table 1. Pore properties of resulting activated carbons (ACs), cow manure-derived activated carbon (CM-AC) and commercial AC (Filtrasorb-300).

Property	CM-AC-500	CM-AC-600	CM-AC-700	Filtrasorb-300
Single point surface area (m^2/g) [a]	385.0	634.0	912.6	643.0
BET surface area (m^2/g) [b]	383.3	627.9	946.3	660.0
Langmuir surface area (m^2/g)	576.0	935.9	1297.6	934.6
Micropore surface area (m^2/g) [c]	212.8	425.5	673.8	499.4
External surface area (m^2/g) [d]	170.5	202.4	272.5	160.6
Total pore volume (cm^3/g) [e]	0.371	0.560	0.719	0.382
Micropore volume (cm^3/g) [c]	0.113	0.223	0.330	0.250
Pore diameter (Å) [f]	38.7	35.7	30.4	23.2

[a] Obtained at relative pressure (P/P_0) of about 0.30. [b] Calculated in the range 0.05 < relative pressure (P/P_0) < 0.30. [c] Micropore area by *t*-plot method. [d] Obtained by subtracting micropore surface area from Brunauer–Emmett–Teller (BET) surface area. [e] Total pore volume obtained at relative pressure (P/P_0) of about 0.99. [f] Obtained from the values of total pore volume and BET surface area.

To describe pore structures of the AC studied, the adsorption-desorption isotherms of nitrogen (N_2) at liquid nitrogen temperature (i.e., −196 °C) is the commonly used method [41]. Figure 1 showed the N_2 adsorption-desorption isotherms of CM-AC-700 and Filtrasorb-300. Obviously, these ACs belonged to type I isotherms due to their microporous feature (i.e., pore width < 2 nm) [46]. Because of its high potential for adsorption, the micropore filling occurs significantly at P/P_0 of < 0.05 [46]. On the other hand, the hysteresis loops occurred at the isotherms from the adsorption to desorption route, especially for CM-AC-700. It is well known that the loop is associated with mesoporous solids, where capillary condensation occurs at higher relative pressures. As listed in Table 1, these isotherms

(Figure 1) also confirmed that CM-AC-700 has larger pore properties with wider pore size in comparison with Filtrasorb-300. Furthermore, the hysteresis loop pattern can be correlated with corresponding pore shapes. According to the classification of adsorption hysteresis loops recommended by the International Union of Pure and Applied Chemistry (IUPAC), the isotherms in Figure 1 belong to the type H4 loops [41], which are given by many ACs and some other microporous adsorbents with slit-shaped pores.

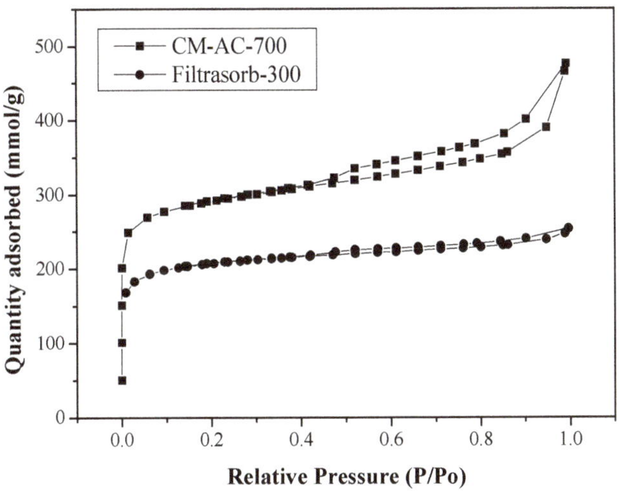

Figure 1. N_2 adsorption-desorption isotherms of cow manure-derived activated carbon (CM-AC) and commercial activated carbon (i.e., Filtrasorb-300).

3.2. Adsorption Performances of AC

Figures 2 and 3 show the adsorption performances of CM-AC-700 for MB removal at the specific adsorption conditions under various initial MB concentrations and AC dosages, respectively. Obviously, the fast decrease in residual MB concentrations (C_t/C_0) occurred at a limited time. It implies that strong interactions between the cationic dye and the resulting AC exist. In this work, the pseudo-second order model was used to fit the adsorption system [47]. In order to fit the adsorption kinetic data, its linear form can be expressed as:

$$t/q_t = 1/(k \times q_e^{?}) + (1/q_e) \times t, \tag{1}$$

where q_t is the adsorbed amount of MB at time t (mg/g), q_e is the adsorbed amount of MB fitted at equilibrium (mg/g), k is the rate constant (g/(mg·min)), and t is the time (min). Furthermore, the time required for the adsorbent (i.e., CM-AC) to take up half as much adsorbate (i.e., MB) as occurs at equilibrium (i.e., $t = t_{1/2}$ as $q_t = q_e/2$) can be obtained by the following equation.

$$t_{1/2} = 1/(k \times q_e) \tag{2}$$

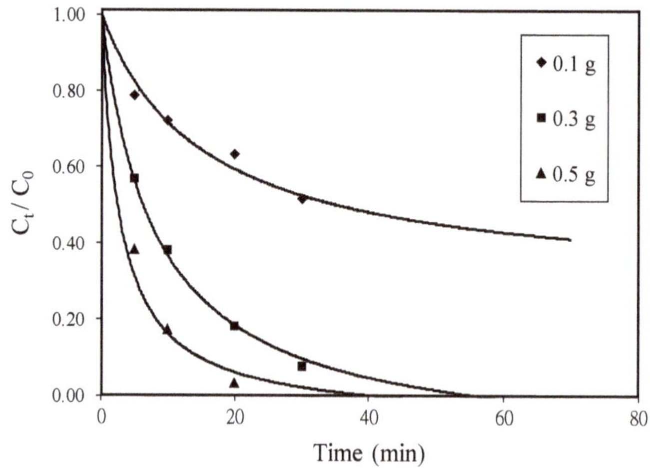

Figure 2. Effect of adsorbent CM-AC-700 dosage on methylene blue (MB) adsorption (initial MB concentrations = 10 mg/L, agitation speed = 200 rpm, and temperature = 25 °C); symbols: measurement data, full lines: calculated from the fitted values (Table 2).

Table 2. Kinetic parameters for methylene blue (MB) adsorption onto adsorbent CM-AC-700 at various adsorbent dosages based on pseudo-second order model [a].

Adsorbent Dosage (g/2 L)	k (g/mg·min)	q_e (mg/g)	Correlation Coefficient	$t_{1/2}$ (min)
0.1	0.0005	142.86	0.987	14.0
0.3	0.0016	76.34	0.999	8.2
0.5	0.0084	42.74	0.999	2.8

[a] Adsorption conditions: initial MB concentration = 10 mg/L, agitation speed = 200 rpm, and temperature = 25 °C.

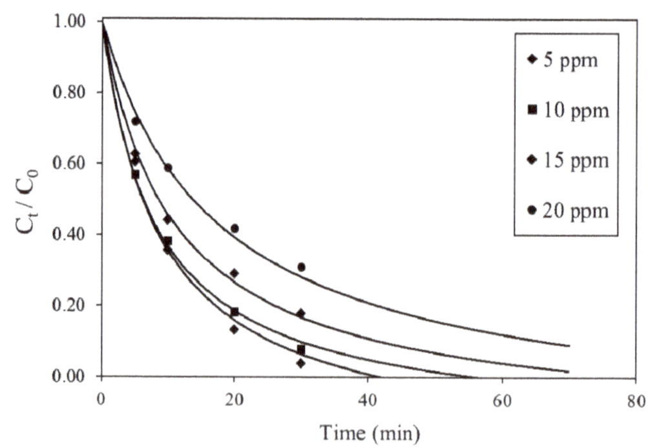

Figure 3. Effect of initial MB concentration on MB adsorption; adsorption conditions: adsorbent CM-AC-700 dosage = 0.3 g/2 L, adsorption temperature = 25 °C, and agitation speed = 200 rpm; symbols: measurement data, full lines: calculated from the fitted values (Table 3).

Table 3. Kinetic parameters for methylene blue (MB) adsorption onto adsorbent CM-AC-700 at initial MB concentrations based on pseudo-second order model [a].

Initial MB Concentration (mg/L)	k (g/mg·min)	q_e (mg/g)	Correlation Coefficient	$t_{1/2}$ (min)
5	0.0028	40.32	0.993	8.9
10	0.0016	76.34	0.999	8.2
15	0.0008	113.64	0.999	11.0
20	0.0004	151.52	0.996	16.5

[a] Adsorption conditions: adsorbent MC-AC dosage = 0.3 g/2 L, agitation speed = 200 rpm, and temperature = 25 °C.

The fitted values of kinetic parameters for the adsorption system were listed in Tables 2 and 3. Obviously, the kinetics of MB adsorption into CM-AC followed this model with high correlation coefficients (>0.99). As listed in Table 2, the fitted adsorption capacity (i.e., q_e) decreased as the initial MB concentration (i.e., C_0) increased. The decrease of q_e with increasing adsorbent dosage implies the exhaustion of adsorption onto excessive adsorbent due to the limited MB molecules. On the other hand, the rate constant values significantly increased at higher adsorbent dosage because of the increase in more adsorption sites provided. This result agrees with previous studies with relation to the adsorption of dyes onto activated carbon and clay [45,48,49]. On the other hand, the fitted adsorption capacity (seen in Table 3) indicated an increasing trend as the initial MB concentration (i.e., C_0) increased. This result can be expected because more adsorbate molecules were adsorbed at the equilibrium condition onto the adsorbent without exhausting its adsorption.

Figure 4 shows the adsorption kinetic profiles of CM-AC-700 in comparison with that of the commercial AC at the same conditions, that is, initial MB concentration of 10 mg/L, CM-AC-700 dosage of 0.3 g/2 L, agitation speed of 200 rpm, and temperature of 25 °C. As listed in Table 4, CM-AC-700 exhibited a better adsorption performance for the removal of MB from the aqueous solution as compared to the commercial AC (Filtrasorb-300). This difference between the fitted values of adsorption capacities (i.e., q_e) can be associated with the interactions between the adsorbate and the adsorbent, including the pore properties of the adsorbent for physical adsorption and the electrostatic attraction interaction between the cationic dye and the negatively charged surface [44]. The observations using energy dispersive X-ray spectroscopy (Figure 5) showed a higher content of oxygen on the CM-AC-700 surface in comparison with the adsorbent Filtrasorb-300, thus enhancing its polar nature (i.e., hydrophilicity, acidity) due to more oxygen complexes or groups on the surface [50]. These surface oxygen complexes and oxygen-containing functional groups (e.g., carboxylic, carbonyl, phenolic groups) could arise from its lignocellulosic constituents and ashes (consisting mainly of silica, alumina, alkaline and alkaline earth metals) in the starting materials (i.e., CM). The carbon contents of the resulting carbon materials were further analyzed using an elemental analyzer and showed a range from 60 to 65 wt%, while that of the commercial activated carbon (Filtrasorb-300) demonstrated a range up to about 90 wt%. These results were consistent with the energy dispersive X-ray spectroscopy analyses (Figure 5). Therefore, the interaction between the cationic dye (MB molecule) and the adsorbent CM-AC-700 via stronger van der Waals forces or other binding forms also plays a vital role in the liquid-phase adsorption. On the other hand, Figure 6 depicted the Fourier transform infrared spectroscopy (FTIR) spectra of CM-AC700 and Filtrasorb-300. The peaks of CM-AC-700 at about 3440, 2400, 1630, 1380, 1180 and 620 cm^{-1} could be mostly associated with functional groups containing oxygen (i.e., O–H, C=O, C–O) [51], which were more significant than those of Filtrasorb-300. For example, there is a broad absorption peak at 3440 cm^{-1} in the CM-AC-700, which is assigned to the stretching vibration of O–H. In addition, a sharp peak at about 3440 cm^{-1} may correspond to the bending band of O–H. The interaction between the cationic dye and the adsorbent CM-AC-700 via stronger van der Waals forces or other binding forms also plays a vital role in the adsorption process.

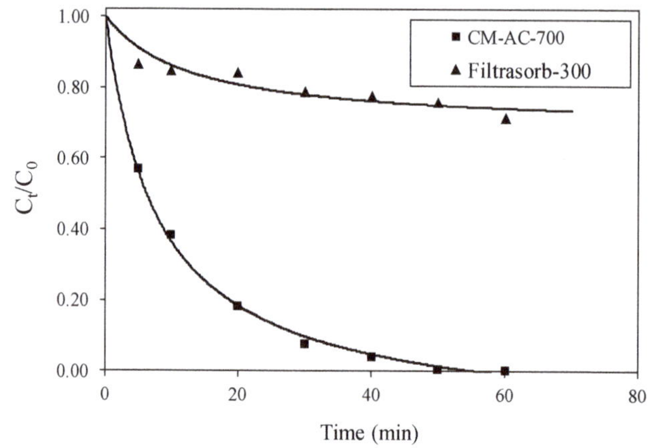

Figure 4. Effect of different ACs on MB adsorption; adsorption conditions: initial MB concentration = 10 mg/L, adsorbent dosage = 0.2 g/2 L, adsorption temperature = 25 °C, and agitation speed = 200 rpm; symbols: experimental data, full lines: calculated from Equation (1) and Table 4.

Table 4. Kinetic parameters for methylene blue (MB) adsorption onto resulting AC (CM-AC-700) and commercial AC (Filtrasorb-300) based on pseudo-second order model [a].

Adsorbent	k (g/mg·min)	q_e (mg/g)	Correlation Coefficient	$t_{1/2}$ (min)
CM-AC-700	0.0016	76.34	0.999	8.2
Filtrasorb-300	0.0038	20.88	0.944	12.6

[a] Adsorption conditions: initial MB concentration = 10 mg/L, adsorbent dosage = 0.3 g/2 L, agitation speed = 200 rpm, and temperature = 25 °C.

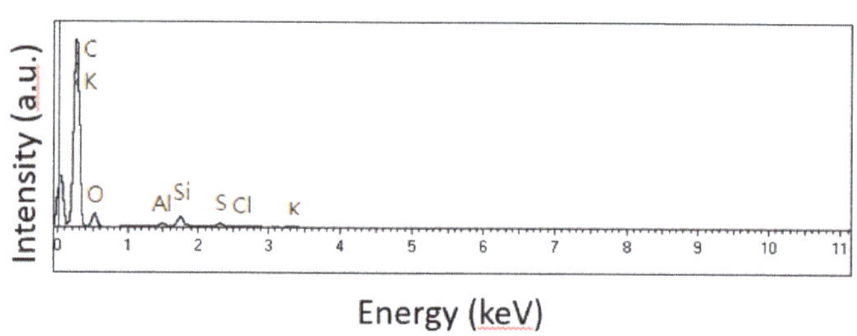

Element	Weight%	Atomic%
Carbon	59.705	68.236
Oxygen	33.389	28.648
Aluminum	0.719	0.366
Silicon	3.386	1.655
Sulfur	1.209	0.518
Chlorine	0.523	0.202
Potassium	1.069	0.375

(**a**) cow manure-derived activated carbon (CM-AC-700)

Figure 5. *Cont.*

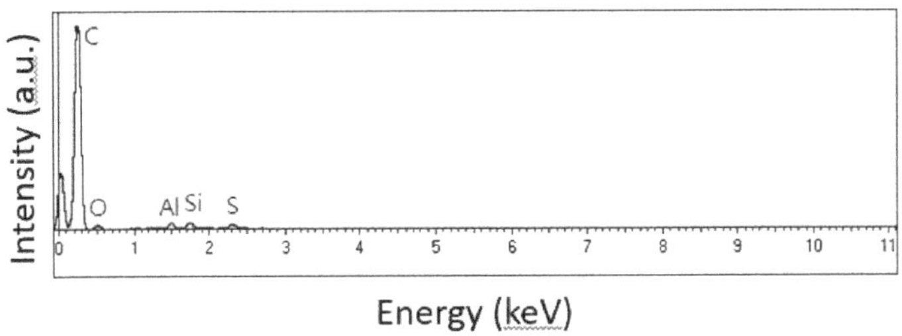

Element	Weight%	Atomic%
Carbon	89.628	93.069
Oxygen	7.319	5.706
Aluminum	0.821	0.379
Silicon	0.937	0.416
Sulfur	0.846	0.329
Iron	0.449	0.100

(**b**) Commercial activated carbon (i.e., Filtrasorb-300)

Figure 5. Energy dispersive X-ray spectroscopy spectra of (**a**) CM-AC-700 and (**b**) Filtrasorb-300.

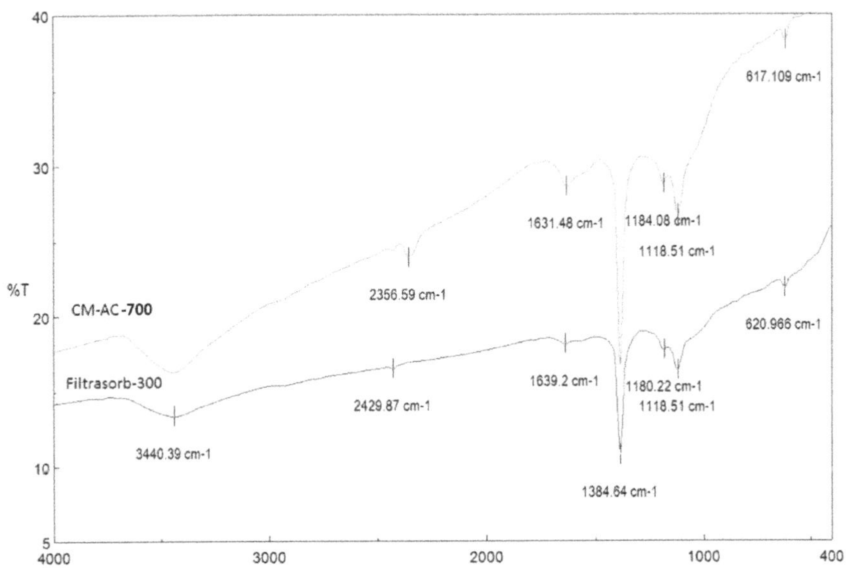

Figure 6. Fourier transform infrared spectroscopy (FTIR) spectra of CM-AC-700 and Filtrasorb-300.

4. Conclusions

The thermochemical conversion of cattle manure can be positively related to climate change mitigation (reduction of greenhouse gas emissions) and waste and wastewater management (improvement of environmental quality). In this work, based on the experimental results of the chemical, pore and adsorption properties for the resulting and commercial ACs, cattle (cow) manure could be an excellent precursor for producing AC at different temperatures (i.e., 500, 600 and 700 °C) using a KOH activation process. The conclusions can be itemized as follows:

- The nitrogen adsorption-desorption isotherms of the resulting ACs showed that an increase in the activation temperature created the higher pore properties due to the development of more micropores.
- The optimal AC product (i.e., CM-AC-700) showed its microporous feature with the pore properties of a BET surface area of 946 m^2/g and total pore volume of about 0.72 cm^3/g, which are close to those of common commercial ACs.
- The resulting AC could be an effective low-cost adsorbent for the removal of cationic dye (i.e., methylene blue, MB) from aqueous solution based on the fittings of the pseudo-second order model.
- The resulting AC exhibited a better adsorption performance for the removal of MB from the aqueous solution as compared to the commercial AC.
- The cow manure-derived adsorbent has a stronger interaction due to its larger pore properties and more oxygen-containing complex on the surface based on the observations of energy dispersive X-ray spectroscopy and Fourier transform infrared spectroscopy (FTIR).

Author Contributions: Conceptualization, W.-T.T.; methodology, P.-C.H.; validation, P.-C.H.; data curation, Y.-Q.L.; formal analysis, Y.-Q.L; Writing—Original draft preparation, W.-T.T.; Writing—Review and editing, W.-T.T.

Funding: This research received no external funding.

Acknowledgments: The authors express sincere appreciation to the Instrument Center of National Pingtung University of Science and Technology for assistance with the energy dispersive X-ray spectroscopy (EDS) observation.

Conflicts of Interest: The authors declare no conflicts of interest.

References

1. Teodosiu, C.; Gilca, A.F.; Barjoveanu, G.; Fiore, S. Emerging pollutants removal through advanced drinking water treatment: A review on processes and environmental performances assessment. *J. Clean. Prod.* **2018**, *197*, 1210–1221. [CrossRef]
2. Hodaifa, G.; Callardo, P.A.R.; Garcia, C.A.; Kowalska, M.; Seyedsalehi, M. Chemical oxidation methods for treatment of real industrial olive oil mill wastewater. *J. Taiwan Inst. Chem. Eng.* **2019**, *97*, 247–254. [CrossRef]
3. Garcia, J.R.; Sedran, U.; Zaini, M.A.A.; Zakaria, Z.A. Preparation, characterization, and dye removal study of activated carbon prepared from palm kernel shell. *Environ. Sci. Pollut. Res.* **2018**, *25*, 5076–5085. [CrossRef] [PubMed]
4. Jeirani, Z.; Niu, C.H.; Soltan, J. Adsorption of emerging pollutants on activated carbon. *Rev. Chem. Eng.* **2017**, *33*, 491–522. [CrossRef]
5. Al-Ghouti, M.A.; Al-Kaabi, M.A.; Ashfag, M.Y.; Da'na, D.A. Produced water characteristics, treatment and reuse: A review. *J. Water Process Eng.* **2019**, *28*, 222–239. [CrossRef]
6. Loannidou, O.; Zabaniotou, A. Agricultural residues as precursors for activated carbon production—A review. *Renew. Sustain. Energy Rev.* **2007**, *11*, 1966–2005. [CrossRef]
7. Demirbas, A. Agricultural based activated carbons for the removal of dyes from aqueous solutions: A review. *J. Hazard. Mater.* **2009**, *167*, 1–9. [CrossRef]
8. Alslaibi, T.M.; Abustan, I.; Ahmad, M.A.; Foul, A.A. A review: Production of activated carbon from agricultural byproducts via conventional and microwave heating. *J. Chem. Technol. Biotechnol.* **2013**, *88*, 1183–1190. [CrossRef]
9. Abioye, A.M.; Ani, F.N. Recent development in the production of activated carbon electrodes from agricultural waste biomass for supercapacitors: A review. *Renew. Sustain. Energy Rev.* **2015**, *52*, 1282–1293. [CrossRef]
10. Yahya, M.A.; Al-Qodah, Z.; Ngah, C.W.Z. Agricultural bio-waste materials as potential sustainable precursors used for activated carbon production: A review. *Renew. Sustain. Energy Rev.* **2015**, *46*, 218–235. [CrossRef]
11. Gonzalez-Garcia, P. Activated carbon from lignocellulosics precursors: A review of the synthesis methods, characterization techniques and applications. *Renew. Sustain. Energy Rev.* **2018**, *82*, 1393–1414. [CrossRef]
12. Lu, Y.; Shan, G.; Huang, J.; Li, Q. Insights into characteristics of dissolved organic matter fractions in co-composted dairy manure and Chinese herbal residues. *Waste Biomass Valor.* **2018**, *9*, 777–782. [CrossRef]

13. Cantrell, K.B.; Ducey, T.; Ro, K.S.; Hunt, P.G. Livestock waste-to-bioenergy generation opportunities. *Bioresour. Technol.* **2008**, *99*, 7941–7953. [CrossRef] [PubMed]
14. Dagnall, S.; Hill, J.; Pegg, D. Resource mapping and analysis of farm livestock manures—assessing the opportunities for biomass-to-energy schemes. *Bioresour. Technol.* **2000**, *71*, 225–234. [CrossRef]
15. Raven, R.P.J.M. Implementation of manure digestion and co-combustion in the Dutch electricity regime: A multi-level analysis of market implementation in the Netherlands. *Energy Policy* **2004**, *32*, 29–39. [CrossRef]
16. Carlin, N.; Annamalai, K.; Sweeten, J.; Mukhtar, S. Thermo-chemical conversion analysis on dairy manure-based biomass through direct combustion. *Int. J. Green Energy* **2007**, *4*, 133–159. [CrossRef]
17. Cantrell, K.B.; Hunt, P.G.; Uchimiya, M.; Novak, J.M.; Ro, K.S. Impact of pyrolysis temperature and manure source on physicochemical characteristics of biochar. *Bioresour. Technol.* **2012**, *107*, 419–428. [CrossRef]
18. Fernandez-Lopez, M.; Puig-Gamero, M.; Lopez-Gonzalez, D.; Avalos-Ramirez, A.; Valverde, J.; Sanchez-Silva, L. Life cycle assessment of swine and dairy manure: Pyrolysis and combustion processes. *Bioresour. Technol.* **2015**, *182*, 184–192. [CrossRef]
19. Kim, J.K.; Kim, H.S.; Park, S.U.; Lee, H.D.; Kim, S.C. Combustion possibility of cattle manure as a blended fuel of pulverized coal fired power plant using TGA and 100 kg/h test furnace. *J. Chem. Eng. Jpn.* **2015**, *48*, 307–319. [CrossRef]
20. Yi, B.J.; Yuan, Q.X.; Cao, H.L.; Niu, W.J.; Wang, M.; Zhu, Y.; Yan, S.P. Effect of alkali and alkaline earth metal species on the combustion characteristics of cattle manures. *RSC Adv.* **2018**, *8*, 11705–11713. [CrossRef]
21. Xu, X.Y.; Cao, X.D.; Zhao, L.; Wang, H.L.; Yu, H.R.; Gao, B. Removal of Cu, Zn, and Cd from aqueous solutions by the dairy manure-derived biochar. *Environ. Sci. Pollut. Res.* **2013**, *20*, 358–368. [CrossRef] [PubMed]
22. Cao, H.L.; Xin, Y.; Yuan, Q.X. Prediction of biochar yield from cattle manure pyrolysis via least squares support vector machine intelligent approach. *Bioresour. Technol.* **2016**, *202*, 158–164. [CrossRef] [PubMed]
23. Ma, F.F.; Zhao, B.W.; Diao, J.R.; Qiao, H.T.; Zhong, J.K. Sorption of cadmium by biochar produced from pyrolysis of cattle manure in aqueous solution. *Fresen. Environ. Bull.* **2016**, *25*, 5226–5236.
24. Zhu, Y.; Yi, B.J.; Yuan, Q.X.; Wang, M.; Yan, S.P. Removal of methylene blue from aqueous solution by cattle manure-derived low temperature biochar. *RSC Adv.* **2018**, *8*, 19917–19929. [CrossRef]
25. Chen, Z.L.; Zhang, J.Q.; Huang, L.; Yuan, Z.H.; Li, Z.J.; Liu, M.C. Removal of Cd and Pb with biochar made from dairy manure at low temperature. *J. Integr. Agric.* **2019**, *18*, 201–210. [CrossRef]
26. Tsai, T.W.; Hsu, C.H.; Lin, Y.Q. Highly porous and nutrients-rich biochar derived from dairy cattle manure and its potential for removal of cationic compound from water. *Agriculture* **2019**, *9*, 114. [CrossRef]
27. Klass, D.J. *Biomass for Renewable Energy, Fuels, and Chemicals*; Academic Press: San Diego, CA, USA, 1998.
28. Qian, Q.R.; Machida, M.; Tatsumoto, H. Preparation of activated carbons from cattle-manure compost by zinc chloride activation. *Bioresour. Technol.* **2007**, *98*, 353–360. [CrossRef]
29. Qian, Q.R.; Machida, M.; Aikawa, M.; Tatsumoto, H. Effect of $ZnCl_2$ impregnation ratio on pore structure of activated carbons prepared from cattle manure compost: Application of N_2 adsorption-desorption isotherms. *J. Mater. Cycles Waste Manag.* **2008**, *10*, 53–61. [CrossRef]
30. Qian, Q.R.; Machida, M.; Tatsumoto, H. Textural and surface chemical characteristics of activated carbons prepared from cattle manure compost. *Waste Manag.* **2008**, *28*, 1064–1071. [CrossRef]
31. Qian, Q.R.; Sunohara, S.; Kato, Y.; Zaini, M.A.A.; Machida, M.; Tatsumoto, H. Water vapor adsorption onto activated carbons prepared from cattle manure compost. *Appl. Surf. Sci.* **2008**, *254*, 4868–4874. [CrossRef]
32. Qian, Q.R.; Chen, Q.H.; Machida, M.; Tatsumoto, H.; Mochidzuki, K.; Sakoda, A. Removal of organic contaminants from aqueous solution by cattle manure compost (CMC) derived activated carbons. *Appl. Surf. Sci.* **2009**, *255*, 6107–6114. [CrossRef]
33. Zaini, M.A.A.; Amano, Y.; Machida, M. Removal of lead (II) binding properties of heat-treated cattle-manure-compost-activated carbons. *Desalin. Water Treat.* **2014**, *52*, 6420–6429. [CrossRef]
34. Zaini, M.A.A.; Okayama, R.; Machida, M. Adsorption of aqueous metal ions on cattle-manure-compost based activated carbons. *J. Hazard. Mater.* **2009**, *170*, 1119–1124. [CrossRef] [PubMed]
35. Watanabe, T.; Zaini, M.A.A.; Amano, Y.; Machida, M. Removal of 2-methylisoborneol from aqueous solution by cattle manure compost (CMC) derived activated carbons. *J. Water Supply Res. Technol.-AQUA* **2014**, *63*, 239–247. [CrossRef]
36. Li, H.C.; Yang, S.S.; Sun, H.Z.; Liu, X.H. Production of activated carbon from cow manure for wastewater treatment. *BioResources* **2018**, *13*, 3135–3143. [CrossRef]

37. Marsh, H.; Rodriguez-Reinoso, F. *Activated Carbon*; Elsevier: Amsterdam, The Netherlands, 2006.
38. Ge, C.; Lian, D.; Cui, S.; Gao, J.; Lu, J. Highly selective CO_2 capture on waste polyurethane foam-based activated carbon. *Processes* **2019**, *7*, 592. [CrossRef]
39. Liu, S.C.; Tsai, W.T. Thermochemical characteristics of dairy manure and its derived biochars from a fixed-bed pyrolysis. *Int. J. Green Energy* **2016**, *13*, 963–968. [CrossRef]
40. Tsai, T.W.; Chang, C.Y.; Wang, S.Y.; Chang, C.F.; Chien, S.F.; Sun, H.F. Preparation of activated carbons from corn cob catalyzed by potassium salts and subsequent gasification with CO_2. *Bioresour. Technol.* **2001**, *78*, 203–208. [CrossRef]
41. Lowell, S.; Shields, J.E.; Thomas, M.A.; Thommes, M. *Characterization of Porous Solids and Powders: Surface Area, Pore Size and Density*; Springer: Dordrecht, The Netherlands, 2006.
42. Smith, J.M. *Chemical Engineering Kinetics*, 3rd ed.; McGraw-Hill: New York, NY, USA, 1981.
43. Tsai, W.T.; Hsu, H.C.; Su, T.Y.; Lin, K.Y.; Lin, C.M.; Dai, T.H. The adsorption of cationic dye from aqueous solution onto acid-activated andesite. *J. Hazard. Mater.* **2007**, *147*, 1056–1062. [CrossRef]
44. Alvarez-Mateos, P.; Garcia-Martin, T.F.; Guerrero-Vacas, F.J.; Maranjo-Calderon, C.; Barrios-Sanchez, C.C.; Perez-Camino, M.C. Valorization of a high-acidity residual oil generated in the waste cooking oils recycling industries. *Grasas Aceites* **2019**, *70*, 335. [CrossRef]
45. Tamai, H.; Yoshida, T.; Sasaki, M.; Yasuda, H. Dye adsorption on mesoporous activated carbon fiber obtained from pitch containing yttrium complex. *Carbon* **1999**, *37*, 983–989. [CrossRef]
46. Gregg, S.J.; Sing, K.S.W. *Adsorption, Surface Area, and Porosity*; Academic Press: London, UK, 1982.
47. Ho, Y.S.; Chiang, C.C.; Hsu, Y.C. Sorption kinetics for dye removal from aqueous solution using activated clay. *Sep. Sci. Technol.* **2001**, *36*, 2473–2488. [CrossRef]
48. Demirbas, E.; Kobya, M.; Sulak, M.T. Adsorption kinetics of a basic dye from aqueous solutions onto apricot stone activated carbon. *Bioresour. Technol.* **2008**, *99*, 5368–5373. [CrossRef] [PubMed]
49. Ghaedi, M.; Golestani Nasab, A.; Khodadoust, S.; Rajabi, M.; Azizian, S. Application of activated carbon as adsorbents for efficient removal of methylene blue: Kinetics and equilibrium study. *J. Ind. Eng. Chem.* **2014**, *20*, 2317–2324. [CrossRef]
50. Suzuki, M. *Adsorption Engineering*; Elsevier: Amsterdam, The Netherlands, 1990.
51. Islam, M.S.; Ang, B.C.; Gharehkhani, S.; Afifi, A.B.M. Adsorption capability of activated carbon synthesized from coconut shell. *Carbon Lett.* **2016**, *20*, 1–9. [CrossRef]

processes

MDPI

Article

Effect of Initial Salt Composition on Physicochemical and Structural Characteristics of Zero-Valent Iron Nanopowders Obtained by Borohydride Reduction

Denis Leybo *, Marat Tagirov *, Dmitry Arkhipov, Elizaveta Permyakova, Evgeny Kolesnikov and Denis Kuznetsov

National University of Science and Technology "MISiS", 119049 Moscow, Russia; arhipov.di@misis.ru (D.A.); permyakova.es@misis.ru (E.P.); kolesnikov.ea@misis.ru (E.K.); dk@misis.ru (D.K.)
* Correspondence: leybo.dv@misis.ru (D.L.); m1701562@edu.misis.ru (M.T.); Tel.: +7-499-237-2226 (D.L. & M.T.)

Received: 4 September 2019; Accepted: 18 October 2019; Published: 21 October 2019

Abstract: The effect of initial salt composition on characteristics of zero-valent iron nanopowders produced via borohydride reduction was studied. The samples were characterized by X-ray diffraction, scanning and transmission electron microscopy, and low-temperature nitrogen adsorption. The efficiency of Pb^{2+} ions removal from aqueous media was evaluated. The use of ferric salts led to enhanced reduction kinetics and, consequently, to a smaller size of iron particles in comparison with ferrous salts. A decrease in the ionic strength of the synthesis solutions resulted in a decrease in iron particles. The formation of small highly-reactive iron particles during synthesis led to their oxidation during washing and drying steps with the formation of a ferrihydrite phase. The lead ions removal efficiency was improved by simultaneous action of zero-valent iron and ferrihydrite phases of the sample produced from iron sulphate.

Keywords: nano zero-valent iron; borohydride reduction method; wastewater treatment; iron nanopowders; lead ions

1. Introduction

Wastewater is one of the main challenges of humanity in the 21st century. The fast growth of the population leads to the diminution of clean water reserves both because of increased global utilization and worsening of the ecological situation as a result of a constant increase in production. As mentioned in the United Nations (UN) World Water Development report [1], water demand is continuously increasing by 1% annually, and is expected to reach 6000 km^3 per year in 2050. About 80% of the water used by industry is released without any treatment [2], which puts the ecosystem and human health at risk.

Several chemical species have been identified as potentially harmful to human health. Among them are nitrates, fluorides, arsenic, and selenium, which constitute the main part of inorganic pollutants found in many locations around the world and are considered as the main priority chemicals for wastewater treatment technologies [3]. Other substances threatening human health include recalcitrant organic dyes [4], chlorinated organic compounds [5], phenolic compounds [6], and heavy metals (e.g., Cr, and Pb) [7–9]. Lead ions fall into wastewater mainly from the mining industry, smelters, hydroelectric power plants, among others [10]. The presence of lead in drinking water leads to irreversible health effects that can affect every organ of the human body [11]. The recommended limit concentration was set in the guidelines for drinking water as 0.01 mg/L for Pb(II) [12]; however, in industrial wastewaters, the lead-ion concentrations approach 200–500 mg/L [13]. Even though producers try to lower or exclude lead from manufacturing processes, lead is still used and exerts detrimental impact on the ecosystem [14].

So far different wastewater treatment technologies have been suggested, e.g., adsorption [15], biosorption [16,17], membrane technology [18,19], biodegradation [20,21], ion exchange [22,23], ionic liquid-based treatment [24]. Sorption methods have several advantages, such as high efficiency uptake for heavy metal ions at low concentrations, energy saving and easy sorbent synthesis [25]. Among these technologies are nano zero valent iron (nZVI)-based techniques that have gained much attention of some scientific groups [26]. nZVI particles showed superior efficiency in the removal of antibiotics [27], phenolic compounds [28], heavy metals [29,30], inorganic pollutants [31], and organic dyes [32,33]. Our previous studies showed that nZVI particles are very effective for the decomposition of azo dyes [34] and as a fenton-like zinc removal catalyst [35]. nZVI have a higher reaction rate because of their very high specific surface area and surface energy for the redox process [36]. The small particle size significantly increases the diffusion properties of the materials, reduces the required amount of sorbent [37], and reduces the required contact time between nZVI and water to prevent oxidation during the synthesis [38].

The primary method for the preparation of standalone and supported nZVI particles used by scientific groups is the borohydride reduction of iron ions [39]. Although this method has been used for several decades, it is still insufficiently studied in terms of different parameters that influence the properties of the final product. R. Eljamal et al. [40] reported the results of a comprehensive study of eight relevant experimental parameters, such as concentration, reaction temperature, pH, mixing speed, and reaction time on composition, particles' size and the surface area of the produced nZVI. Their findings resulted in some improvement of nitrate reduction efficiency by 27% and phosphorus adsorption by 9.5%. N. Goldstein et al. [41] studied the influence of the organophosphate stabilizers on nZVI particles' size and zeta potential. H. Woo et. al. [42] investigated the effect of a washing solution and drying condition on the reactivity of nanosized nZVI synthesized by borohydride reduction. However, up to date, there has been no information in the literature on the effect of initial salt composition on the properties of nZVI powders produced by the borohydride reduction method. Although the authors of [41] reported differences in the stabilization mechanisms using different organic additives for ferrous and ferric salts, it is still unclear how these salts affect the borohydride reduction of iron.

Another issue related to the method is the need to use a protective atmosphere to avoid oxidation of the as-prepared nZVI particles. This concern imposes restrictions on the widespread use of the method because of the increased complexity of the overall process. Thus, it is highly desirable to identify conditions that will allow producing nZVI in an ambient atmosphere.

Our research investigated the effect of ferric and ferrous salts choice on structural characteristics and the physicochemical properties of nZVI samples synthesized by borohydride reduction. The obtained samples were characterized by X-ray diffraction, scanning and transmission electron microscopy, and low-temperature nitrogen adsorption techniques. A tentative scheme of the initial salt effect was proposed based on the experimental results obtained in this work. The nZVI samples were evaluated in the process of Pb^{2+} ion removal from aqueous solution.

2. Materials and Methods

2.1. Materials

All the reagents ($FeSO_4{\cdot}7H_2O$, $Fe(NO_3)_3{\cdot}9H_2O$, $FeCl_2{\cdot}4H_2O$, $FeCl_3{\cdot}6H_2O$, sodium carboxymethylcellulose ($C_6H_7O_2(OH)_2CH_2COONa$), and $NaBH_4$) used in this work were of analytical grade and used without any additional purification. MilliQ water and absolute isopropanol were used for the synthesis and water purification experiments.

2.2. *nZVI Synthesis*

All the synthesis procedures were performed in an ambient atmosphere. The nZVI samples were obtained by the reduction of iron ions from different salts according to the following reactions [40,43]:

$$FeSO_4 + 2NaBH_4 + 6H_2O = Fe + Na_2SO_4 + 2B(OH)_3 + 7H_2 \quad (1)$$

$$Fe(NO_3)_3 + 3NaBH_4 + 9H_2O = Fe + 3NaNO_3 + 3B(OH)_3 + 10.5H_2 \quad (2)$$

$$FeCl_2 + 2NaBH_4 + 6H_2O = Fe + 2NaCl + 2B(OH)_3 + 7H_2 \quad (3)$$

$$FeCl_3 + 3NaBH_4 + 9H_2O = Fe + 3NaCl + 3B(OH)_3 + 10.5H_2 \quad (4)$$

In detail, 2.5 g of $FeSO_4{\cdot}7H_2O$/4 g of $Fe(NO_3)_3{\cdot}9H_2O$/2 g of $FeCl_2{\cdot}4H_2O$/2.7 g of $FeCl_3{\cdot}6H_2O$ were dissolved in 30 mL of deionized water. 30 mL of 10 wt. % $NaBH_4$ aqueous solution was added drop-by-drop for 10 min. Sodium borohydride was taken excessively to provide a complete reduction. The suspension was mixed during the next 30 min.

The precipitate was filtered through a 0.22 μm membrane filter; afterwards, it was washed from by-products with 250 mL of water and 250 mL of isopropanol, sequentially. The obtained sample was dried in a drying oven under vacuum at room temperature. For the sample obtained from iron nitrate, a possibility of passivating the nanoparticles' surface under the flow of 1% O_2/N_2 was also studied. For this experiment, the washing of the sample was performed in a Drexel flask with further drying under the flow of O_2/N_2 overnight. The washing of the sample, in this case, was done by decantation using a magnet to increase the sedimentation speed.

The reduction process was also performed in the presence of sodium carboxymethylcellulose (Na-CMC) [44] with Fe:Na-CMC molar ratio equal to 1:0.002, 1:0.0055, and 1:0.0108 to passivate the surface of iron nanoparticles obtained from $Fe(NO_3)_3{\cdot}9H_2O$ by the reaction (2). The precipitate was separated by vacuum filtration followed by washing with isopropanol (250 mL) and drying under vacuum at room temperature as in the previous experiments.

2.3. *Characterization*

The X-ray diffraction analysis (XRD) of the samples was done on Diffray 401 diffractometer (Scientific Instruments, Russia) operating with Cr-Kα radiation (λ = 2.2909 Å). The crystallites' size was estimated using the Scherrer equation, and the Si standard was used for instrumental broadening elimination. Scanning electron microscopy (SEM) and transmission electron microscopy (TEM) were used to study the morphology of the obtained particles. The SEM analysis was done on Tescan Vega 3 SB instrument, and the LEO 912 ab omega was used for the TEM analysis. The specific surface area analysis was performed using a low-temperature N_2 adsorption method with the Nova 1200e analyzer (Quantachrome, Boynton Beach, FL, USA) at 77 K. Before the analysis, the samples were degassed in vacuum at 120 °C for 3 h. The specific surface area was calculated using the Brunauer–Emmett–Teller (BET) equation.

The pH zero charge point (pH_{pzc}) was measured using the pH drift method [45,46]. Samples of 2.5 mg nZVI were introduced into a known volume (10 mL) of a solution of 0.1 M KNO_3. KNO_3 was selected as an inert electrolyte. The initial pH ($pH_{Initial}$) of the KNO_3 solutions was adjusted from 3 to 12 by adding 0.1 M HCl or NaOH solutions. Suspensions were left to balance for 24 h in closed vessels at room temperature. Then, the pH values (pH_{final}) were measured again. Based on the data obtained, we plotted the pH_{final} vs. $pH_{initial}$ dependence. pH_{pzc} was determined by the intersection with the $pH_{initial} = pH_{final}$ line.

2.4. *Pb^{2+} Ions Removal*

The water purification experiments were performed by measuring the concentration of Pb^{2+} ions in an aqueous solution in the presence of the nZVI samples reduced from different salts. In detail,

200 mL of 1.2 mM $Pb(NO_3)_2$ solution was mixed with 50 mg of nZVI. The suspension was processed in an ultrasonic bath for 10 min for homogenization. After the ultrasonic treatment, the suspension was left in a shaker for 230 min. The probes of the solutions were taken in 1-mL portions after 10 min of the ultrasonic treatment and after 10, 20, 50, 80, 110, 140, 170, 200, and 230 min of mechanical mixing for our further analysis of the Pb concentration. The concentration of Pb in the supernatant solutions was determined by inductively coupled plasma mass spectrometry on ICAP Q instrument (Thermoscientific) in triplicates.

The lead removal capacity q ($mg \cdot g^{-1}$) was calculated for the samples after treatment for 240 min by the following equation:

$$q = \frac{(C_0 - C_\tau) \cdot V}{m} \quad (5)$$

where C_0 is the initial concentration of Pb^{2+}, C_τ is the concentration of Pb^{2+} after 240 min of the experiment, V is the volume of the solution, and m is the mass of nZVI.

3. Results and Discussion

3.1. The Effects of Initial Salt on the Phase Composition of the nZVI Samples

The X-ray diffraction patterns of the powders obtained after the reduction of iron with sodium borohydride from $FeSO_4$, $FeCl_2$, and $FeCl_3$ had characteristic peaks at 2θ angles equal to 68.6° and 106.0°, corresponding to metallic iron (Figure 1, PDF card no. 06-0696). The data confirmed that iron precipitated in a zero valence state. The sample obtained from $FeSO_4$ also contained the 2-line ferrihydrite (Fh) phase, $Fe_{10}O_{14}(OH)_2$ [47], which was confirmed by the presence of the broad peaks at 2θ angles equal to 52.6° and 99.5° [48]. nZVI could not be obtained by reducing Fe^{3+} ions from $Fe(NO_3)_3$ with $NaBH_4$. This sample consisted of only the Fh phase. The XRD patterns show that the iron particles reduced from $FeCl_2$ had the highest crystallinity (Figure 1), which was confirmed by the lowest full width at half maximum (FWHM) value.

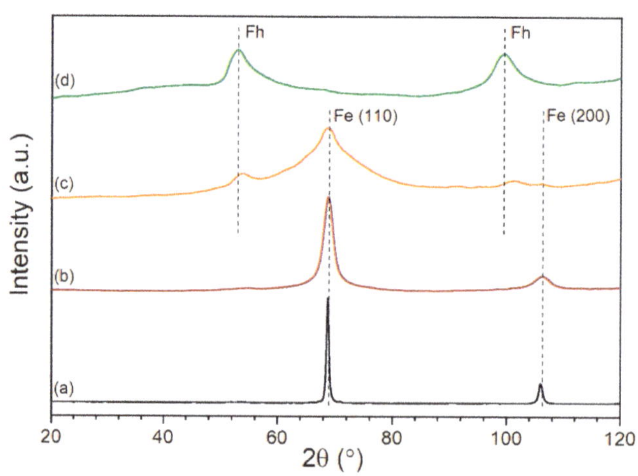

Figure 1. The X-ray diffraction analysis (XRD) patterns of powders obtained by reduction from $FeCl_2$ (**a**); $FeCl_3$ (**b**); $FeSO_4$ (**c**); $Fe(NO_3)_3$ (**d**).

The XRD pattern in Figure 1 shows that the best results were obtained by reducing iron from its chlorides. The content of the oxidized products in the final samples obtained from different salts decreased in the series $Fe(NO_3)_3 > FeSO_4 > FeCl_3 \approx FeCl_2$.

The ferrihydrite phase is usually synthetically produced by hydrolyzing ferric or ferrous salts [48]. Our observations on suspension's color change during the reduction of the iron nitrate salt suggested that the formation of the Fh phase occurred after the end of the reduction process, which means that this was a secondary product of the iron particles oxidation. H. Pullin et al. [49] observed the formation of the Fh phase as a corrosion product of the nZVI particles in aqueous media. The authors stated that the formation of 2-line ferrihydrite, being intermediate to more stable iron oxide and oxohydroxide phases, was detectable during the entire experiment lasting for more than four weeks. Similar results were obtained by A. Liu et al. [50] during their study of the nZVI composition evolution in static water after ten days of the experiment.

The corrosion rate of nZVI is inversely proportional to the size of the particles comprising the powder. The results obtained in our work indicate the dependence of the final product composition on the initial salt used for the production of nZVI, which could be attributed to the different sizes of the particles formed during the synthesis. Thus, the use of iron nitrate led to the formation of extremely reactive small iron particles stable only under reductive conditions. The iron sulphate-derived sample only partially consisted of the Fh phase, which suggested the formation of more stable and larger particles during the borohydride reduction; whereas the use of iron chlorides allowed producing the pure iron phase.

During the reduction of the iron ions from the nitrate salt, the color change of the suspension from black to rusty was observed during the drying step. Thus, an additional experiment on the passivation of the surface of the iron particles reduced from nitrate during drying using an O_2/N_2 mixture was conducted to protect against oxidation. The XRD pattern of this sample is shown in Figure 2a. It can be seen from the results that the only crystalline phase present in the sample was sodium borate. Sodium borate forms as a by-product of the sodium borohydride hydrolysis reaction [51] and are usually removed during the washing procedure. The decantation is less efficient for the sample purification. The rusty color of the sample suggested that it should contain iron-based compounds, probably, in a low crystallinity form. To confirm this, the energy dispersive X-ray (EDX) analysis of the sample was done. The results (Figure 2b) show that the sample contained iron, most probably, in the form of the amorphous Fh phase. Therefore, this route is inappropriate for the production of nZVI.

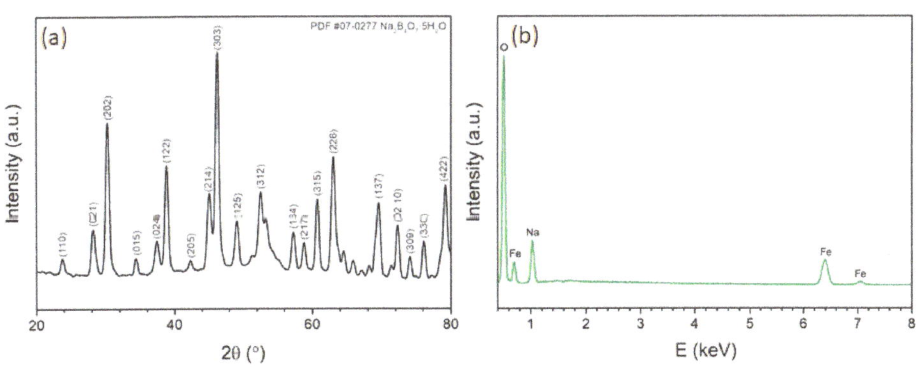

Figure 2. The XRD pattern (**a**) and the energy dispersive X-ray (EDX) analysis (**b**) of the powder obtained by reduction from $Fe(NO_3)_3$ with passivation in a current of 1% O_2/N_2 in a Drexel bottle.

The second route for stabilizing the surface of the iron particles reduced from the nitrate salt in this work was the usage of Na-CMC as an additive. Sodium CMC is known to be soluble in water and poorly soluble in alcohols. Thus, changing the solvent during the washing procedure could result in the formation of the Na-CMC protective coating on the surface of the iron particles. The XRD pattern in Figure 3 shows that all the samples obtained using the sodium CMC as a passivating agent consisted of ferrihydrite phase. The results obtained in this work show that it is impossible to produce nZVI with a sodium borohydride reduction of the nitrate salt using the abovementioned techniques.

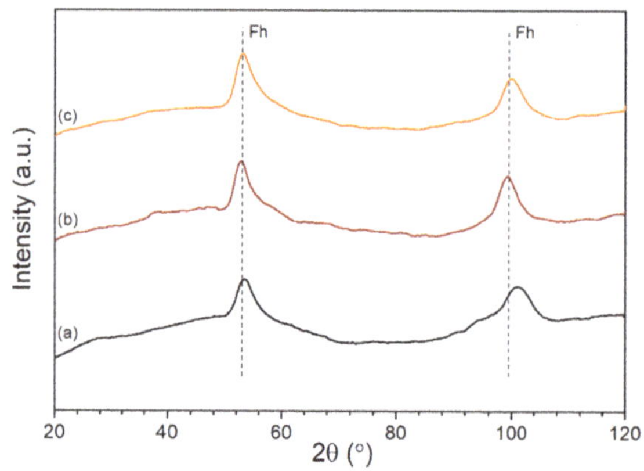

Figure 3. The XRD patterns of the powders obtained by reduction from $Fe(NO_3)_3$ with a concentration of CMC 1:0.002 (**a**), 1:0.0055 (**b**), and 1:0.0108 (**c**).

3.2. *The Effects of Initial Salt on Structural Characteristics of Iron-Based Samples*

3.2.1. SEM Results

The SEM images of nanoparticles are shown in Figure 4. The nZVI particles reduced from $FeSO_4$ formed aggregates of complex shapes (Figure 4a). The SEM images of the powder reduced from $Fe(NO_3)_3$ showed aggregates of the particles with an extended surface (Figure 4b). The image revealed that these aggregates consist of flake-shaped particles. The particles of the iron sample produced from $FeCl_2$ and $FeCl_3$ were arranged in chains (Figure 4c,d). The particle size of the sample produced from $FeSO_4$ was approximately 50–60 nm, the cross-section particle size of the sample produced from $FeCl_2$ was approximately 70–90 nm and of $FeCl_3$ 30–45 nm. The aggregates of the particles obtained from $Fe(NO_3)_3$ had sizes in a range from 500 nm to 1 μm.

Figure 4. The scanning electron microscopy (SEM) images of nZVI produced from $FeSO_4$ (**a**), $Fe(NO_3)_3$ (**b**), $FeCl_2$ (**c**), $FeCl_3$ (**d**).

3.2.2. TEM Results

The TEM images of the nanoparticles are shown in Figure 5. The aggregates of the nZVI particles obtained from $FeSO_4$ had a developed morphology (Figure 5a). The sample consisted of flake-like particles as well as particles of an irregular form with a size of around 20 nm. The TEM images of the powder produced from $Fe(NO_3)_3$ showed that two-dimensional structures with a thickness of about 2 nm were obtained (Figure 5b). The nZVI particles obtained from $FeCl_2$ and $FeCl_3$ were arranged in chains and had a core-shell structure with a shell thickness of about 3–6 nm (Figure 5c,d). The particles' size of the iron samples reduced from $FeCl_2$ was approximately 100 nm, and from $FeCl_3$ was 20–30 nm.

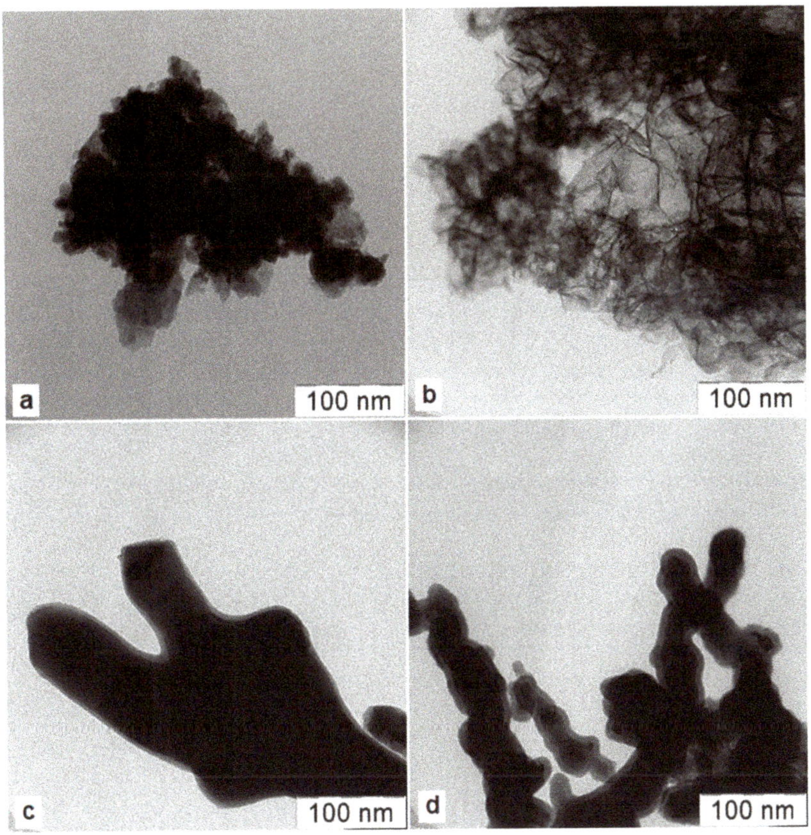

Figure 5. The transmission electron microscopy (TEM) images of nZVI produced from $FeSO_4$ (**a**), $Fe(NO_3)_3$ (**b**), $FeCl_2$ (**c**), $FeCl_3$ (**d**).

The results of the morphology study of the nZVI samples obtained in this work are well correlated with their composition and the results obtained by other authors. The 2D structures produced using the iron nitrate salt are characteristic of ferrihydrite being a poorly crystalline iron compound with the crystallites' size of 2–6 nm [52–54]. The structures similar to the obtained ones via the reduction of iron sulphate in this work were also observed by different researchers [55,56]. Mainly, these structures were assigned to the presence of iron oxides [57] on the surface of the Fe^0 particles. Our results showed that these structures correspond to the formation of ferrihydrite via oxidation of small reactive iron particles. The core-shell structured chains of the nZVI aggregates are the most commonly encountered morphology of iron produced by a borohydride reduction process [30,58,59]. The shell composition according to the XPS analysis was shown to be Fe_2O_3 [59], the most stable iron oxide, comprising the passivation coating preventing their fast oxidation upon exposure to air/water environment.

3.2.3. Specific Surface Area and Crystallites' Size

The specific surface area and the crystallites' sizes of the iron samples reduced from various salts are presented in Table 1.

Table 1. The characteristics of the recovered powders.

Initial Salt	Phase Composition	S_{sp}, m^2/g	Crystallites' Size, nm
$FeSO_4$	Fe, Fh	73.0	5.6
$Fe(NO_3)_3$	Fh	277.3	-
$FeCl_2$	Fe	6.2	108.1
$FeCl_3$	Fe	46.3	11.7

The powder produced from iron nitrate had the largest surface area among the samples. The smallest specific surface area was identified for iron particles reduced from $FeCl_2$. The obtained results are in good agreement with the particles' size and the morphology observations visible on the SEM and TEM images. Thus, the 2D morphology of the particles produced from $Fe(NO_3)_3$ salt resulted in the highest surface area. The presence of such two-dimensional structures in the sample produced from sulphate was also responsible for the high surface area of this sample. Because both samples containing 2D structures consisted of Fh, we could conclude that the particles of such morphology corresponded to the Fh phase. In the case of the iron chloride salts, the largest specific surface area was obtained for the powder produced from $FeCl_3$, which is well correlated to the smaller size of the particles produced from this salt.

The crystallites' size for the iron particles reduced from $FeSO_4$ had the smallest value among the samples although it could be underestimated because of the presence of the amorphous phase in the sample. The powders obtained from $FeCl_2$ had the largest crystallites' size, almost ten times larger than those reduced from the $FeCl_3$ salt. The comparison of these values with a transverse size of the particles measured from the SEM and TEM images led us to the conclusion that the chain aggregates produced from the iron chloride salts consisted of single-crystal domains of iron.

3.3. *The Scheme of Initial Salt Composition Effect on Reduction of Fe Ions*

The sodium borohydride reduction of iron was performed using Fe^{3+} and Fe^{2+} containing salts in present work according to the reactions [55,58]:

$$Fe^{2+} + 2BH_4^- + 6H_2O = Fe^0 + 2B(OH)_3 + 7H_2 \quad (6)$$

$$Fe^{3+} + 3BH_4^- + 9H_2O = Fe^0 + 3B(OH)_3 + 10.5H_2 \quad (7)$$

The standard reduction potentials of these ions are different, with Fe^{2+} having a more negative value of −0.447 V in comparison with −0.037 V for Fe^{3+} ion. Consequently, it is easier to reduce ferric ion [60] from the solution. Since we can expect that iron ions form identical transition complexes [43], the rate constant of the reaction (6) is higher, which leads to the formation of smaller particles using ferric ion as the source of iron.

Among the salts containing identical iron ions the difference in the properties of the produced iron powders originates from the corresponding anions. Although the anions do not participate in the reactions (5) and (6) directly, they do form the ionic atmosphere that influences the reaction rate. According to the Brönsted–Bjerrum equation [61], an increase in the ionic strength of the solution results in a decrease of the reaction rate constant when the reacting ions have different signs. Thus, the formation of smaller iron particles is more favorable in the solutions of lower ionic strength.

Taking into account the abovementioned qualitative relationships, a tentative scheme of the initial salt composition effect on the reduction of Fe ions is proposed in this work (Figure 6). According to the scheme, the use of ferric ion-containing $FeCl_3$ and $Fe(NO_3)_3$ salts leads to the formation of smaller particles during the reduction process as compared to the $FeCl_2$ and $FeSO_4$. The ionic strength of the solutions used in this work was calculated to be equal (in mol/L) to 1.8, 2.0, 2.4, and 2.6 for $FeSO_4$, $FeCl_2$, $Fe(NO_3)_3$, and $FeCl_3$ initial salts, respectively. The iron particle formation rate is higher for the $FeSO_4$ and $Fe(NO_3)_3$ initial salts due to their lower ionic strengths. After the formation of nZVI during the washing and drying steps, small highly reactive particles are oxidized by water and dissolved oxygen with the formation of the Fh phase in the case of $Fe(NO_3)_3$ and, partially, $FeSO_4$ salts. Larger nZVI particles produced from the chlorides agglomerate in the chain structures with the formation of a passivation oxide layer on their surface.

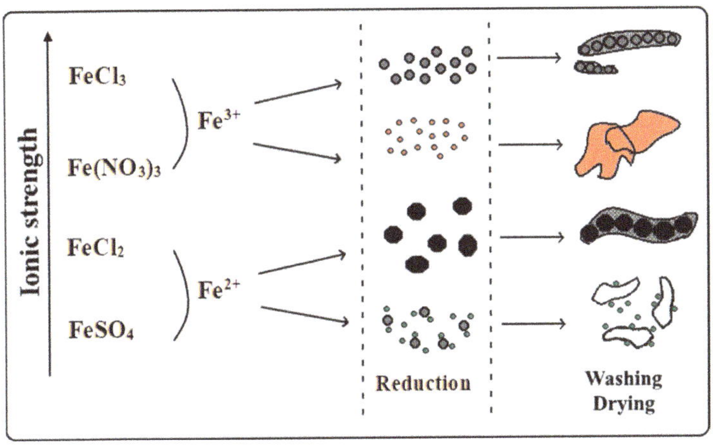

Figure 6. The scheme of the nZVI production via a borohydride reduction of different salts.

3.4. The Effects of Initial Salt Composition to Pb^{2+} Ions Removal Efficiency

The results of the removal of Pb^{2+} ions removal are illustrated in Figure 7. The horizontal axis shows the total processing time as the sum of the ultrasonic and mechanical treatment. The concentration of the Pb^{2+} ions rapidly decreased after the ultrasonic treatment for the nZVI samples obtained from all the salts. Then there was a gradual decrease in the concentration of all the samples within the next 120 min. There was a relatively slight fluctuation after 120 min of the experiment. The lead removal capacity q was equal to 744, 646, 462, and 650 mg·g^{-1} for the nZVI samples obtained from $FeSO_4$, $Fe(NO_3)_3$, $FeCl_2$, and $FeCl_3$, respectively, after the treatment for 240 min. Interestingly, the nZVI sample obtained from $FeSO_4$ showed maximum efficiency in the removal of the Pb^{2+} ions as compared to other nZVI samples.

Table 2 shows the comparison of the lead removal capacity results obtained in the present study and by other authors. Zhang Dongsheng et al. [30] gave q calculations for the removal of lead from the obtained nZVI and nZVI powders after freeze-drying. Chuang Yu et al. [62] reported the results of the adsorption of lead ions by GO-B-nZVI composite. Humera Jabeen et al. [63] in their work compared the adsorption capacity q for the synthesized powders of nZVI and composites based on graphene and nZVI (G-nZVI).

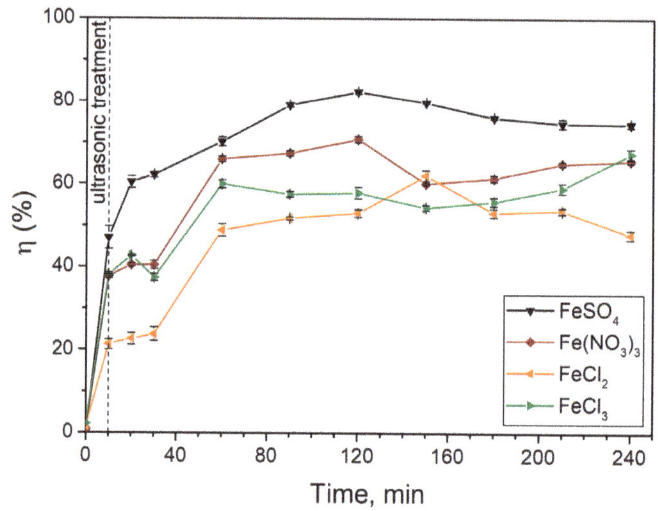

Figure 7. Pb^{2+} ions removal efficiency by the nZVI samples produced from different salts.

Table 2. Comparison of the lead removal capacity results.

Material	q, $mg{*}g^{-1}$	Reference
nZVI	855	[30]
nZVI-drying	758	[30]
GO-B-nZVI	199	[62]
nZVI	133	[63]
G-nZVI	181	[63]
nZVI from $FeSO_4$	744	this work
nZVI from $Fe(NO_3)_3$	646	this work
nZVI from $FeCl_2$	463	this work
nZVI from $FeCl_3$	650	this work

The comparison of the results with the characteristics of the samples obtained in this work reveals that there was a synergy between Fh and zero-valent iron, which led to the increased efficiency of the lead removal. Thus, the removal of the lead ions by the iron chloride originated samples was lower, which can be attributed to lower suspension stability of the nZVI particles in the solution as well as the lower specific surface area of these samples. These samples consisted of iron in a metallic form. The sample obtained from the iron nitrate had moderate efficiency. Although this sample did not contain zero-valent iron, its high specific surface area resulted in high adsorption capacity. The composition of the nZVI sample produced from sulphate ensured its high efficiency as a result of both positive actions of zero-valent iron capable of reducing lead ions [30] and ferrihydrite phase providing high surface area and the suspension stability of this sample.

Three mechanisms leading to the removal of lead ions from an aqueous solution are considered to be the main ones: sorption, reduction, and coprecipitation. During Pb^{2+} ions removal experiments, the initial pH was 4.53 for all samples. The final pH also did not differ significantly and varied in the series 5.54, 5.52, 5.51, 5.48 for $FeSO_4$, $Fe(NO_3)_3$, $FeCl_2$, and $FeCl_3$, respectively. An increase in pH is associated with the oxidation of nZVI in an acid solution. A relatively small change in pH during the removal of lead ions indicates a slight contribution of the coprecipitation mechanism on the overall Pb ions removal.

An important characteristic of nZVI is the point of zero charge (pH_{pzc}). pH_{pzc} is the pH at which the number of positively and negatively charged centers on the surface of the powder is the same. If $pH < pH_{pzc}$, part of the protons from the solution is adsorbed on the surface with the formation of a positive charge, part of the counterions (OH^-) in the surface layer of the solution will be released and the pH of the contacting solution will increase, approaching pH_{pzc}; otherwise, some OH^- groups will sorb, and the surface will acquire a negative charge.

The results of determining pH_{pzc} for nZVI samples are presented in Figure 8. The dashed line shows the position where $pH_{initial} = pH_{final}$. pH_{pzc} was determined by the intersection of the graphs with a dashed line. pHpzc were 7.9, 6.6, 9.1, and 8.8 for the nZVI samples obtained from $FeSO_4$, $Fe(NO_3)_3$, $FeCl_2$, and $FeCl_3$, respectively. The pH_{pzc} of the obtained nZVI samples are not favorable for the electrostatic adsorption of Pb^{2+} cations. Therefore, adsorption also does not make a priority contribution to the removal of the lead ions.

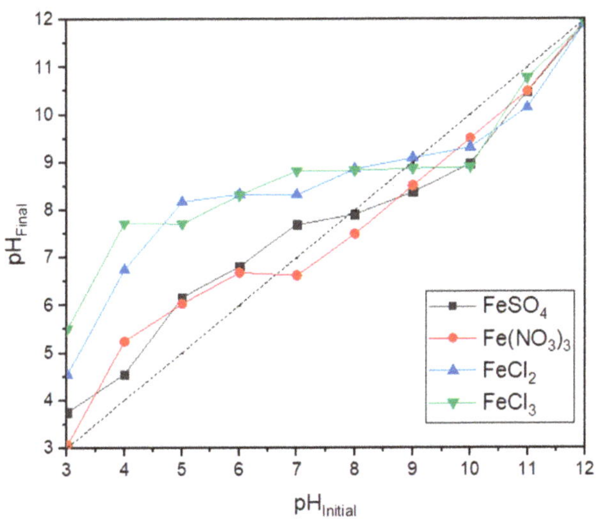

Figure 8. pH_{final} versus $pH_{initial}$ during equilibration of nZVI powders in 0.1 M KNO_3 solution.

The obtained results of measuring the pH of the solution and pH_{pzc} of nZVI samples show that the processes of coprecipitation and electrostatic adsorption were not the main ones during the removal of Pb^{2+} ions from the aqueous environment. Apparently, the reduction of Pb^{2+} ions by nZVI made a predominant contribution to the process of water purification from lead ions.

4. Conclusions

Iron-based samples with different phase compositions, particles' size, and morphology can be obtained by a borohydride reduction of various ferric and ferrous iron salts. The use of ferric salts leads to the formation of smaller iron particles during the synthesis because of a higher value of the reduction potential and the enhanced reaction kinetics. The formation of such highly reactive particles results in their further oxidation by water and dissolved oxygen with the formation of 2D ferrihydrite. The oxidation cannot be avoided by O_2/N_2 passivation and Na-CMC assisted synthesis. Among the salts containing iron ions with the same oxidation state the difference in the corresponding nZVI samples' characteristics originated from different solution's ionic strength, which influences the reduction rate.

The synergistic action of zero-valent iron and ferrihydrite leads to superior Pb^{2+} ion removal from water. This effect is easily achieved by the borohydride reduction of iron from sulphate salt.

Pure ferrihydrite shows moderate lead ion removal efficiency despite the highest specific surface area among the samples since the adsorption is the only action mechanism. The phase pure zero-valent iron samples have the lowest Pb^{2+} ions removal efficiency because of the lowest suspension stability and the adsorption capacity.

Author Contributions: Conceptualization, D.K.; formal analysis, D.L.; funding acquisition, D.A.; investigation, D.L. and M.T.; methodology, D.L., M.T., E.P., and E.K.; project administration, D.A.; supervision, D.K.; visualization, M.T.; writing–original draft, D.L. and M.T.; writing–review and editing, D.L., M.T., and D.K.

Funding: RFBR funded the reported study according to the research project No. 18-29-25051.

Conflicts of Interest: The authors declare no conflict of interest.

References

1. UN. *The United Nations World Water Development Report 2018: Nature-Based Solutions for Water*; UN: Paris, France, 2018.
2. UN. *The United Nations World Water Development Report 2017. Wastewater: The Untapped Resource*; UN: Paris, France, 2017.
3. Thompson, T.; Fawell, J.; Kunikane, S.; Jackson, D.; Appleyard, S.; Callan, P.; Bartram, J.; Kingston, P. *Chemical Safety of Drinking Water: Assessing Priorities for Risk Management*; WHO: Geneva, Switzerland, 2007.
4. Rovira, J.; Domingo, J.L. Human health risks due to exposure to inorganic and organic chemicals from textiles: A review. *Environ. Res.* **2019**, *168*, 62–69. [CrossRef]
5. Huang, B.; Lei, C.; Wei, C.; Zeng, G. Chlorinated volatile organic compounds (Cl-VOCs) in environment—Sources, potential human health impacts, and current remediation technologies. *Environ. Int.* **2014**, *71*, 118–138. [CrossRef] [PubMed]
6. Raza, W.; Lee, J.; Raza, N.; Luo, Y.; Kim, K.H.; Yang, J. Removal of phenolic compounds from industrial waste water based on membrane-based technologies. *J. Ind. Eng. Chem.* **2019**, *71*, 1–18. [CrossRef]
7. Sharma, S.; Nagpal, A.K.; Kaur, I. Appraisal of heavy metal contents in groundwater and associated health hazards posed to human population of Ropar wetland, Punjab, India and its environs. *Chemosphere* **2019**, *227*, 179–190. [CrossRef] [PubMed]
8. Mao, C.; Song, Y.; Chen, L.; Ji, J.; Li, J.; Yuan, X.; Yang, Z.; Ayoko, G.A.; Frost, R.L.; Theiss, F. Human health risks of heavy metals in paddy rice based on transfer characteristics of heavy metals from soil to rice. *Catena* **2019**, *175*, 339–348. [CrossRef]
9. Rai, P.K.; Lee, S.S.; Zhang, M.; Tsang, Y.F.; Kim, K.H. Heavy metals in food crops: Health risks, fate, mechanisms, and management. *Environ. Int.* **2019**, *125*, 365–385. [CrossRef]
10. Gorchev, H.G.; Ozolins, G. *Guidelines for Drinking-Water Quality*; WHO: Geneva, Switzerland, 2011; Volume 1, p. 564.
11. Ahamed, M.; Akhtar, M.J.; Alhadlaq, H.A. Preventive effect of TiO2 nanoparticles on heavy metal Pb-induced toxicity in human lung epithelial (A549) cells. *Toxicol. Vitr.* **2019**, *57*, 18–27. [CrossRef]
12. Pandey, R.; Dwivedi, M.K.; Singh, P.K.; Patel, B.; Pandey, S.; Patel, B.; Patel, A.; Singh, B. Effluences of Heavy Metals, Way of Exposure and Bio-toxic Impacts: An Update. *J. Chem. Chem. Sci.* **2016**, *66*, 2319–7625.
13. Arbabi, M.; Hemati, S.; Amiri, M. Removal of lead ions from industrial wastewater: A review of Removal methods. *Int. J. Epidemiol. Res.* **2015**, *2*, 105–109.
14. Wani, A.L.; Ara, A.; Usmani, J.A. Lead toxicity: A review. *Interdiscip. Toxicol* **2015**, *8*, 55–64. [CrossRef]
15. Sharma, A.; Syed, Z.; Brighu, U.; Gupta, A.B.; Ram, C. Adsorption of textile wastewater on alkali-activated sand. *J. Clean. Prod.* **2019**, *220*, 23–32. [CrossRef]
16. Lapo, B.; Demey, H.; Zapata, J.; Romero, C.; Sastre, A. Sorption of Hg (II) and Pb (II) ions on chitosan-iron (III) from aqueous solutions: Single and binary systems. *Polymers* **2018**, *10*, 367. [CrossRef] [PubMed]
17. Demey, H.; Melkior, T.; Chatroux, A.; Attar, K.; Thiery, S.; Miller, H.; Marchand, M. Evaluation of torrefied poplar-biomass as a low-cost sorbent for lead and terbium removal from aqueous solutions and energy co-generation. *Chem. Eng. J.* **2019**, *361*, 839–852. [CrossRef]
18. Abdel-Fatah, M.A. Nanofiltration systems and applications in wastewater treatment: Review article. *Ain Shams Eng. J.* **2018**, *9*, 3077–3092. [CrossRef]

19. Sunil, K.; Karunakaran, G.; Yadav, S.; Padaki, M.; Zadorozhnyy, V.; Pai, R.K. Al-Ti_2O_6 a mixed metal oxide based composite membrane: A unique membrane for removal of heavy metals. *Chem. Eng. J.* **2018**, *348*, 678–684. [CrossRef]
20. Yang, L.; Sheng, M.; Zhao, H.; Qian, M.; Chen, X.; Zhuo, Y.; Cao, G. Treatment of triethyl phosphate wastewater by Fenton oxidation and aerobic biodegradation. *Sci. Total Environ.* **2019**, *678*, 821–829. [CrossRef]
21. Goswami, L.; Kumar, R.V.; Pakshirajan, K.; Pugazhenthi, G.A. A novel integrated biodegradation—Microfiltration system for sustainable wastewater treatment and energy recovery. *J. Hazard. Mater.* **2019**, *365*, 707–715. [CrossRef]
22. Mohammadtabar, F.; Pillai, R.G.; Khorshidi, B.; Hayatbakhsh, A.; Sadrzadeh, M. Efficient treatment of oil sands produced water: Process integration using ion exchange regeneration wastewater as a chemical coagulant. *Sep. Purif. Technol.* **2019**, *221*, 166–174. [CrossRef]
23. Muhammad, A.; Soares, A.; Jefferson, B. The impact of background wastewater constituents on the selectivity and capacity of a hybrid ion exchange resin for phosphorus removal from wastewater. *Chemosphere* **2019**, *224*, 494–501. [CrossRef]
24. Isosaari, P.; Srivastava, V.; Sillanpää, M. Ionic liquid-based water treatment technologies for organic pollutants: Current status and future prospects of ionic liquid mediated technologies. *Sci. Total Environ.* **2019**, *690*, 604–619. [CrossRef]
25. Attar, K.; Demey, H.; Bouazza, D.; Sastre, A.M. Sorption and desorption studies of Pb (II) and Ni (II) from aqueous solutions by a new composite based on alginate and magadiite materials. *Polymers* **2019**, *11*, 340. [CrossRef] [PubMed]
26. Crane, R.A.; Scott, T.B. Nanoscale zero-valent iron: Future prospects for an emerging water treatment technology. *J. Hazard. Mater.* **2012**, *211*, 112–125. [CrossRef] [PubMed]
27. Pirsaheb, M.; Moradi, S.; Shahlaei, M.; Wang, X.; Farhadian, N. A new composite of nano zero-valent iron encapsulated in carbon dots for oxidative removal of bio-refractory antibiotics from water. *J. Clean. Prod.* **2019**, *209*, 1523–1532. [CrossRef]
28. Bavandi, R.; Emtyazjoo, M.; Saravi, H.N.; Yazdian, F.; Sheikhpour, M. Study of capability of nanostructured zero-valent iron and graphene oxide for bioremoval of trinitrophenol from wastewater in a bubble column bioreactor. *Electron. J. Biotechnol.* **2019**, *39*, 8–14. [CrossRef]
29. Zhang, W.; Oswal, H.; Renew, J.; Ellison, K.; Huang, C.H. Removal of heavy metals by aged zero-valent iron from flue-gas-desulfurization brine under high salt and temperature conditions. *J. Hazard. Mater.* **2019**, *373*, 572–579. [CrossRef] [PubMed]
30. Zhang, D.; Gao, W.; Chang, G.; Luo, S.; Jiao, W.; Liu, Y. Removal of heavy metal lead (II) using nanoscale zero-valent iron with different preservation methods. *Adv. Powder Technol.* **2019**, *30*, 581–589. [CrossRef]
31. Yan, W.; Herzing, A.A.; Kiely, C.J.; Zhang, W.X. Nanoscale zero-valent iron (nZVI): Aspects of the core-shell structure and reactions with inorganic species in water. *J. Contam. Hydrol.* **2010**, *118*, 96–104. [CrossRef]
32. Xu, J.; Wang, X.; Pan, F.; Qin, Y.; Xia, J.; Li, J.; Wu, F. Synthesis of the mesoporous carbon-nano-zero-valent iron composite and activation of sulfite for removal of organic pollutants. *Chem. Eng. J.* **2018**, *353*, 542–549. [CrossRef]
33. Zhang, M.H.; Dong, H.; Zhao, L.; Wang, D.X.; Meng, D. A review on Fenton process for organic wastewater treatment based on optimization perspective. *Sci. Total Environ.* **2019**, *670*, 110–121. [CrossRef]
34. Mikhailov, I.; Levina, V.; Leybo, D.; Masov, V.; Tagirov, M.; Kuznetsov, D. ; Synthesis, Characterization and Reactivity of Nanostructured Zero-Valent Iron Particles for Degradation of Azo Dyes. *Int. J. Nanosci.* **2017**, *16*, 1750017. [CrossRef]
35. Mikhailov, I.; Komarov, S.; Levina, V.; Gusev, A.; Issi, J.P.; Kuznetsov, D. Nanosized zero-valent iron as Fenton-like reagent for ultrasonic-assisted leaching of zinc from blast furnace sludge. *J. Hazard. Mater.* **2017**, *321*, 557–565. [CrossRef] [PubMed]
36. Turabik, M.; Simsek, U.B. Effect of synthesis parameters on the particle size of the zero valent iron particles. *Inorg. Nano-Met. Chem.* **2017**, *47*, 1033–1043. [CrossRef]
37. Hwang, Y.H.; Kim, D.G.; Shin, H.S. Effects of synthesis conditions on the characteristics and reactivity of nano scale zero valent iron. *Appl. Catal. B Environ.* **2011**, *105*, 144–150. [CrossRef]
38. Han, Y.; Yang, M.D.; Zhang, W.; Yan, W. Optimizing synthesis conditions of nanoscale zero valent iron (nZVI) through aqueous reactivity assessment. *Front. Environ. Sci. Eng.* **2015**, *9*, 813–822. [CrossRef]

39. Ezzatahmadi, N.; Ayoko, G.A.; Millar, G.J.; Speight, R.; Yan, C.; Li, J.; Li, S.; Zhu, J.; Xi, Y. Clay-supported nanoscale zero-valent iron composite materials for the remediation of contaminated aqueous solutions: A review. *Chem. Eng. J.* **2017**, *312*, 336–350. [CrossRef]
40. Eljamal, R.; Eljamal, O.; Khalil, A.M.; Saha, B.B.; Matsunaga, N. Improvement of the chemical synthesis efficiency of nano-scale zero-valent iron particles. *J. Environ. Chem. Eng.* **2018**, *6*, 4727–4735. [CrossRef]
41. Goldstein, N.; Greenlee, L.F. Influence of synthesis parameters on iron nanoparticle size and zeta potential. *J. Nanopart. Res.* **2012**, *14*, 760. [CrossRef]
42. Woo, H.; Park, J.; Lee, S.; Lee, S. Effects of washing solution and drying condition on reactivity of nano-scale zero valent irons (nZVIs) synthesized by borohydride reduction. *Chemosphere* **2014**, *97*, 146–152. [CrossRef]
43. Glavee, G.N.; Klabunde, K.J.; Sorensen, C.M.; Hadjipanayis, G.C. Chemistry of borohydride reduction of iron (II) and iron (III) ions in aqueous and nonaqueous media. Formation of nanoscale Fe, FeB, and Fe2B powders. *Inorg. Chem.* **1995**, *34*, 28–35. [CrossRef]
44. Ngo, Q.B.; Dao, T.H.; Nguyen, H.C.; Tran, X.T.; Van Nguyen, T.; Khuu, T.D.; Huynh, T.H. Effects of nanocrystalline powders (Fe, Co and Cu) on the germination, growth, crop yield and product quality of soybean (Vietnamese species DT-51). *Adv. Nat. Sci. Nanosci. Nanotechnol.* **2014**, *5*, 015016. [CrossRef]
45. Habish, A.J.; Lazarević, S.; Janković-Častvan, I.; Jokić, B.; Kovač, J.; Rogan, J.; Janaćković, Đ.; Petrović, R.; Petrović, R. Nanoscale zerovalent iron (nZVI) supported by natural and acid-activated sepiolites: The effect of the nZVI/support ratio on the composite properties and Cd 2+ adsorption. *Environ. Sci. Pollut. Res.* **2017**, *24*, 628–643. [CrossRef] [PubMed]
46. Jia, Y.F.; Xiao, B.; Thomas, K.M. Adsorption of metal ions on nitrogen surface functional groups in activated carbons. *Langmuir* **2002**, *18*, 470–478. [CrossRef]
47. Michel, F.M.; Ehm, L.; Antao, S.M.; Lee, P.L.; Chupas, P.J.; Liu, G.; Strongin, D.R.; Schoonen, M.A.; Phillips, B.L.; Parise, J.B. The structure of ferrihydrite, a nanocrystalline material. *Science* **2007**, *316*, 1726–1729. [CrossRef] [PubMed]
48. Drits, V.A.; Sakharov, B.A.; Salyn, A.L.; Manceau, A. Structural model for ferrihydrite. *Clay Miner.* **1993**, *28*, 185–207. [CrossRef]
49. Pullin, H.; Springell, R.; Parry, S.; Scott, T. The effect of aqueous corrosion on the structure and reactivity of zero-valent iron nanoparticles. *Chem. Eng. J.* **2017**, *308*, 568–577. [CrossRef]
50. Liu, A.; Liu, J.; Zhang, W.X. Transformation and composition evolution of nanoscale zero valent iron (nZVI) synthesized by borohydride reduction in static water. *Chemosphere* **2015**, *119*, 1068–1074. [CrossRef]
51. Li, Z.; Sun, T.; Jia, J. An extremely rapid, convenient and mild coal desulfurization new process: Sodium borohydride reduction. *Fuel Process. Technol.* **2010**, *91*, 1162–1167. [CrossRef]
52. Li, F.Y.; Koopal, L.; Tan, W.F. Effect of citrate on the species and levels of Al impurities in ferrihydrite. *Colloids Surf. A Physicochem. Eng. Asp.* **2018**, *539*, 140–147. [CrossRef]
53. Pariona, N.; Camacho-Aguilar, K.I.; Ramos-González, R.; Martinez, A.I.; Herrera-Trejo, M.; Baggio-Saitovitch, E. Magnetic and structural properties of ferrihydrite/hematite nanocomposites. *J. Magn. Magn. Mater.* **2016**, *406*, 221–227. [CrossRef]
54. Xu, H.; Ji, L.; Kong, M.; Jiang, H.; Chen, J. Molecular weight-dependent adsorption fractionation of natural organic matter on ferrihydrite colloids in aquatic environment. *Chem. Eng. J.* **2019**, *363*, 356–364. [CrossRef]
55. Sawafta, R.; Shahwan, T. A comparative study of the removal of methylene blue by iron nanoparticles from water and water-ethanol solutions. *J. Mol. Liq.* **2019**, *273*, 274–281. [CrossRef]
56. Eljamal, O.; Mokete, R.; Matsunaga, N.; Sugihara, Y. Chemical pathways of nanoscale zero-valent iron (NZVI) during its transformation in aqueous solutions. *J. Environ. Chem. Eng.* **2018**, *6*, 6207–6220. [CrossRef]
57. Ma, X.; He, D.; Jones, A.M.; Collins, R.N.; Waite, T.D. Reductive reactivity of borohydride-and dithionite-synthesized iron-based nanoparticles: A comparative study. *J. Hazard. Mater.* **2016**, *303*, 101–110. [CrossRef] [PubMed]
58. Zhao, R.; Zhou, Z.; Zhao, X.; Jing, G. Enhanced Cr (VI) removal from simulated electroplating rinse wastewater by amino-functionalized vermiculite-supported nanoscale zero-valent iron. *Chemosphere* **2019**, *218*, 458–467. [CrossRef]
59. Zhang, S.H.; Wu, M.F.; Tang, T.T.; Xing, Q.J.; Peng, C.Q.; Li, F.; Liu, H.; Luo, X.B.; Zou, J.P.; Luo, J.M.; et al. Mechanism investigation of anoxic Cr (VI) removal by nano zero-valent iron based on XPS analysis in time scale. *Chem. Eng. J.* **2018**, *335*, 945–953. [CrossRef]

60. Joseyphus, R.J.; Shinoda, K.; Kodama, D.; Jeyadevan, B. Size controlled Fe nanoparticles through polyol process and their magnetic properties. *Mater. Chem. Phys.* **2010**, *123*, 487–493. [CrossRef]
61. Logan, S.R. Effects of ionic strength on the rates of reaction between ions in solution. *Trans. Faraday Soc.* **1967**, *63*, 3004–3008. [CrossRef]
62. Yu, C.; Shao, J.; Sun, W.; Yu, X. Treatment of lead contaminated water using synthesized nano-iron supported with bentonite/graphene oxide. *Arab. J. Chem.* **2018**. [CrossRef]
63. Jabeen, H.; Kemp, K.C.; Chandra, V. Synthesis of nano zerovalent iron nanoparticles—Graphene composite for the treatment of lead contaminated water. *J. Environ. Manag.* **2013**, *130*, 429–435. [CrossRef]

processes

MDPI

Article

Mesoporous Mn-Doped Fe Nanoparticle-Modified Reduced Graphene Oxide for Ethyl Violet Elimination: Modeling and Optimization Using Artificial Intelligence

Yu Hou [1], Jimei Qi [1], Jiwei Hu [1,2,*], Yiqiu Xiang [1], Ling Xin [1] and Xionghui Wei [3]

[1] Guizhou Provincial Key Laboratory for Information Systems of Mountainous Areas and Protection of Ecological Environment, Guizhou Normal University, Guiyang 550001, China; hyhyhouyu@163.com (Y.H.); qqijimei@163.com (J.Q.); xxiangyiqiu@163.com (Y.X.); xinling901231@163.com (L.X.)

[2] Cultivation Base of Guizhou National Key Laboratory of Mountainous Karst Eco-Environment, Guizhou Normal University, Guiyang 550001, China

[3] Department of Applied Chemistry, College of Chemistry and Molecular Engineering, Peking University, Beijing 100871, China; xhwei@pku.edu.cn

* Correspondence: jiweihu@yahoo.com or jwhu@gznu.edu.cn; Tel.: +86-851-8670-2710

Received: 16 February 2020; Accepted: 16 April 2020; Published: 22 April 2020

Abstract: Mesoporous Mn-doped Fe nanoparticle-modified reduced graphene oxide (Mn-doped Fe/rGO) was prepared through a one-step co-precipitation method, which was then used to eliminate ethyl violet (EV) in wastewater. The prepared Mn-doped Fe/rGO was characterized by X-ray diffraction, X-ray photoelectron spectroscopy, Raman spectroscopy, high-resolution transmission electron microscopy, scanning electron microscopy, energy dispersive spectroscopy, N_2-sorption, small angle X-ray diffraction and superconducting quantum interference device. The Brunauer–Emmett–Teller specific surface area of Mn-doped Fe/rGO composites was 104.088 m^2/g. The EV elimination by Mn-doped Fe/rGO was modeled and optimized by artificial intelligence (AI) models (i.e., radial basis function network, random forest, artificial neural network genetic algorithm (ANN-GA) and particle swarm optimization). Among these AI models, ANN-GA is considered as the best model for predicting the removal efficiency of EV by Mn-doped Fe/rGO. The evaluation of variables shows that dosage gives the maximum importance to Mn-doped Fe/rGO removal of EV. The experimental data were fitted to kinetics and adsorption isotherm models. The results indicated that the process of EV removal by Mn-doped Fe/rGO obeyed the pseudo-second-order kinetics model and Langmuir isotherm, and the maximum adsorption capacity was 1000.00 mg/g. This study provides a possibility for synthesis of Mn-doped Fe/rGO by co-precipitation as an excellent material for EV removal from the aqueous phase.

Keywords: ethyl violet; Mn-doped Fe/rGO nanocomposites; mesoporous materials; artificial intelligence; gradient boosted regression trees

1. Introduction

Decolorization of organic dyes in industrial wastewater is an essential process for achieving a pollution free environment [1]. Since these dyes are toxic substances and produce unpleasant odors and non-biodegradable wastes, they pose significant hazards to the environment and the health of humans. For example, it may affect the growth of plants and pose a carcinogenic threat to humans and other mammals [1–3]. Triphenylmethane (TPM) dyes are the third largest category in synthetic dyes after azo and anthraquinone dyes, and they are mainly used in food, paper, cosmetic, leather and textile industries [4–7]. It is estimated that about 15% of the total world production of dyes is lost during the dyeing process, and this quantity is then released into the wastewater [8].

Ethyl violet (EV) ($C_{31}H_{42}ClN_3$, MW = 492.15 g/mol) is a typical cationic dye of triphenylmethane (Figure 1), which is toxic and has strong coloring ability and is difficult to degrade in the natural environment. Current approaches to detoxify the dye wastewater include Fenton or photo-Fenton oxidation, reduction by zero-valence metals, coagulation/flocculation, electrochemical oxidation, biological treatment, membrane filtration, ozonation, electrochemical degradation and adsorption [1,9–15]. Among these options, zero-valence metals have attracted considerable attention in industries due to their ease of synthesis and operation, low cost and high adsorption capacity [9].

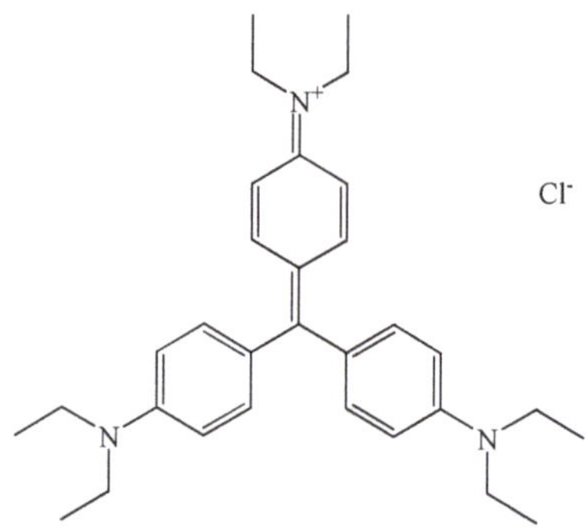

Figure 1. Chemical structure of cationic ethyl violet dye.

Nanoscale zero-valence iron (nZVI) is widely used in the removal of dyes, heavy metals, trichloroethylene and nitrate wastewater because of its large specific surface area and high reactivity [16–20]. However, the bulk of the synthesized nZVI tends to agglomerate over time, which ultimately reduces the specific surface area and reactivity of these iron nanoparticles [21,22]. Bimetallic nanoparticles are formed by the combination of Fe and Mn to improve the reactivity and function of nZVI, but Mn/Fe nanoparticles are also prone to agglomeration. Graphene oxide (GO) is a good substrate material for Mn/Fe nanoparticles that can disperse nanoparticles and avoid agglomeration. GO has a large specific surface area (theoretical value is about 2600 m^2/g), good chemical stability and abundant functional groups and is easily dispersed in water to produce stable suspensions [23–26]. In addition, it can be obtained from inexpensive bulk graphite and has been proved to be biodegradable and non-toxic [27,28]. Due to these unique characteristics, this material has great application prospects in sensors, membrane material, catalysis, electrochemistry, energy storage devices, cell supercapacitors and other fields [29–36].

At present, artificial intelligence (AI) has led to enormous breakthroughs in big data, automatic driving, pattern recognition, speech recognition, human-computer games, automatic programming, computer vision, robots and intelligent searches, which will have a far-reaching impact on human society [37,38]. Artificial neural network (ANN) is one of the major AI tools, which is inspired by human brain recognition and can be used for predicting and modeling phenomena [39]. ANN is a non-linear dynamical system, which can accurately express the complex correlation between inputs and outputs. The trained ANN models have already worked for the prognosis and optimal routes in a variety of fields based on appointed parameter settings [40]. The back propagation (BP) neural network algorithm is a multilayer feedforward network and can be trained according to the forward propagation of operating signal and the back propagation of error [41]. It is one of the most widely used neural networks. In recent years, the radial basis function (RBF) neural network has attracted much attention because of its ability to approximate nonlinear behavior. RBF-NN has the advantages of simple network structure, fast learning ability and strong approximation ability, and does not encounter local minimal problems [42]. Genetic algorithm (GA) and particle swarm optimization (PSO) have a parallel search strategy and global optimization characteristics, which can facilitate the ANN to achieve fast convergence and high prediction accuracy [43,44]. As an ensemble learning technology, random forest (RF) can improve the accuracy of single-model classification methods and solve the problem of over-fitting [45,46]. The goal of RF is to reduce the correlation among the separate trees by randomizing and bootstrapping variable selection methods, which results in the reduction of variance in tree aggregation [42].

The overall objectives of this study were to synthesize mesoporous Mn-doped Fe/rGO nanocomposites by a co-precipitation method and investigate ultrasonic assisted fast removal of EV from simulated wastewater. The prepared nanocomposites were characterized by X-ray diffraction (XRD), X-ray photoelectron spectroscopy (XPS), Raman spectroscopy, high-resolution transmission electron microscopy (HRTEM), scanning electron microscopy (SEM), energy dispersive spectroscopy (EDS), N_2-sorption, small angle X-ray diffraction (SA-XRD) and superconducting quantum interference device (SQUID). AI tools (ANN-GA, ANN-PSO, RF and RBF) were combined with response surface methodology (RSM) to optimize the removal efficiency of Mn-doped Fe/rGO nanocomposites for EV. Various parameters (initial EV concentration, sonication time, pH and amount of adsorbent (Mn-doped Fe/rGO) on the removal efficiency of EV were examined through batch experiments. The importance of four factors in the removal process was evaluated by the Garson method, RF, variance analysis and gradient boosting regression tree (GBRT). The adsorption kinetics and isotherm of the adsorbent for EV were studied.

2. Materials and Methods

2.1. Chemicals

All chemicals used in this work were of analytical grade. $FeSO_4$ $7H_2O$ and H_2O_2 were obtained from Chengdu Jinshan Chemical Reagent Co., Ltd. (Sichuan, China). $MnCl_2$ $4H_2O$ was supplied by Tianjin Bodi Chemical Co., Ltd. (Tianjin, China). $NaBH_4$ was purchased from Tianjin Kermel Chemical Reagent Co., Ltd. (Tianjin, China). HCl, H_2SO_4 and NaOH were obtained from Sinopharm Chemical Reagent Co., Ltd. (Shanghai, China). C_2H_5OH was procured from Tianjin Fuyu Fine Chemistry Engineering Co., Ltd. (Tianjin, China). Graphite powder was provided by Sinopharm Chemical Reagent (Beijing, China). Ethyl violet (purity > 90%, $C_{31}H_{42}ClN_3$, molecular weight = 492.14 g/mol) was purchased from Shanghai Acmec Biochemical Co. Ltd. (Shanghai, China), and 1000 mg/L stock solution of EV was prepared with deionized water.

2.2. Fabrication of Fe–Mn and Mn-doped Fe/rGO

GO was synthesized following the improved Hummers method by Shi et al. [47]. The Mn-doped Fe/rGO nanocomposites were prepared by a one-step synthesis approach (Figure 2). $FeSO_4{\cdot}7H_2O$ (10 g/100 mL), $MnCl_2{\cdot}4H_2O$ (3.60 g/50 mL) and GO suspensions (1.0 g/300 mL) were mixed with ultrasound for 2 h, which was stirred at room temperature for 12 h. Then, $NaBH_4$ (5.4 g/50 mL) solution was slowly added under a nitrogen atmosphere. The Mn-doped Fe/rGO nanocomposites (Mn:Fe = 1:2) were treated with deionized water and ethanol several times and dried in a vacuum oven at 50 °C. Additionally, bimetallic Fe–Mn nanoparticles were synthesized by the same method without adding GO.

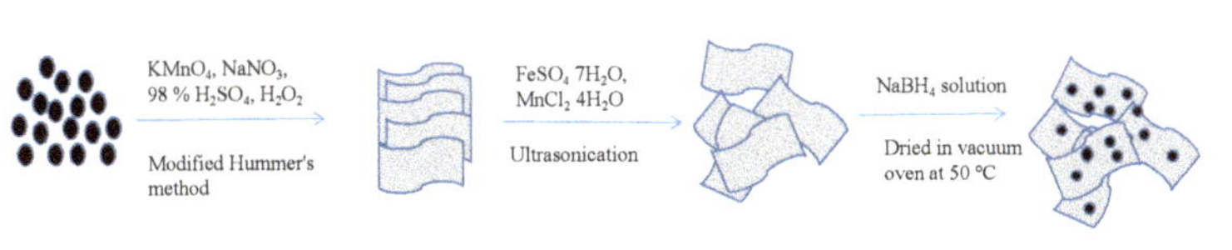

Figure 2. Schematic of Mn-doped Fe/rGO nanocomposite synthesis.

2.3. Characterization

Phase structures of GO, Fe–Mn and Mn-doped Fe/rGO were measured by X-ray diffraction (XRD) using a Philips Analytical X-ray (Lelyweg 1 7602, EA, Almelo, The Netherlands) with a Cu-Kα X-ray source at 40 KV and 40 mA within the 2θ angle range from 5 to 90°. The X-ray photoelectron spectroscopy (XPS) was observed by the surface chemical states of the samples using an ESCALAB 250Xi spectrometer (Thermo Electron Corporation, Waltham, MA, USA). Raman spectra were recorded using a Raman spectrometer (LabRAM HR800, Horiba Jobin Yvon, Paris, France) with a 532 nm laser. The morphologies of samples were studied by a high-resolution transmission electron microscope (HR-TEM) and scanning electron microscope (SEM) with an energy dispersive spectrometer (EDS). A nitrogen adsorption–desorption apparatus at 77 K (Quadrasorb SI, Quantachrome Instruments, Boynton Beach, Florida, USA) was employed to determine the Brunauer–Emmet–Teller (BET) specific surface area. The ordered structure of mesoporous Mn-doped Fe/rGO was evaluated using small angle X-ray diffraction (SA-XRD). The magnetic properties of nanocomposites were estimated by a superconducting quantum interference device (SQUID).

2.4. Dye Removal Procedure

The EV adsorption experiments were conducted using solutions with varying pH, sonication time (min), EV concentration (mg/L) and amount of adsorbent (mg) in 100 mL conical flasks, and reactions were carried out at room temperature using an ultrasonic agitation. Then, initial pH of EV solutions was adjusted to expected values by 0.1 mol/L HCl and 0.1 mol/L NaOH. The mixed solution was separated by a magnet and supernatant was obtained. The supernatant liquid was determined using a UV–VIS spectrophotometer at a wavelength of 595 nm. The dye removal percentage was calculated by the following equation:

$$Y(\%) = (C_0 - C_t)/C_0 \times 100\,\% \tag{1}$$

where Y is the removal percentage of EV, C_0 is the initial dye concentration (mg/L) and C_t is the concentration of unadsorbed dye residual in the solution. The amount of EV removal at equilibrium was calculated as follows:

$$q_e = (C_0 - C_t) \times V/m_s \tag{2}$$

where q_e is the removal equilibrium of EV (mg/g), V is the volume of EV solution and m_s is the adsorbent dosage (g).

2.5. Response Surface Methodology

Response surface methodology (RSM) consists of the following three steps: experimental statistical design, estimation of variable coefficients in empirical formulas and final prediction of response, model validation and adequacy study [48]. The parameter contributions of Mn-doped Fe/rGO nanocomposites to the elimination of ethyl violet dye were studied by central composite design (CCD). According to the principle of central combination design, there are five levels of code values ($-\alpha$, -1, 0, 1, α) in each coefficient, which represent the actual operating parameters as presented in Table 1.

$$Y = \beta_0 + \sum_{i=1}^{k} \beta_i X_i + \sum_{i=1}^{k} \beta_{ii} X_i^2 + \sum_{i=1}^{k} \sum_{i \neq j=1}^{k} \beta_{ij} X_i X_j + \varepsilon \tag{3}$$

where Y is the response variable; β_0, β_i, β_{ii} and β_{ij} are the constant, the linear, the quadratic and the interaction coefficients, respectively; ε is error; X_i and X_j are the independent variables.

Table 1. Four factor, five level central composite design.

Factors	Coded Variable Levels				
	$-\alpha$	-1	0	1	α
Initial concentration (mg/L)	250	300	350	400	450
Initial pH	3	4	5	6	7
Sonication time (min)	7	10	13	16	19
Dosage (mg)	10	13	16	19	22

Analysis of variance (ANOVA) was the statistical analysis method used to explain the complex relationship between two dependent variables and four independent variables in the whole set of data.

2.6. Artificial Neural Network (ANN)

BP-ANN is one of the most widely used ANN methods. Generally, computational neural networks are superior to rule-based and knowledge-based expert systems because they have better fault tolerance and generalization capabilities [49]. In this study, MATLAB R2016a software was used for all computations related to neural networks. A non-linear prediction model based on RSM data was established by using three layers (input layer, hidden layer and output layer) of BP-ANN. There were four inputs (i.e., initial pH, dosage, initial EV concentration and sonication time) and one output (i.e., elimination efficiency), as demonstrated in Figure 3. In the 30 experimental data groups of RSM, network training was conducted with groups 1–24, and network testing was conducted with groups 25–30. In the process of neural network training, 30 samples were normalized in the range of 0.1 to 0.9 (Equation (13)) [39]. According to the model results, the calculation equations for mean square error (MSE) and the R^2 correlation coefficient can be written as follows [50]:

$$y = 2\times\left(\frac{x - x_{\min}}{x_{\max} - x_{\min}}\right) - 1 \tag{4}$$

$$MSE = \frac{1}{N}\sum_{i=1}^{N}\left(\left|y_{prd,i} - y_{\exp,i}\right|\right)^2 \tag{5}$$

$$R^2 = 1 - \frac{\sum_{i=1}^{N}\left(y_{prd,i} - y_{exp,i}\right)}{\sum_{i=1}^{N}\left(y_{prd,i} - y_m\right)} \tag{6}$$

where x is the input variable in a group of variables to be scaled, y is the normalized value of x, x_{max} and x_{min} are the maximum and minimum value of x, respectively, $y_{prd,i}$ is the predicted value by the ANN model, $y_{exp,i}$ is the experimental value, N is the number of data and y_m is the average of the experimental values.

GBRT is one of the most effective machine learning models, which is suitable for complex nonlinear relations. For regression problems, GBRT is a combination of gradient boosting and regression trees for solving problems that uses ensembles of regression trees to reduce the error over a large single-tree model [51]. The GBRT model (developed with R, version 2.9.2) was used to evaluate the importance of four parameters, which were measured according to the co-occurrence frequency of these characteristics in all splits of decision trees [52]. The relative influence of the individual variable was calculated by the following Garson equation [53,54]:

$$I_{ab} = \frac{\sum_e^n\left(\frac{|w_{ae}|}{\sum_g^m |w_{ge}|}|w_{eb}|\right)}{\sum_z^n\left(\sum_l^n\left(\frac{|w_{al}|}{\sum_g^m |w_{ge}|}|w_{eb}|\right)\right)} \tag{7}$$

where I_{ab} is the relative importance of the j_{th} input variable on the output variable; w_x is the connection weight; a, e and b are the number of neurons in the input layer, hidden layer and output layer, respectively.

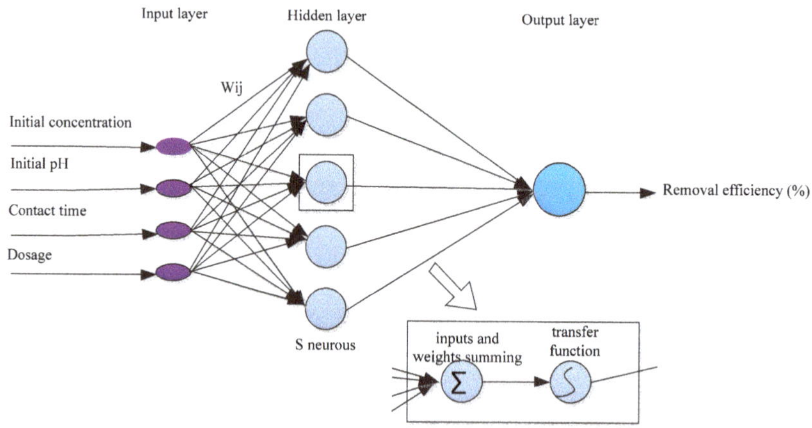

Figure 3. Back propagation artificial neural network (BP-ANN) structure.

2.7. Optimization Using ANN-GA and ANN-PSO Models

A suitable combination of variables was selected to check the aptness of the model. Experiments were carried out with the optimal four variables to verify the prediction models as proposed by ANN-GA and ANN-PSO [55]. GA is a stochastic optimization technology developed based on genetic mechanisms and Darwin's theory of evolution [56,57]. Simply, GA involves a randomly generated initial population and uses genetic operators (selection, mutation and crossover) to modify the population [56]. The algorithm is used to explore different regions of parameter space and to determine optimal conditions for removal of EV [58]. Kennedy and Eberhart have proposed the PSO method, which is an evolutionary computational algorithm [59]. PSO is similar to GA, which initializes the system with the population of random solutions and also uses the concept of population and evolutionary iteration to achieve the purpose of optimization [59,60]. In the first step of PSO, the random position and velocity of each particle were given. The fitness values of each particle was calculated at the initial location by contrasting the properties of known parameters (i.e., initial pH, sonication time, initial concentration, adsorbent dosage) and was obtained with the prediction results. The closeness between simulated and known properties was defined as the fitness value of particles at any given position, and particles with the highest fitness value were selected as the global optimum particles in the whole operation [61]. On the basis of evolutionary mechanisms, all particles guided by the global optimum eventually converged to some system optimum of the optimized problem by successively executing a certain number of iterations [62].

2.8. Random Forest

Random forest (RF), as an ensemble classifier, consists of multiple decision trees [45,63], which combines the classification tree and regression tree by a bagging algorithm. In random forests, decision trees are independent of each other and have different classification results. Decision trees are also called classification trees, in which leaves and branches are represented in class labels and the connection of feature vectors leading to class labels, respectively [63]. Decision trees are derived from a decision tree learning algorithm in which data sets are divided into different subsets according to attribute value tests. This segmentation process is called recursive partitioning. Each internal node delegates an input trait, and each node has a child of another input characteristic [63]. The advantages of RF are as follows: (1) random variable selection seeks to minimize the correlation among trees in the set, which gives a lower error rate; (2) in order to obtain fast learning and prevent over-fitting, random selection of bootstrap samples can be guided to construct bootstrap data sets (sub-training sample sets) [64,65]. The RF model was established using R studio (version 3.4.4.).

2.9. Radial Basis Function Neural Network

Radial basis function (RBF) neural network is a typical feed-forward network, which consists of three layers, namely the input layer, hidden layer and output layer [66]. The hidden layer contains many nodes, and each node uses a non-linear activation function (φ(r)) [67,68]. The input layer takes the input parameters as input vectors, from input layer to hidden layer, nonlinear transformation of input vectors as activation functions of neurons (i.e., radial basis function) and the conversion from hidden layer to output layer is a linear transfer function [67]. In this work, we investigated four main parameters (as input data) including initial dye concentration (250–450 mg/L), initial pH (3–7), sonication time (7–19 min) and dosage (10–22 mg) and EV removal percentage (as output data) with RBF. The radial basis function is expressed by a Gaussian function as follows [69]:

$$\varphi_{ij} = \exp\left(\frac{||x_j - c_i||^2}{\sigma_j^2}\right) \tag{8}$$

where x_j is the input vector, $|| \, ||$ is a measure of Euclidean distance and c_i and σ_j are the center and the spread of j_{th} the RBF, respectively. The output node $f_k(x)$ is calculated as follows [69]:

$$f_k(x) = \sum_{j=1}^{m} (w_{kj}\varphi_j(x)) \tag{9}$$

where w_{kj} is the weight connection between the hidden and output layers.

3. Results and Discussion

3.1. Surface Characterization of Fe–Mn Nanoparticles and Mn-doped Fe/rGO Nanocomposites

X-ray diffraction (XRD) patterns of GO, Fe–Mn and Mn-doped Fe/rGO are illustrated in Figure 4. GO had a characteristic peak at 11° (002), while the peak disappeared for Mn-doped Fe/rGO, indicating that GO was completely reduced to rGO in the synthesis process. In the Fe–Mn and Mn-doped Fe/rGO composites, no obvious diffraction peaks of Fe and Mn were observed by XRD, suggesting that Fe and Mn in the composites existed in an amorphous form [70].

The surface morphology of Fe–Mn nanoparticles and Mn-doped Fe/rGO nanocomposites are discussed based on SEM images (Figure 5a,b). Figure 5a shows that Fe–Mn spherical particles were dispersed on the surface of graphene. The average particle size of Mn-doped Fe/rGO (Figure 5c) and Fe–Mn (Figure 5d) were 82.56 and 119.57 nm, respectively. The EDS measurement confirmed the composition of Fe–Mn and Mn-doped Fe/rGO (Figure 6). The elemental compositions of C, O, Mn and Fe in Fe–Mn were 1.50, 2.38, 1.11 and 95.01 wt%, respectively. The elemental composition of C, O, Mn and Fe in Mn-doped Fe/rGO were 16.67, 8.83, 1.59 and 72.91 wt%, respectively. EDS is a qualitative and semi-quantitative analysis tool [71], which measures a certain point on the surface of the sample, and thus its result cannot be used for the whole sample. The ratio of Mn to Fe measured by EDS was only 1:50 in Mn-doped Fe/rGO, much lower than the theoretical value of Mn-doped Fe/rGO (Mn:Fe = 1:2). The possible reason is that Mn and Fe were not evenly distributed on the rGO and nanoscale Fe agglomerated on the surface of rGO [72].

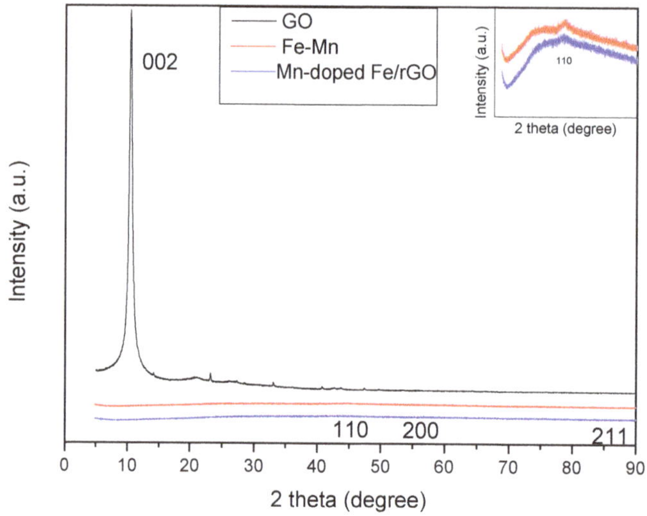

Figure 4. XRD spectra of graphene oxide (GO), Fe–Mn and Mn-doped Fe/rGO.

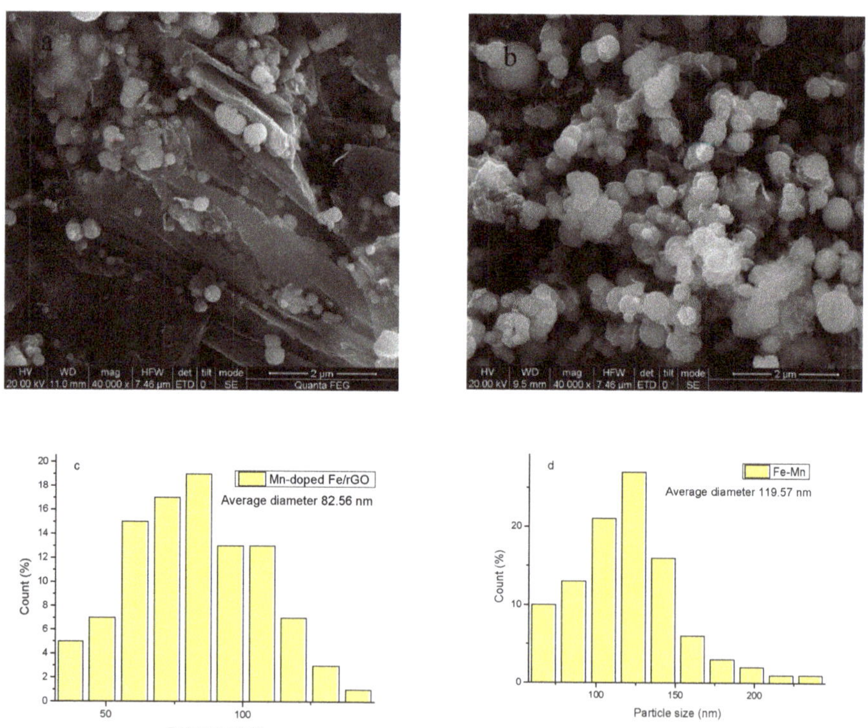

Figure 5. SEM image and the corresponding particle size of Mn-doped Fe/rGO (**a**,**c**) and Fe–Mn (**b**,**d**).

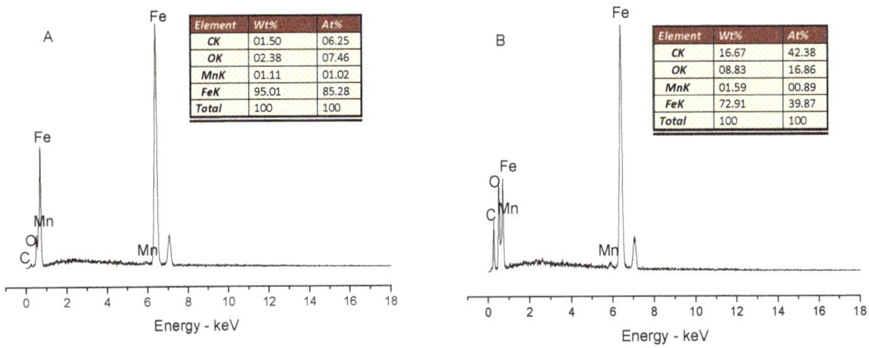

Figure 6. EDS of Fe–Mn (**A**) and Mn-doped Fe/rGO (**B**).

Figure 7 displays the Raman spectra of GO and Mn-doped Fe/rGO nanocomposites. The characteristic D and G peaks appeared at 1354 and 1570 cm^{-1}, respectively. The D peak is related to the defects and disorder in the lattice structure of graphite materials, while the G peak is related to the vibration of sp^2-bonded carbon and is a doubly degenerate (iTO and LO) phonon mode (E_{2g} symmetry) at the BZ center [73–76]. The intensity ratio (I_D/I_G) of Mn-doped Fe/rGO (I_D/I_G = 1.5) was higher than that of GO (I_D/I_G = 1.2), which indicated the creation of smaller sp^2-bonded carbon domains after the reduction process and the increase of disorder in the graphene sheets [77,78]. The surface survey spectra of Fe–Mn and Mn-doped Fe/rGO were measured by XPS, and the spectra of C1s, O1s, Fe2p and Mn2p are presented in Figure 8. The bonding energies at 284 eV, 531 eV, 641 eV and 711 eV correspond to the C1s, O1s, Mn2p and Fe2p, respectively. The contents of C, O, Mn and Fe in Mn-doped Fe/rGO were 76.19, 19.68, 0.826 and 3.305 at%, respectively. The atomic ratio of Fe and Mn measured by XPS was about Mn:Fe ≈ 1:3, significantly lower than the theoretical value of the nanomaterials (Mn:Fe = 1:2). The possible reason is that most of the Fe and Mn were loaded in the pores of nanomaterials, and only a small amount of iron and manganese were exposed on the outer surface of the nanomaterials [79]. Since XPS can only measure elements in the depth range of about 5 nm on the surface, most of the iron and manganese in the pores of adsorbent were not detected. Figure 9c,d shows the Mn2p peak of the Fe–Mn and the Mn-doped Fe/rGO. The characteristic peaks (Figure 9a,b) of Fe^0, Fe2p3/2 and Fe2p1/2 at 707, 711.6 and 725.6 eV, respectively, were observed in the Fe2p XPS spectra. The Fe2p spectra exhibited four peaks at 710.6 eV, 711.9, 723.8 eV and 725.4 eV, which confirmed the Fe^{3+} and Fe^{2+} states in the nanocomposites and revealed the core–shell structure of nZVI.

The nitrogen adsorption–desorption isotherms of Fe–Mn nanoparticles and Mn-doped Fe/rGO nanocomposites are shown in Figure 10. The specific surface areas (SBET) were 19.166 m^2/g (Fe–Mn) and 104.088 m^2/g (Mn-doped Fe/rGO), respectively, which were due to the high specific surface area of reduced graphene sheets. The Fe–Mn and Mn-doped Fe/rGO composites of pore size distribution are shown in Figure 11, and narrow pore size distributions were centered at 1.76 nm and 3.93 nm, respectively. Mn-doped Fe/rGO exhibited type IV isotherms with H_3 hysteresis loops, revealing that the nanocomposites were mesoporous (2 nm < pore size < 50 nm) materials [80]. Nanocomposites with large specific surface area can provide more adsorption and reaction sites, improving the dye removal ability from wastewater.

Figure 12A represents the magnetization curves of the Mn-doped Fe/rGO nanocomposites, which are typical soft magnetic materials. The soft magnetic properties of nanocomposites are associated with the very small area encircled by a closed curve [81]. Magnetization increased linearly with rising magnetic field and then reached saturation; the value of saturation magnetization (Ms) was 47.0514 emu/g for Mn-doped Fe/rGO. The magnetic properties of nanocomposites allow them to be easily and rapidly segregated from dye wastewater. Small angle X-ray diffraction (SAXRD) and high-resolution transmission electron microscopy (HR-TEM) of Mn-doped Fe/rGO nanocomposites are given in Figure 12B,E, respectively. Mn-doped Fe/rGO showed a well-resolved diffraction peak at 0.71° of 2θ angles, a large and repetitive lattice commonly conducive to the presence of diffractions at small angles, indicating that the nanocomposites may have been composed of an ordered structure [82]. From the HR-TEM image of Figure 12C–E, it can be observed that the regularly ordered structure appeared in Figure 12A, while the disordered structure emerged in Figure 12B. Therefore, the Mn-doped Fe/rGO nanocomposites were determined to be partially ordered mesoporous materials.

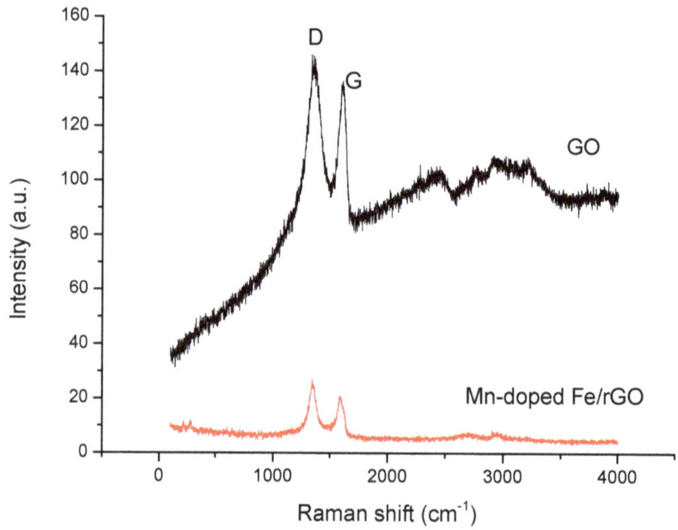

Figure 7. Raman spectra of GO and Mn-doped Fe/rGO.

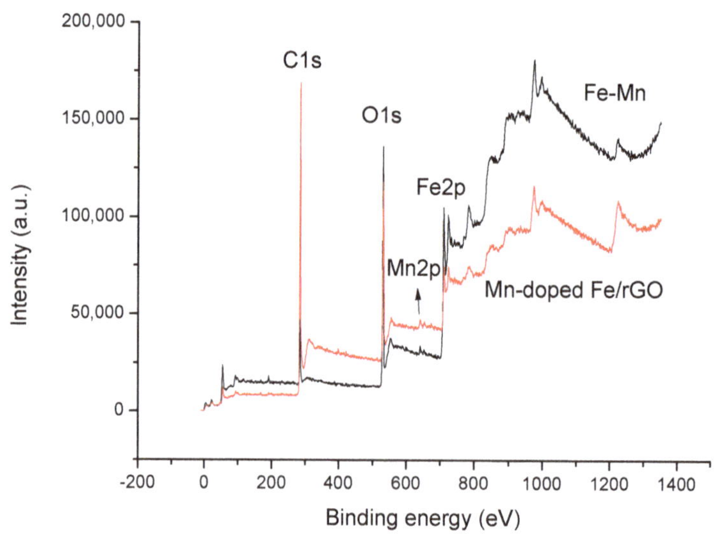

Figure 8. XPS survey spectra of Fe–Mn NPs and Mn-doped Fe/rGO composites.

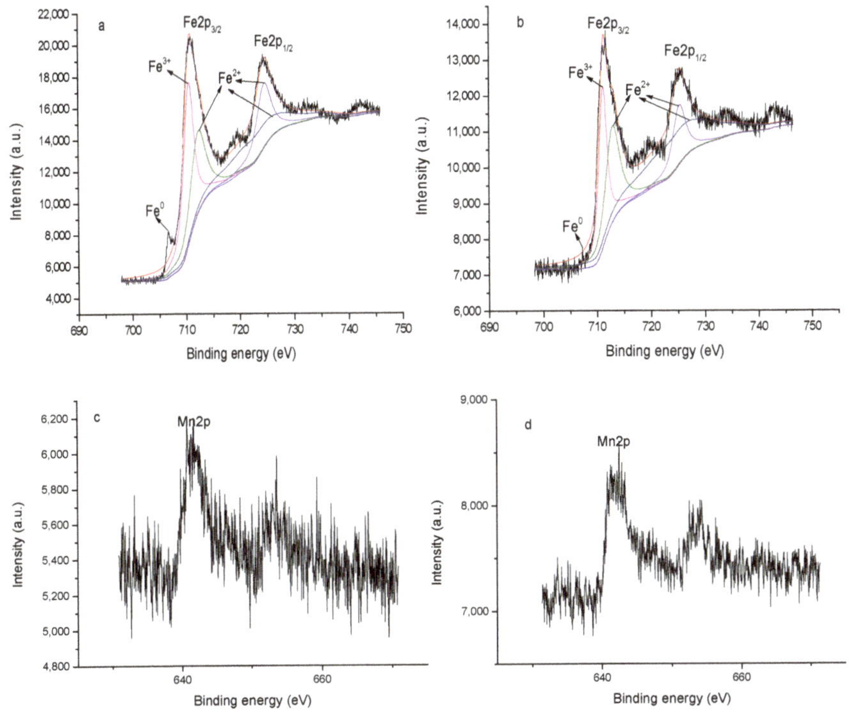

Figure 9. XPS analyses of high resolution spectra of Fe2p for Fe–Mn nanoparticles (NPs) (**a**) and Mn-doped Fe/rGO composites (**b**), Mn2p for Fe–Mn NPs (**c**) and Mn-doped Fe/rGO composites (**d**).

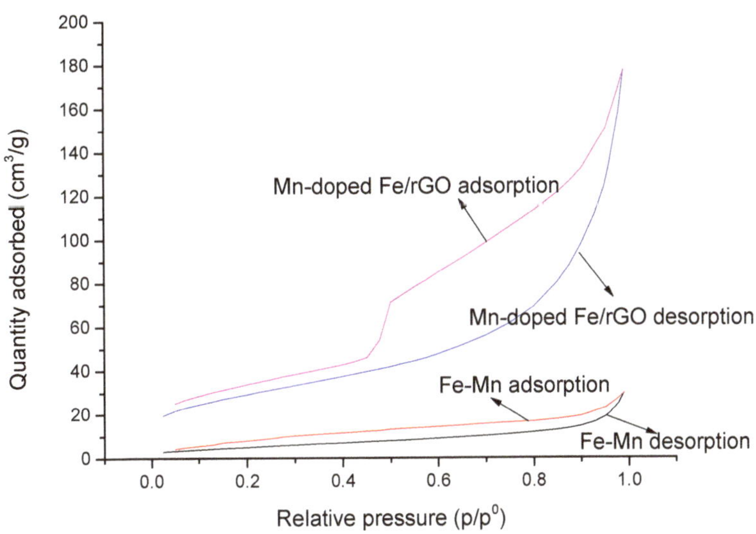

Figure 10. Nitrogen adsorption–desorption isotherm of Fe–Mn and Mn-doped Fe/rGO.

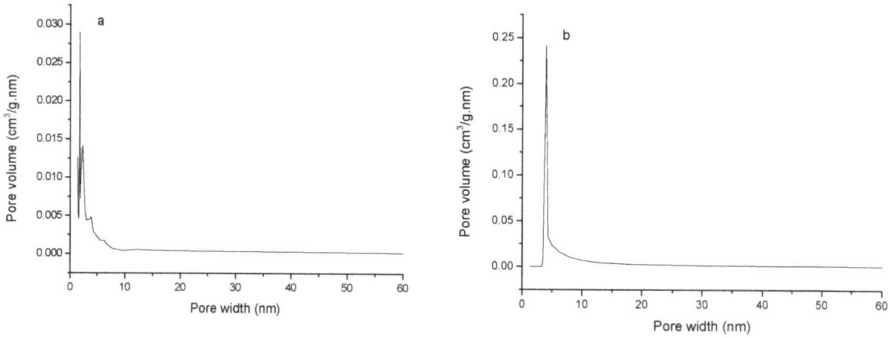

Figure 11. Pore size distribution of Fe/Mn (**a**) and Mn-doped Fe/rGO (**b**).

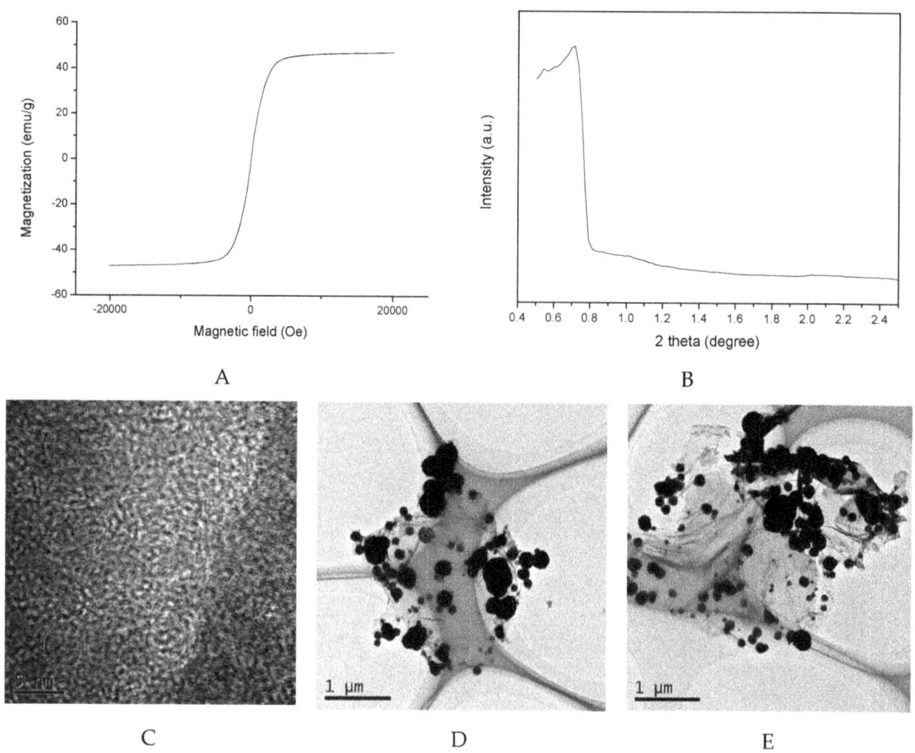

Figure 12. Magnetization curves of Mn-doped Fe/rGO (**A**), small angle X-ray diffraction (SAXRD) patterns of Mn-doped Fe/rGO (**B**), HR-TEM (**C**–**E**) images with different magnifications of Mn-doped Fe/rGO nanocomposites.

3.2. Central Composite Design (CCD)

Four variables were studied at initial concentration (250–450 mg/L), initial pH (3–7), sonication time (7–19 min) and dosage (10–22 mg). Removal efficiency was measured as response, and the maximum and minimum were 92.46% and 60.09%, respectively. In the CCD and ANOVA programs used, 30 sets of data were used to predict the four independent variables (X_1 = A, X_2 = B, X_3 = C, X_4 = D), and the experimental data were analyzed by quadratic multiple regression analysis. Table 2 compares the predicted values with the experimental results, and the relationship between removal efficiency and variables factors are given as follows:

$$Y = 87.20 - 3.68\ A + 1.92\ B + 3.32\ C + 5.07\ D + 0.12\ AB - 0.097\ AC + 0.21\ AD + 0.084\ BC + 0.73\ BD - 0.27\ CD - 1.37\ A^2 - 0.34\ B^2 - 3.75\ C^2 - 3.14\ D^2 \quad (10)$$

Table 2. Central composite design matrix for ethyl violet (EV) removal.

Run No.	Initial Concentration (mg/L)	Initial pH	Sonication Time (min)	Dosage (mg)	Removal Efficiency (%)	Prediction Efficiency (%)
1	350	5	13	16	88	87
2	300	6	10	13	76	75
3	350	3	13	16	83	82
4	300	4	10	19	80	82
5	400	4	10	13	67	65
6	300	4	16	19	85	88
7	400	4	16	13	72	72
8	350	5	13	16	86	87
9	300	6	16	13	82	82
10	300	4	16	13	80	80
11	400	6	10	19	82	80
12	350	5	13	22	87	85
13	350	5	13	16	86	87
14	400	4	10	19	75	75
15	300	6	10	19	87	86
16	400	6	16	19	85	86
17	400	6	16	13	77	75
18	350	5	13	16	90	87
19	300	6	16	19	92	93
20	450	5	13	16	71	74
21	350	7	13	16	86	90
22	350	5	19	16	80	78
23	400	6	10	13	70	67
24	400	4	16	19	80	80
25	250	5	13	16	90	89
26	350	5	13	16	87	87
27	350	5	13	10	60	64
28	300	4	10	13	74	73
29	350	5	13	16	86	87
30	350	5	7	16	61	66

The sufficiency of the model was tested by ANOVA, and the fitting of the second-order polynomial equation with the experimental data was tested (Table 3). The value of "Prob. > F" was used to test the null hypothesis ($p < 0.05$), which implied that the model was significant. The regression model established for removal of EV by Mn-doped Fe/rGO based on A, B, C and D was significant ($p < 0.05$) and the lack of fit values was not significant ($F = 4.4$, $p = 0.0578$) showing that the regression model was valid for the removal of EV. The values of "Prob. > F" for linear effect of A, B, C, D and quadratic effect of A^2, D^2, C^2 were less than 0.05; thus, these terms were significant. The results in Table 3 display a high R^2 (0.9379) for EV, thus demonstrating dependability in the evaluation of EV removal efficiency by Mn-doped Fe/rGO.

Table 3. Results of ANOVA for response surface quadratic model.

Source	Sum of Squares	df	Mean Square	F Value	p-Value	
Model	1899.59	14	135.68	16.19	<0.0001	significant
A	325.24	1	325.24	38.8	<0.0001	
B	88.05	1	88.05	10.5	0.0055	
C	265.14	1	265.14	31.63	<0.0001	
D	617.02	1	617.02	73.6	<0.0001	
AB	0.24	1	0.24	0.028	0.8685	
AC	0.15	1	0.15	0.018	0.8953	
AD	0.68	1	0.68	0.081	0.7802	
BC	0.11	1	0.11	0.014	0.9087	
BD	8.63	1	8.63	1.03	0.3264	
CD	1.2	1	1.2	0.14	0.71	
A^2	51.24	1	51.24	6.11	0.0259	
B^2	3.09	1	3.09	0.37	0.553	
C^2	385.56	1	385.56	45.99	<0.0001	
D^2	271.17	1	271.17	32.35	<0.0001	
Residual	125.75	15	8.38			
Lack of Fit	112.92	10	11.29	4.4	0.0578	not significant
Pure Error	12.83	5	2.57			
Cor Total	2025.33	29				
			$R^2 = 0.9379$			

Cor total: total of all information corrected for the mean.

Figure 13 illustrates the correlation between the experimental and predicted values. The distribution of data points was relatively close and had a linear behavior, which indicated that there was sufficient consistency between predicted and experimental values. Figure 14 displays the normal probability plot of the residual. The maximum number of points fell on a straight line, which indicates that the residual obeyed a normal distribution and the prediction of the RSM model was accurate. The effects of four factors on the EV removal efficiency were compared by using the perturbation plot (Figure 15). The results show that the elimination of EV was more sensitive to adsorbent dosage and initial concentration than sonication time and initial pH. The interaction between the two dependent and the four independent variables can be represented by the three-dimensional response surface and contour diagrams (Figure 16). It can be seen from Figure 16A,a that the maximum decolorization of EV was in the pH range of 4 to 6. Within the scale of experiment, the removal efficiency of dye increased with the increasing amount of adsorbent and decreased with increasing initial ethyl violet concentration (Figure 16C,c).

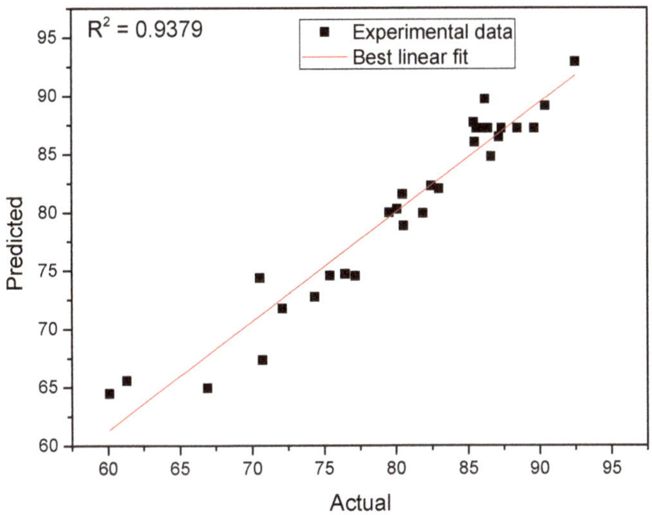

Figure 13. Predicted versus experimental values plot for EV removal.

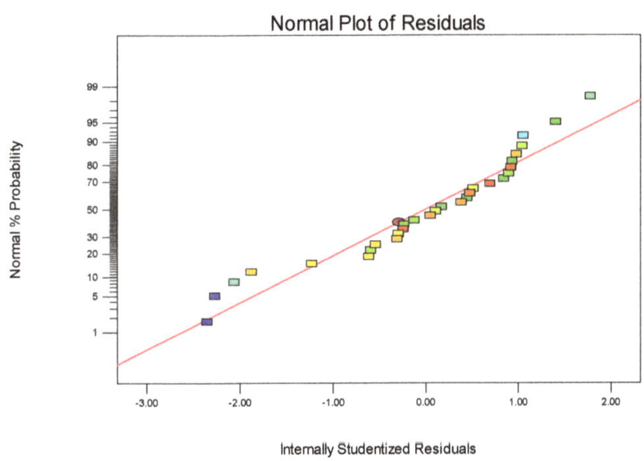

Figure 14. Normal probability plot of residuals.

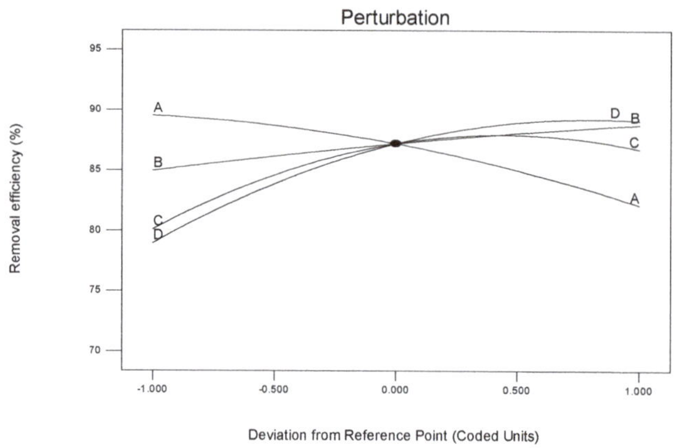

Figure 15. Perturbation plot of EV removal efficiency.

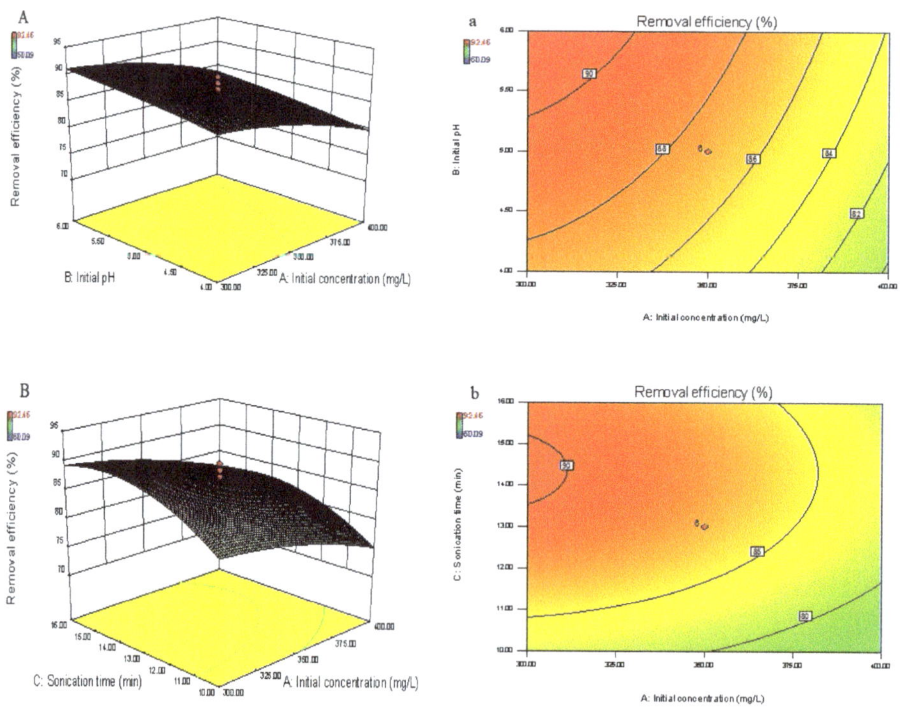

Figure 16. *Cont.*

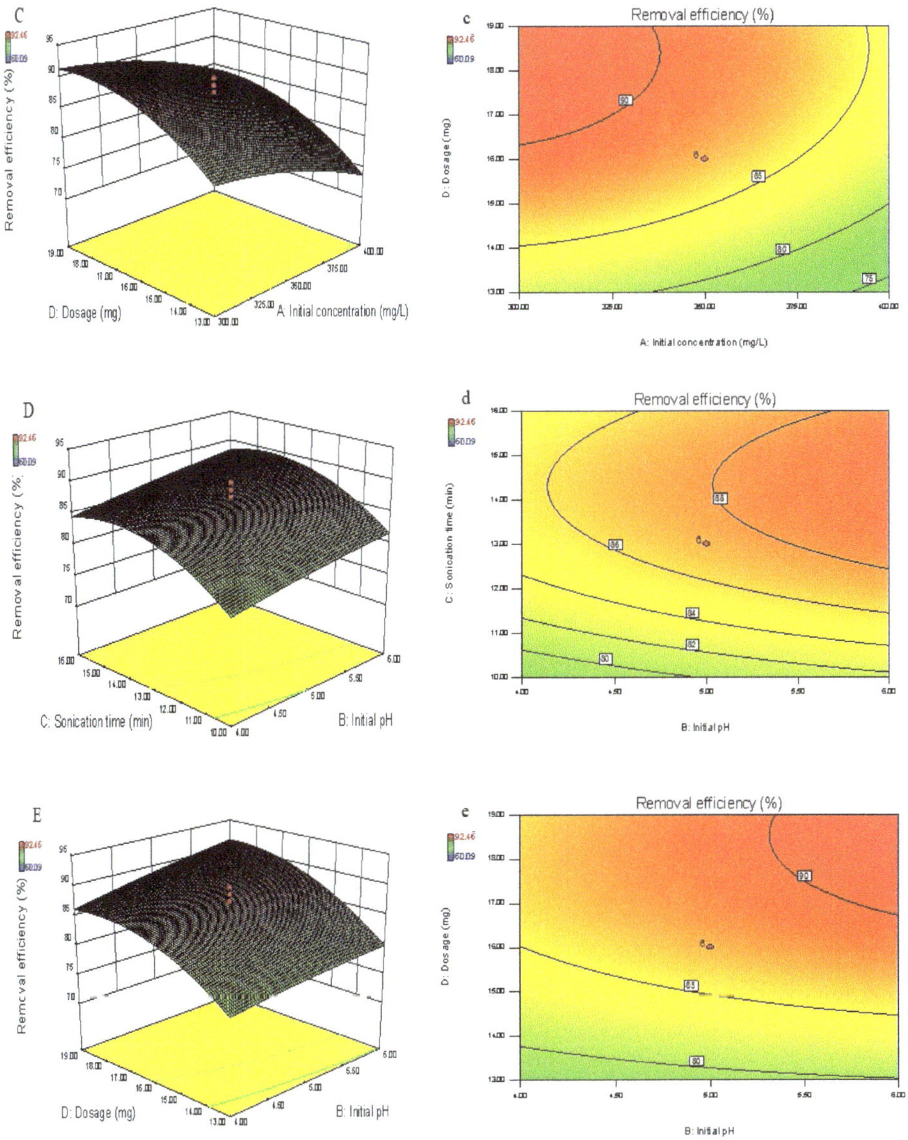

Figure 16. *Cont.*

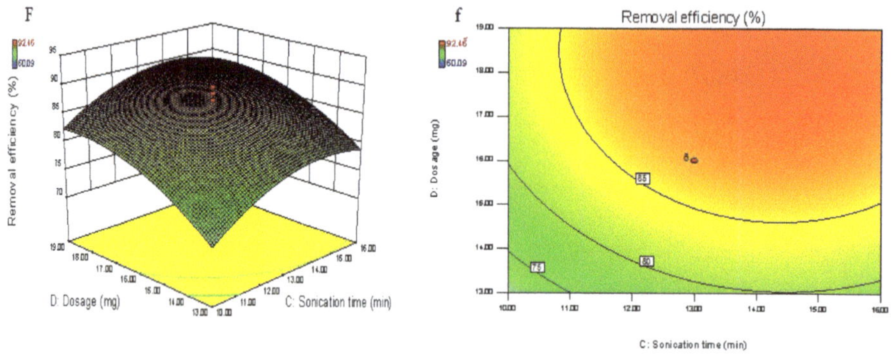

Figure 16. The three-dimensional response surface and contour plots: interactive effects of (**A**) and (**a**) X_1–X_2; (**B**) and (**b**) X_1–X_3; (**C**) and (**c**) X_1–X_4 (**D**) and (**d**) X_2–X_3; (**E**) and (**e**) X_2–X_4; (**F**) and (**f**) X_3–X_4 on decolorization efficiency (%) of dye.

3.3. *Back Propagation Artificial Neural Network (BP-ANN)*

Under the ANN modeling method, it is an important step to determine the optimal number of hidden layer neurons. The optimal number of neurons was determined by MSE, and the minimum MSE value was 0.0005, as presented in Figure 17. Additionally, Figure 18 shows that the BP-ANN training was stopped when the epoch reached 1453, the gradient reached 5.5214×10^{-6}, the best training performance was 0.0030709 and the correlation coefficient (R^2) was 0.99403 (Figure 19). The high R^2 value and low MSE value indicated goodness of the ANN model performance.

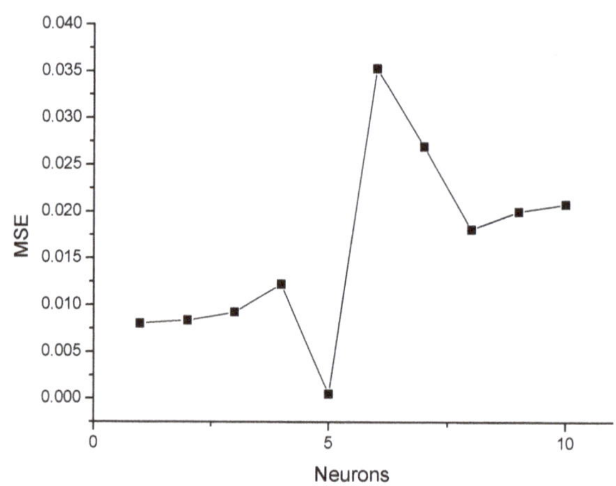

Figure 17. Correlation between MSE and the number of hidden neurons.

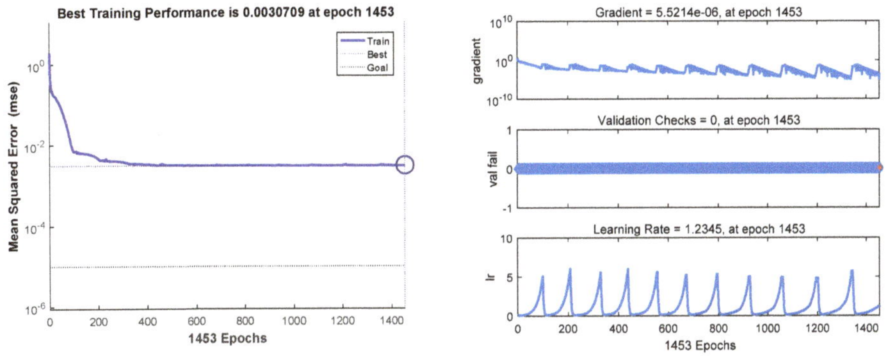

Figure 18. Training process of BP-ANN.

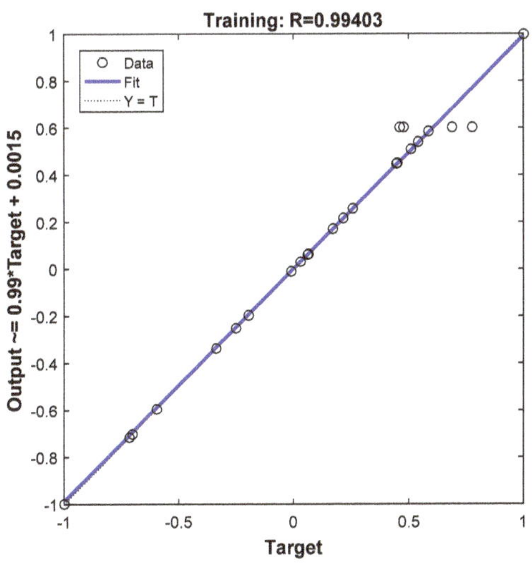

Figure 19. Predicted vs. experimental values of the normalized removal of EV from the BP-ANN.

3.4. Optimization for the Removal of EV by RSM, ANN-GA and ANN-PSO

Both ANN-GA and ANN-PSO can obtain global optimal solutions through iterative optimization processes. Figure 20 shows for ANN-GA and ANN-PSO gradual convergence; when the iteration values were 5 and 8 the maximum predicted values were 92.75% and 90.11%, respectively. The optimization conditions for the four parameters were found to be dosage (GA 17.55 mg and PSO 17.20 mg), sonication time (GA 13.19 min and PSO 13.95 min), initial pH (GA 6.1 and PSO 5) and initial concentration (GA 299.22 mg/L and PSO 371.23 mg/L). The developed models from ANN-GA, ANN-PSO and RSM were compared and validated using the predicted optimal conditions (Table 4). The results show that the average values of absolute errors of ANN-GA, ANN-PSO and RSM models were 0.72%, 1.28% and 1.53%, respectively. Compared with the RSM model, ANN-based models had a high R^2 value. In a word, the ANN-GA model was considered to be the best one for predicting the removal of EV by Mn-doped Fe/rGO.

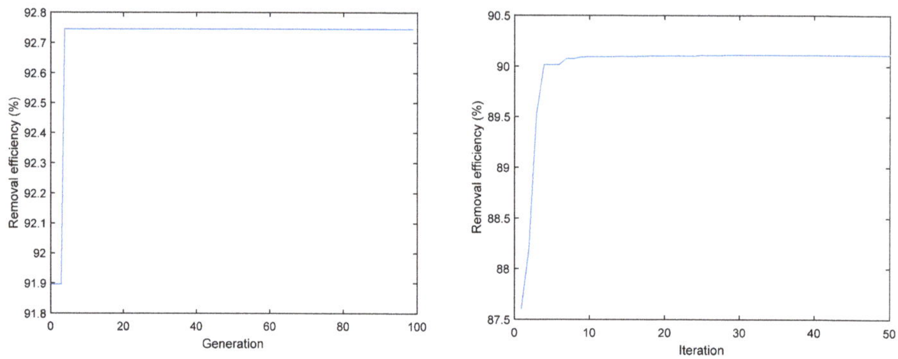

Figure 20. Genetic algorithm (GA) and particle swarm optimization (PSO) chart predicted optimum value of EV removal efficiency.

Table 4. Comparison of the confirmatory results with predicted results.

Process Parameters	Optimization		
	ANN-GA	ANN-PSO	Response Surface Methodology (RSM)
Initial concentration (mg/L)	299.22	371.23	300
Initial pH	6.1	5	6
Sonication time (min)	13.19	13.95	14.32
Dosage (mg)	17.55	17.20	18.62
Decolorization efficiency of model (%)	92.75%	90.11%	94.18%
Experimental verification values (%)	92.03%	88.83%	92.65%
Average values of absolute errors (%)	0.72%	1.28%	1.53%
R^2	0.99403		0.9379
MSE	0.0005		

3.5. Comparison of RF, RBF and BP-ANN

Combined with RSM data, 30 sets of experimental values were predicted by RF and BP-ANN (Figure 21). The mean absolute errors of BP-ANN and RF were 1.55% and 4.56%, respectively. In this study, BP-ANN was more suitable than RF to optimize the parameters for the removal of EV by Mn-doped Fe/rGO nanocomposites. Figure 22 demonstrates that the network had a high R^2 (0.95415) value, and the R^2 value of BP-ANN was 0.99403. According to the results, the RBF-NN and BP-ANN models can be used to predict the elimination of EV from the aqueous phase by Mn-doped Fe/rGO.

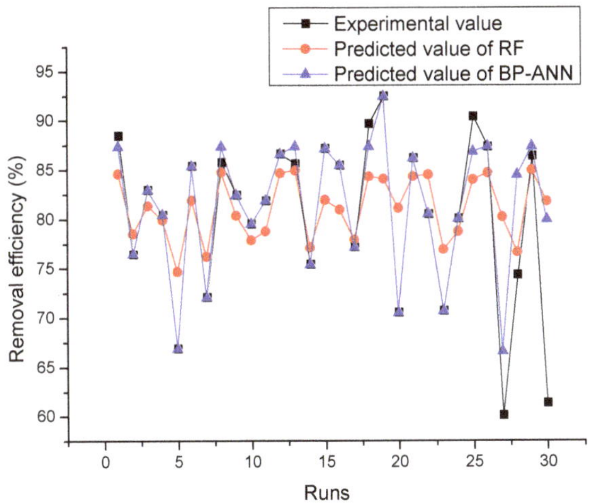

Figure 21. Comparison between experimental and predicted values of random forest (RF) and BP-ANN.

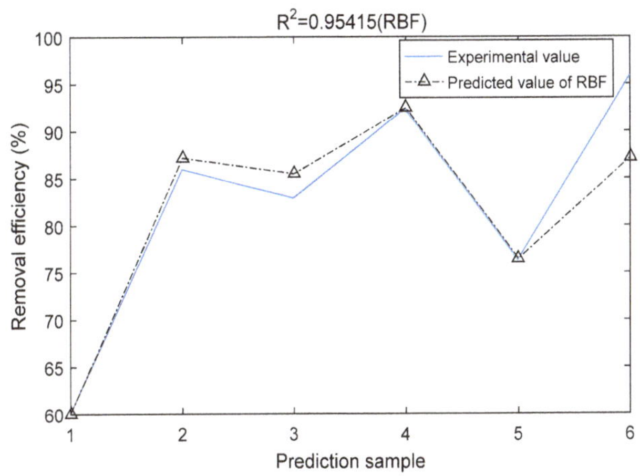

Figure 22. Comparison between experimental and predicted values of RBF.

3.6. Comparison among Analysis of Variance, GBRT, Garson Method and RF

Relative variable importance analysis is one of the advantages of GBRT and RF models (Figures 23 and 24). The Garson method for calculating the influence of each input variable on the output variable using weight and bias is shown in Table 5. The high importance score of variables indicates that the contribution of variables is significant to dye removal prediction [83]. Table 6 gives the factor importance obtained from F value, Garson method, GBRT and RF, which shows that the dosage had the maximum importance to the EV removal by Mn-doped Fe/rGO.

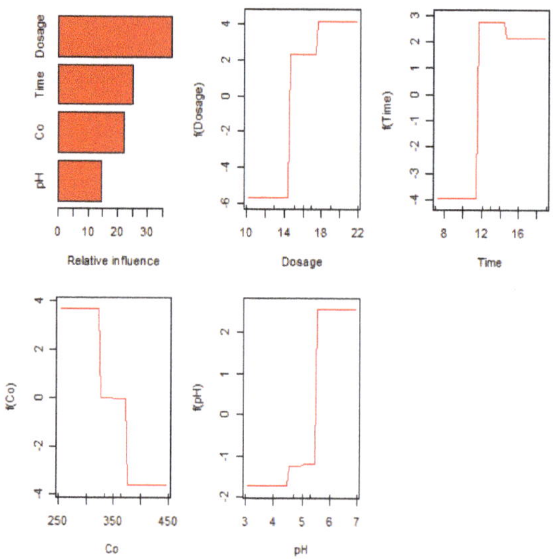

Figure 23. Feature importance for gradient boosting regression tree (GBRT).

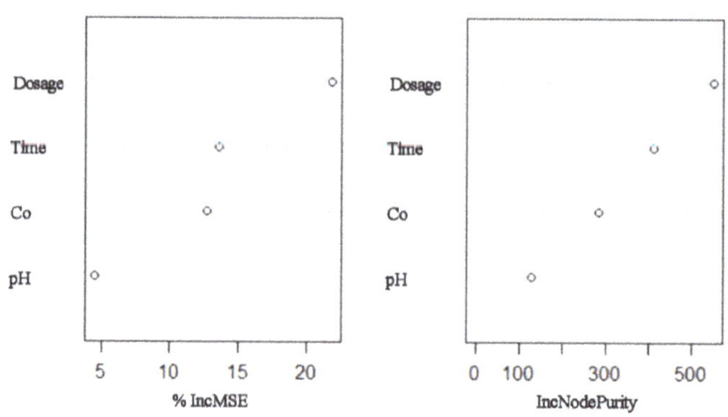

Figure 24. Feature importance for RF.

Table 5. The weights and biases of input layers (wi and bi) and hidden layers (wj and bj).

Number of Neurons	wi Input Weights				bi Input Bias	wj Hidden Weights	bj Hidden Bias
	A	B	C	D			
1	1.901923065	0.366456135	−0.653955781	1.420846441	−2.489591174	−0.930289741	
2	1.309510183	1.345913921	0.27228768	−1.611714128	−1.936348691	−0.26791453	
3	−1.358393495	−1.004381258	−0.631064845	1.716335276	1.383106208	0.145250088	
4	0.835468192	−0.017592712	−1.268234245	1.972645308	−0.829863725	−0.328025674	
5	0.491800507	−1.471722414	1.875969756	0.520544991	−0.276621242	−0.496795294	0.149904727
6	−0.213896428	1.474611835	−1.322999207	−1.492483021	−0.276621242	0.127936272	
7	0.978450096	−0.885530436	1.591737282	−1.386689545	0.829863725	0.612899558	
8	1.812375002	0.731347696	1.323745446	−0.791321431	1.383106208	0.861938074	
9	−2.371460889	−0.698563644	−0.177167065	0.234218125	−1.936348691	0.669754604	
10	−1.325902143	−0.668065352	1.592629016	1.207174065	−2.489591174	0.812121353	

Table 6. Comparison among F value, Garson method, GBRT and RF of factors importance.

	F test		Garson		GBRT		RF			
Factors	F Value	Order	Relative Influence (%)	Order	Relative Influence (%)	Order	% IncMSE	Order	IncNodePurity	Order
Initial concentration	38.8	2	29.96	2	22.34	3	12.76	3	282.69	3
Initial pH	10.5	4	16.51	4	14.64	4	4.54	4	127.76	4
Sonication time	31.63	3	22.64	3	25.01	2	13.63	2	411.46	2
Dosage	73.6	1	30.89	1	38.00	1	21.81	1	551.58	1

3.7. Adsorption Isotherms

The adsorption isotherm for adsorption EV onto Mn doped Fe/rGO was studied under the conditions of different dye concentrations (300–1000 mg/L), sonication time of 23 min and pH of 5. The data were fitted to the Langmuir, Freundlich and Temkin equations to find a model suitable for explaining the adsorption mechanism [84]. The Langmuir isotherm model is given by linear Equation (11) and nonlinear Equation (12) [85–88] as follows:

$$\frac{C_e}{q_e} = \frac{1}{k_L q_m} + \frac{C_e}{q_m} \tag{11}$$

$$q_e = \frac{q_m k_L c_e}{1 + k_L c_e} \tag{12}$$

where C_e (mg/g) is the equilibrium EV concentration in solution (mg/L); q_e is the amount of EV adsorbed at equilibrium (mg/g); q_m is the maximum adsorption capacity of dye per gram of adsorbent (mg/g); and K_L is a constant related to adsorption rate (L/mg). The Langmuir equation was further analyzed and the dimensionless equilibrium parameter (R_L) was calculated, which is given by Equation (13) [89] as follows:

$$R_L = \frac{1}{1 + K_L C_0} \tag{13}$$

where C_0 (mg/L) is the initial EV concentration; the R_L value indicates the adsorption properties of the dye with the sorbent. If the R_L value is equal to 0 or 1, the adsorption is irreversible and linear, respectively; if the value is >1, the adsorption process is unfavorable; and if the value lies between 0 and 1, this indicates the adsorption process is favorable [89].

The Freundlich equilibrium isotherm equation is an empirical relationship describing the multi-layer and heterogenous adsorption of adsorbed intermolecular interactions [84]. The linear and nonlinear forms of the Freundlich Equation (14) and Equation (15) are expressed as follows [90,91]:

$$Inq_e = Ink_F + \frac{1}{n} InC_e \tag{14}$$

$$q_e = k_F (c_e)^{1/n} \tag{15}$$

where k_F = Freundlich isotherm constant (mg/g) and $1/n$ = adsorption intensity. The linear Temkin equation is expressed as follows [86]:

$$q_e = B_T \ln K_T + B_T \ln C_e \tag{16}$$

where K_T and B_T are the Temkin constants.

The adsorption isotherm of EV is illustrated in Figure 25, and Table 7 lists the values of Langmuir, Freundlich and Temkin constants and the correlation coefficient (R^2) values. The Langmuir R_L values were 0.01585–0.05093, indicating that the adsorption of EV by Mn-doped Fe/rGO was favorable, and the value of maximum adsorption capacity was 1000.00 mg/g. The constant of 1/n in the Freundlich model is related with the adsorption intensity, which varies with the heterogeneity of materials [92]. The value of 1/n was 0.0442 (<0.5), which indicates that the adsorption was favorable. The experimental data were fitted with linear and nonlinear models with the Freundlich and Langmuir models. The R^2 given in Table 7 shows that the linear model was better than nonlinear fitting, and compared with the other two models the Langmuir model could better describe the adsorption of EV onto Mn-doped Fe/rGO.

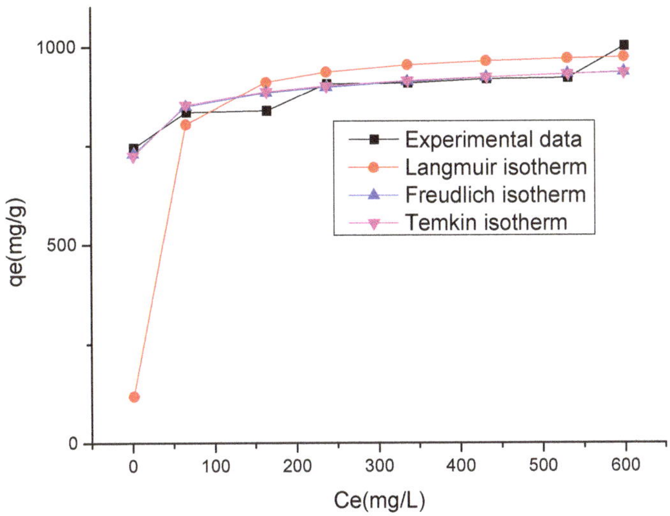

Figure 25. Adsorption isotherm of EV by Mn-doped Fe/rGO (sonication time = 23 min; initial pH = 5.0; dosage = 20 mg).

Table 7. Freundlich, Langmuir and Temkin isotherm parameters for the removal of EV by Mn-doped Fe/rGO.

Isotherms Models	Parameters	Value of Parameters Obtained by the Linear Fitting	Value of Parameters Obtained by the Nonlinear Fitting
Freundlich	K_f (mg/g)	705.7775	698.97
	1/n	0.0442	0.0462
	R^2	0.8503	0.8049
Langmuir	K_L (L/mg)	0.06211	0.05494
	q_m (mg/g)	1000.00	13,417.92
	R^2	0.9945	0.7642
	R_L	0.01585–0.05093	
Temkin	K_T (L/g)	1.2×10^8	
	B_T	37.395	
	R^2	0.7491	

3.8. Kinetics Studies

Four models (pseudo-first-order [93], pseudo-second-order [94,95], intraparticle diffusion [96,97] and Elovich [98]) were investigated in the adsorption of EV by Mn-doped Fe/rGO. The parameters of the four models and calculated values of qt versus t are presented in Table 8 and Figure 26, respectively. It can be seen from this figure that the adsorption of EV onto the nanocomposites increased with sonication time until the equilibrium was attained after 23 min. The adsorption system adsorbed EV rapidly within 7 min. During the rapid adsorption process, the adsorbent diffused from the bulk phase to the outer membrane around the adsorbent surface and then was gradually adsorbed until it reached equilibrium [89]. The R^2 values for pseudo-first-order, pseudo-second-order, intraparticle diffusion and Elovich were 0.9509, 0.9969, 0.7926 and 0.9246, respectively. The results indicated that the pseudo-second-order equation ($R^2 = 0.9969$) could fit the experimental data better than the other models.

Table 8. Kinetic parameters for the removal of EV on Mn-doped Fe/rGO.

Kinetic Models	Equations	Parameters	Values of Parameters
Pseudo-first-order	$\log(q_e - q_t) = \log q_e - \frac{k_1 t}{2.303}$	k_1 (1/min) qe (mg/g) R^2	0.2034 807.4209 0.9509
Pseudo-second-order	$\frac{t}{q_t} = \frac{1}{k_2 q_e^2} + \frac{t}{q_e}$	k_2 (g/mg/min) qe (mg/g) R^2	0.37×10^{-3} 909.0909 0.9969
Intraparticle diffusion	$q_t = k_d t^{0.5} + C$	k_d (mg/g/min$^{1/2}$) C (mg/g) R^2	115.09 305.97 0.7926
Elovich	$q_t = \frac{1}{\beta}\ln(\alpha\beta) + \frac{1}{\beta}\ln t$	α (mg/g/min) β (g/mg) R^2	715.2812 5.209×10^{-3} 0.9246

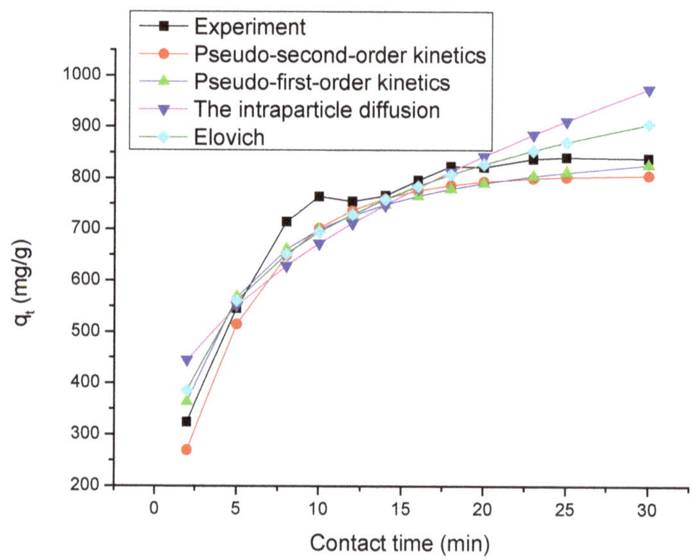

Figure 26. Time dependent study of EV removal by Mn-doped Fe/rGO (initial pH = 5.0; Mn-doped Fe/rGO dosage = 20 mg; EV concentration = 500 mg/L).

4. Conclusions

Mesoporous Mn-doped Fe/rGO nanocomposites were successfully synthesized in this work with a one-step co-precipitation method, characterized by XRD, XPS, Raman, HR-TEM, SEM, EDS, N_2-sorption, SA-XRD and SQUID techniques and used for EV elimination. The SBET of Mn-doped Fe/rGO composites was 104.088 m^2/g, and narrow pore size distributions centered at 3.93 nm. Mn-doped Fe/rGO demonstrated a well-resolved peak at 0.71° of 2θ angles of SA-XRD, and partially ordered pores were clearly observed in HR-TEM images. The results show that the Mn-doped Fe/rGO nanocomposites were partially ordered mesoporous materials. The effects of experimental parameters (initial EV concentration, sonication time, initial pH and amount of adsorbent (Mn-doped Fe/rGO)) on the elimination efficiency of EV were studied by using AI tools (ANN-GA, ANN-PSO, RF and RBF). The mean absolute errors of BP-ANN and RF were 1.55% and 4.56%, respectively, and the R^2 values of RBF and BP-ANN were 0.95415 and 0.99403, respectively. According to the results, random forest optimization is not as effective as BP-ANN, while the RBF and BP-ANN models were suitable for predicting the EV removal. The developed models from ANN-GA, ANN-PSO and RSM were compared and validated using the predicted optimal conditions. The results indicated that the absolute errors of ANN-GA, ANN-PSO and RSM models were 0.72%, 1.28% and 1.53%, respectively. Therefore, ANN-GA is considered as the best model for the prediction of the EV elimination by Mn-doped Fe/rGO.

The factor importance was analyzed by F test, Garson method, GBRT and RF. It can be seen that dosage gives the maximum importance to Mn-doped Fe/rGO EV elimination. The isotherms and kinetic models were investigated in the adsorption of EV onto Mn-doped Fe/rGO. The results illustrated that the adsorption of EV accords with Langmuir isotherm and pseudo-second-order models, respectively. The R_L values were 0.01585–0.05093, indicating that the adsorption of EV by Mn-doped Fe/rGO is favorable, and the value of maximum adsorption capacity is 1000.00 mg/g. The Mn-doped Fe/rGO composites is an effective adsorbent for removal of dyes in water because of its easy synthesis, large adsorption capacity, high specific surface area and good magnetic property (saturation magnetization was 47.0514 emu/g).

Author Contributions: Conceptualization, Y.H.; methodology, Y.H., J.H. and X.W.; software, Y.H. and J.Q.; formal analysis, J.Q. and Y.X.; data curation, Y.H.; validation, Y.X. and L.X.; supervision, J.Q. and L.X.; Writing the draft and data preparation, Y.H. and J.H. All authors have read and agreed to the published version of the manuscript.

Funding: This research was funded by the National Natural Science Foundation of China (Grant No. 21667012), the Government of Guizhou Province (Project No. [2017] 5726-42) and the National 111 Project of China (Grant No. D17016).

Conflicts of Interest: The authors declare no conflict of interest.

References

1. Velusamy, P.; Pitchaimuthu, S.; Rajalakshmi, S.; Kannan, N. Modification of the photocatalytic activity of TiO_2 by β-Cyclodextrin in decoloration of ethyl violet dye. *J. Adv. Res.* **2014**, *5*, 19–25. [CrossRef] [PubMed]
2. Tsai, W.T.; Chang, Y.M.; Lai, C.W.; Lo, C.C. Adsorption of ethyl violet dye in aqueous solution by regenerated spent bleaching earth. *J. Colloid Interface Sci.* **2005**, *289*, 333–338. [CrossRef] [PubMed]
3. Khan, H.; Ahmad, N.; Yasar, A.; Shahid, R. Advanced oxidative decolorization of red Cl-5B: Effects of dye concentration, process optimization and reaction kinetics. *Pol. J. Environ. Stud.* **2010**, *19*, 83–92.
4. Lee, W.W.; Chung, W.H.; Lu, C.S.; Lin, W.Y.; Chen, C.C. A study on the degradation efficiency and mechanisms of ethyl violet by HPLC—PDA—ESI—MS and GC—MS. *Sep. Purif. Technol.* **2012**, *98*, 488–496. [CrossRef]
5. Duxbury, D.F. The photochemistry and photophysics of triphenylmethane dyes in solid and liquid media. *Chem. Rev.* **1993**, *93*, 381–433. [CrossRef]
6. Zhang, C.; Li, H.; Li, C.; Li, Z. Fe-loaded MOF-545 (Fe): Peroxidase-like activity for dye degradation dyes and high adsorption for the removal of dyes from wastewater. *Molecules* **2020**, *25*, 168. [CrossRef]
7. Ruan, W.; Shi, X.; Hu, J.; Hou, Y.; Fan, M.; Cao, R.; Wei, X. Modeling of malachite green removal from aqueous solutions by nanoscale zerovalent zinc using artificial neural network. *Appl. Sci.* **2018**, *7*, 3. [CrossRef]
8. Chen, Y.; Wang, K.; Lou, L. Photodegradation of dye pollutants on silica gel supported TiO_2 particles under visible light irradiation. *J. Photochem. Photobiol. A* **2004**, *163*, 281–287. [CrossRef]
9. Wang, J.Q.; Liu, Y.H.; Chen, M.W.; Xie, G.Q.; Louzguine-Luzgin, D.V.; Inoue, A.; Perepezko, J.H. Rapid degradation of azo dye by Fe-based metallic glass powder. *Adv. Funct. Mater.* **2012**, *22*, 2567–2570. [CrossRef]
10. Zhang, W. Nanoscale iron particles for environmental remediation: An overview. *J. Nanopart. Res.* **2003**, *5*, 323–332. [CrossRef]
11. Kalme, S.D.; Parshetti, G.K.; Jadhav, S.U.; Govindwar, S.P. Biodegradation of benzidine based dye direct blue-6 by pseudomonas desmolyticum NCIM 2112. *Bioresour. Technol.* **2007**, *98*, 1405–1410. [CrossRef] [PubMed]
12. Amin, N.K. Removal of direct blue-106 dye from aqueous solution using new activated carbons developed from pomegranate peel: Adsorption equilibrium and kinetics. *J. Hazard. Mater.* **2009**, *165*, 52–62. [CrossRef] [PubMed]
13. Hmani, E.; Samet, Y.; Abdelhédi, R. Electrochemical degradation of auramine-O dye at boron-doped diamond and lead dioxide electrodes. *Diam. Relat. Mater.* **2012**, *30*, 1–8. [CrossRef]

14. Basiri, F.; Ravandi, S.A.H.; Feiz, M.; Moheb, A. Recycling of direct dyes wastewater by nylon-6 nanofibrous membrane. *Curr. Nanosci.* **2011**, *7*, 633–639. [CrossRef]
15. Kamal, S.; Pan, G.T.; Chong, S.; Yang, T.C.K. Ultrasonically induced sulfur-doped carbon nitride/cobalt ferrite nanocomposite for efficient sonocatalytic removal of organic dyes. *Processes* **2020**, *8*, 104. [CrossRef]
16. Dong, H.; Zhang, C.; Deng, J.; Jiang, Z.; Zhang, L.; Cheng, Y.; Hou, K.; Tang, L.; Zeng, G. Factors influencing degradation of trichloroethylene by sulfide-modified nanoscale zero-valent iron in aqueous solution. *Water Res.* **2018**, *135*, 1–10. [CrossRef]
17. Sun, Y.P.; Li, X.Q.; Cao, J.; Zhang, W.X.; Wang, H.P. Characterization of zero-valent iron nanoparticles. *Adv. Colloid Interface* **2006**, *120*, 47–56. [CrossRef]
18. Zhang, X.; Lin, S.; Lu, X.Q.; Chen, Z.l. Removal of Pb(II) from water using synthesized kaolin supported nanoscale zero-valent iron. *Chem. Eng. J.* **2010**, *163*, 243–248. [CrossRef]
19. Ziajahromi, S.; Mehrdad, M.; Khanizadeh, M. Nitrate Removal from Water Using Synthesis Nanoscale Zero-Valent Iron (NZVI). In Proceedings of the International Conference on Applied Life Sciences (ICALS2012), Konya, Turkey, 10–12 September 2012; pp. 105–110.
20. Arabi, S.; Sohrabi, M.R. Experimental design and response surface modelling for optimization of vat dye from water by nano zero valent iron (NZVI). *Acta Chim. Slov.* **2013**, *60*, 853–860.
21. Huber, D.L. Synthesis, properties, and applications of iron nanoparticles. *Small* **2005**, *1*, 482–501. [CrossRef]
22. Lu, H.; Wang, J.; Ferguson, S.; Wang, T.; Bao, Y.; Hao, H. Mechanism, synthesis and modification of nano zerovalent iron in water treatment. *Nanoscale* **2016**, *8*, 9962–9975. [CrossRef] [PubMed]
23. Diagboya, P.N.; Olu-Owolabi, B.I.; Zhou, D.; Han, B.H. Graphene oxide–tripolyphosphate hybrid used as a potent sorbent for cationic dyes. *Carbon* **2014**, *79*, 174–182. [CrossRef]
24. Loh, K.P.; Bao, Q.; Ang, P.K.; Yang, J. The chemistry of graphene. *J. Mater. Chem.* **2010**, *20*, 2277–2289. [CrossRef]
25. Gao, W.; Alemany, L.B.; Ci, L.; Ajayan, P.M. New insights into the structure and reduction of graphite oxide. *Nat. Chem.* **2009**, *1*, 403–408. [CrossRef]
26. Boukhvalov, D.W.; Katsnelson, M.I. Modeling of graphite oxide. *J. Am. Chem. Soc.* **2008**, *130*, 10697–10701. [CrossRef]
27. Akhavan, O.; Ghaderi, E. Toxicity of graphene and graphene oxide nanowalls against bacteria. *ACS Nano* **2010**, *4*, 5731–5736. [CrossRef]
28. Tang, J.; Huang, Y.; Gong, Y.; Lyu, H.; Wang, Q.; Ma, J. Preparation of a novel graphene oxide/Fe-Mn composite and its application for aqueous Hg(II) removal. *J. Hazard. Mater.* **2016**, *316*, 151–158. [CrossRef]
29. Tan, W.C.; Hofmann, M.; Hsieh, Y.P.; Lu, M.L.; Chen, Y.F. A graphene-based surface plasmon sensor. *Nano Res.* **2012**, *5*, 695–702. [CrossRef]
30. Joshi, R.K.; Alwarappan, S.; Yoshimura, M.; Sahajwalla, V.; Nishina, Y. Graphene oxide: The new membrane material. *Appl. Mater. Today* **2015**, *1*, 1–12. [CrossRef]
31. Huang, C.; Bai, H.; Li, C.; Shi, G. A graphene oxide/hemoglobin composite hydrogel for enzymatic catalysis in organic solvents. *Chem. Commun.* **2011**, *47*, 4962. [CrossRef]
32. Li, W.; Zhang, Z.; Tang, Y.; Bian, H.; Ng, T.W.; Zhang, W.; Lee, C.S. Graphene-nanowall-decorated carbon felt with excellent electrochemical activity toward VO\r 2+\r /VO\r 2+\r couple for all vanadium redox flow battery. *Adv. Sci.* **2016**, *3*, 1500276. [CrossRef] [PubMed]
33. Islam, M.M.; Aboutalebi, S.H.; Cardillo, D.; Liu, H.K.; Konstantinov, K.; Dou, S.X. Self-assembled multifunctional hybrids: Toward developing high-performance graphene-based architectures for energy storage devices. *ACS Cent. Sci.* **2015**, *1*, 206–216. [CrossRef] [PubMed]
34. Bissett, M.A.; Kinloch, I.A.; Dryfe, R.A.W. Characterization of MoS_2-graphene composites for high-performance coin cell supercapacitors. *ACS Appl. Mater. Inter.* **2015**, *7*, 17388. [CrossRef] [PubMed]
35. He, Q.; Wu, S.; Yin, Z.; Zhang, H. Graphene-based electronic sensors. *Chem. Sci.* **2012**, *3*, 1764–1772. [CrossRef]
36. Wan, L.; Luo, T.; Wang, S.; Wang, X.; Guo, Z.; Xiong, H.; Dong, B.; Zhao, L.; Xu, Z.; Zhang, X.; et al. Pt/graphene nanocomposites with low Pt-loadings: Synthesis through one- and two-step chemical reduction methods and their use as promising counter electrodes for DSSCs. *Compos. Sci. Technol.* **2015**, *113*, 46–53. [CrossRef]
37. Fan, M.; Hu, J.; Cao, R.; Ruan, W.; Wei, X. A review on experimental design for pollutants removal in water treatment with the aid of artificial intelligence. *Chemosphere* **2018**, *200*, 330. [CrossRef]

38. Mocanu, D.C.; Mocanu, E.; Stone, P.; Nguyen, P.H.; Gibescu, M.; Liotta, A. Scalable training of artificial neural networks with adaptive sparse connectivity inspired by network science. *Nat. Commun.* **2018**, *9*, 2383–2395. [CrossRef]
39. Vakili, M.; Karami, M.; Delfani, S.; Khosrojerdi, S.; Kalhor, K. Experimental investigation and modeling of thermal conductivity of CuO–water/EG nanofluid by FFBP-ANN and multiple regressions. *J. Therm. Anal. Calorim.* **2017**, *129*, 629–637. [CrossRef]
40. Hua, J.; Xiao, C.; Ke, Z.; Wang, Y. Construct Drawing Man-Hour Forecasting based on GA-BP in Chemical Equipment Design Process. In Proceedings of the 2016 22nd International Conference on Automation and Computing (ICAC), Colchester, UK, 7–8 September 2016; pp. 1–6.
41. Li, J.; Cheng, J.; Shi, J.; Huang, F. Brief Introduction of Back Propagation (BP) Neural Network Algorithm and its Improvement. In *Advances in Computer Science and Information Engineering-Volume 2*; Jin, D., Lin, S., Eds.; Vol. 169 of Advances in Intelligent and Soft Computing; Springer: Berlin/Heidelberg, Germany, 2012; pp. 553–558.
42. Ahmadi Azqhandi, M.H.; Ghaedi, M.; Yousefi, F.; Jamshidi, M. Application of random forest, radial basis function neural networks and central composite design for modeling and/or optimization of the ultrasonic assisted adsorption of brilliant green on ZnS-NP-AC. *J. Colloid Interface Sci.* **2017**, *505*, 278–292. [CrossRef]
43. Kayarvizhy, N.; Kanmani, S.; Rhymend, U.V. Improving fault prediction using ANN-PSO in object oriented systems. *Int. J. Comput. Appl.* **2014**, *73*, 18–25.
44. Che, Z.H. PSO-based back-propagation artificial neural network for product and mold cost estimation of plastic injection molding. *Comput. Ind. Eng.* **2010**, *58*, 625–637. [CrossRef]
45. Breiman, L. Random Forests. *Mach. Learn.* **2001**, *45*, 5–32. [CrossRef]
46. Zhang, M.; Liu, Y. Signal Sorting Using Teaching-Learning-Based Optimization and Random Forest. In Proceedings of the 2018 17th International Symposium on Distributed Computing and Applications for Business Engineering and Science (DCABES), Wuxi, China, 19–23 October 2018; pp. 258–261.
47. Shi, X.; Ruan, W.; Hu, J.; Fan, M.; Cao, R.; Wei, X. Optimizing the removal of Rhodamine B in aqueous solutions by reduced graphene oxide-supported nanoscale zerovalent iron (nZVI/rGO) using an artificial neural network-genetic algorithm (ANN-GA). *Nanomaterials* **2017**, *7*, 134. [CrossRef]
48. Parsaee, Z.; Karachi, N.; Abrishamifar, S.M.; Kahkha, M.R.R.; Razavi, R. Silver-choline chloride modified graphene oxide: Novel nano-bioelectrochemical sensor for Celecoxib detection and CCD-RSM model. *Ultrason. Sonochem.* **2018**, *45*, 106–115. [CrossRef] [PubMed]
49. Wang, Y.X.; Liu, B.; Gao, J.X.; Zhang, X.F.; Li, S.L.; Liu, J.Q.; Tian, Z.P. Auto recognition of carbonate microfacies based on an improved back propagation neural network. *J. Cent. South Univ.* **2015**, *22*, 3521–3535. [CrossRef]
50. Ghaedi, M.; Zeinali, N.; Ghaedi, A.M.; Teimuori, M.; Tashkhourian, J. Artificial neural network-genetic algorithm based optimization for the adsorption of methylene blue and brilliant green from aqueous solution by graphite oxide nanoparticle. *Spectrochim. Acta A* **2014**, *125*, 264–277. [CrossRef]
51. Huang, Y.; Liu, Y.; Li, C.; Wang, C. GBRTVis: Online analysis of gradient boosting regression tree. *J. Visual.* **2019**, *22*, 125–140. [CrossRef]
52. Li, X.; Bai, R. Freight Vehicle Travel Time Prediction Using Gradient Boosting Regression Tree. In Proceedings of the 2016 15th IEEE International Conference on Machine Learning and Applications (ICMLA), Anaheim, CA, USA, 18–20 December 2016; pp. 1–7.
53. Montaño, J.J.; Palmer, A. Numeric sensitivity analysis applied to feedforward neural networks. *Neural Comput. Appl.* **2003**, *12*, 119–125. [CrossRef]
54. Ruan, W.; Hu, J.; Qi, J.; Hou, Y.; Cao, R.; Wei, X. Removal of crystal violet by using reduced-graphene-oxide-supported bimetallic Fe/Ni nanoparticles (rGO/Fe/Ni): Application of artificial intelligence modeling for the optimization process. *Materials* **2018**, *11*, 865. [CrossRef]
55. Dhingra, S.; Dubey, K.K.; Bhushan, G. A polymath approach for the prediction of optimized transesterification process variables of polanga biodiesel. *J. Am. Oil Chem. Soc.* **2014**, *91*, 641–653. [CrossRef]
56. Gulati, T.; Chakrabarti, M.; Sing, A.; Duvuuri, M.; Banerjee, R. Comparative study of response surface methodology, artificial neural network and genetic algorithms for optimization of soybean hydration. *Food Technol. Biotech.* **2010**, *48*, 11–18
57. Sarkar, D.; Modak, J.M. Optimisation of fed-batch bioreactors using genetic algorithms. *Chem. Eng. Sci.* **2003**, *58*, 2283–2296. [CrossRef]

58. Lee, K.H.; Jun, S.O.; Pak, K.H.; Lee, D.H.; Lee, K.W.; Park, J.P. Numerical optimization of site selection for offshore wind turbine installation using genetic algorithm. *Curr. Appl. Phys.* **2010**, *10*, S302–S306. [CrossRef]
59. Eberhart, R.; Kennedy, J. A New Optimizer Using Particle Swarm Theory. In Proceedings of the MHS'95. Proceedings of the Sixth International Symposium on Micro Machine and Human Science, Nagoya, Japan, 4–6 October 1995; pp. 39–43.
60. Mitchell, M.L.D. Davis, handbook of genetic algorithms. *Artif. Intell.* **1998**, *100*, 325–330. [CrossRef]
61. Bejagam, K.K.; Singh, S.; An, Y.; Berry, C.; Deshmukh, S.A. PSO assisted development of new transferable coarse-grained water models. *J. Phys. Chem. B* **2018**, *122*, 1–37. [CrossRef]
62. Wang, D.; Liu, Y.; Wu, Z.; Fu, H.; Shi, Y.; Guo, H. Scenario analysis of natural gas consumption in China based on wavelet neural network optimized by particle swarm optimization algorithm. *Energies* **2018**, *11*, 825. [CrossRef]
63. Sami, M.; Hassanien, A.E.; Nashwa, E.B.; Berwick, R.C. Incorporating Random Forest Trees with Particle Swarm Optimization for Automatic Image Annotation. In Proceedings of the 2012 Federated Conference on Computer Science and Information Systems (FedCSIS), Wroclaw, Poland, 9–12 September 2012; pp. 791–797.
64. Ming, D.; Zhou, T.; Min, W.; Tian, T. Land cover classification using random forest with genetic algorithm-based parameter optimization. *J. Appl. Remote Sens.* **2016**, *10*, 035021. [CrossRef]
65. Gislason, P.O.; Benediktsson, J.A.; Sveinsson, J.R. Random forests for land cover classification. *Pattern Recogn. Lett.* **2006**, *27*, 294–300. [CrossRef]
66. Li, J.; Liu, X. Melt index prediction by RBF neural network optimized with an adaptive new ant colony optimization algorithm. *J. Appl. Polym. Sci.* **2011**, *119*, 3093–3100. [CrossRef]
67. Najafi-Marghmaleki, A.; Khosravi-Nikou, M.R.; Barati-Harooni, A. A new model for prediction of binary mixture of ionic liquids+water density using artificial neural network. *J. Mol. Liq.* **2016**, *220*, 232–237. [CrossRef]
68. Tatar, A.; Shokrollahi, A.; Mesbah, M.; Rashid, S.; Arabloo, M.; Bahadori, A. Implementing radial basis function networks for modeling CO_2-reservoir oil minimum miscibility pressure. *J. Nat. Gas Sci. Eng.* **2013**, *15*, 82–92. [CrossRef]
69. Rasouli, Z.; Hassanzadeh, Z.; Ghavami, R. Application of a new version of GA-RBF neural network for simultaneous spectrophotometric determination of Zn(II), Fe(II), Co(II) and Cu(II) in real samples: An exploratory study of their complexation abilities toward MTB. *Talanta* **2016**, *160*, 86–98. [CrossRef] [PubMed]
70. Zhang, K.; Dwivedi, V.; Chi, C.; Wu, J. Graphene oxide/ferric hydroxide composites for efficient arsenate removal from drinking water. *J. Hazard. Mater.* **2010**, *182*, 162–168. [CrossRef] [PubMed]
71. Trejos, T.; Corzo, R.; Subedi, K.; Almirall, J. Characterization of toners and inkjets by laser ablation spectrochemical methods and Scanning Electron Microscopy-Energy Dispersive X-ray Spectroscopy. *Spectrochim. Acta B* **2014**, *92*, 9–22. [CrossRef]
72. Su, J.; Lin, S.; Chen, Z.; Megharaj, M.; Naidu, R. Dechlorination of p-chlorophenol from aqueous solution using bentonite supported Fe/Pd nanoparticles: Synthesis, characterization and kinetics. *Desalination* **2011**, *280*, 167–173. [CrossRef]
73. Casiraghi, C.; Pisana, S.; Novoselov, K.S.; Geim, A.K.; Ferrari, A.C. Raman fingerprint of charged impurities in graphene. *Appl. Phys. Lett.* **2007**, *91*, 183. [CrossRef]
74. Kotutha, I.; Swatsitang, E.; Meewassana, W.; Maensiri, S. One-pot hydrothermal synthesis, characterization, and electrochemical properties of rGO/$MnFe_2O_4$ nanocomposites. *Jpn. J. Appl. Phys.* **2015**, *54*, 06FH10. [CrossRef]
75. Pimenta, M.; Dresselhaus, G.; Dresselhaus, M.; Cancado, L.; Jorio, A.; Saito, R. Studying disorder in graphite-based systems by Raman spectroscopy. *Phys. Chem. Chem. Phys.* **2007**, *9*, 1276–1291. [CrossRef]
76. Fontana, M.D.; Bourson, P. Microstructure and defects probed by Raman spectroscopy in lithium niobate crystals and devices. *Appl. Phys. Rev.* **2015**, *2*, 040602. [CrossRef]
77. Srivastava, R.K.; Xavier, P.; Gupta, S.N.; Kar, G.P.; Bose, S.; Sood, A. Excellent electromagnetic interference shielding by graphene-$MnFe_2O_4$-multiwalled carbon nanotube hybrids at very low weight percentage in polymer matrix. *ChemistrySelect* **2016**, *1*, 5995–6003. [CrossRef]
78. Stankovich, S.; Dikin, D.A.; Piner, R.D.; Kohlhaas, K.A.; Kleinhammes, A.; Jia, Y.; Wu, Y.; Nguyen, S.T.; Ruoff, R.S.J.C. Synthesis of graphene-based nanosheets via chemical reduction of exfoliated graphite oxide. *Carbon* **2007**, *45*, 1558–1565. [CrossRef]

79. Liu, G.; Chen, Q.; Oyunkhand, E.; Ding, S.; Yamane, N.; Yang, G.; Yoneyama, Y.; Tsubaki, N. Nitrogen-rich mesoporous carbon supported iron catalyst with superior activity for Fischer-Tropsch synthesis. *Carbon* **2018**, *130*, 304–314. [CrossRef]
80. Leofanti, G.; Padovan, M.; Tozzola, G.; Venturelli, B. Surface area and pore texture of catalysts. *Catal. Today* **1998**, *41*, 207–219. [CrossRef]
81. Lin, X.; Lv, X.; Wang, L.; Zhang, F.; Duan, L. Preparation and characterization of $MnFe_2O_4$ in the solvothermal process: Their magnetism and electrochemical properties. *Mater. Res. Bull.* **2013**, *48*, 2511–2516. [CrossRef]
82. Hung, I.M.; Hung, D.T.; Fung, K.Z.; Hon, M.H. Synthesis and characterization of highly ordered mesoporous YSZ by tri-block copolymer. *J. Porous Mat.* **2006**, *13*, 225–230. [CrossRef]
83. Solaymani, E.; Ghaedi, M.; Karimi, H.; Azqhandi, M.H.A.; Asfaram, A. Intensified removal of Malachite green by AgOH-AC nanoparticles combined with ultrasound: Modeling and optimization. *Appl. Organomet. Chem.* **2017**, *31*, e3857. [CrossRef]
84. El-Gamal, S.; Amin, M.; Ahmed, M. Removal of methyl orange and bromophenol blue dyes from aqueous solution using Sorel's cement nanoparticles. *J. Environ. Chem. Eng.* **2015**, *3*, 1702–1712. [CrossRef]
85. Forouz, F.S.; Ravandi, S.A.H.; Allafchian, A.R. Removal of Ag and Cr heavy metals using nanofiber membranes functionalized with aminopropyltriethoxysilane (APTES). *Curr. Nanosci.* **2016**, *12*, 266–274. [CrossRef]
86. Chowdhury, S.; Misra, R.; Kushwaha, P.; Das, P. Optimum sorption isotherm by linear and nonlinear methods for safranin onto alkali-treated rice husk. *Bioremediat. J.* **2011**, *15*, 77–89. [CrossRef]
87. Zhang, J.; Lin, S.; Han, M.; Su, Q.; Xia, L.; Hui, Z. Adsorption properties of magnetic magnetite nanoparticle for coexistent Cr (VI) and Cu (II) in mixed solution. *Water* **2020**, *12*, 446. [CrossRef]
88. Chen, C.; Cheng, T.; Zhang, X.; Wu, R.; Wang, Q. Synthesis of an efficient Pb adsorption nano-crystal under strong alkali hydrothermal environment using a gemini surfactant as directing agent. *J. Chem. Soc. Pak.* **2019**, *41*, 1034–1038.
89. Altaher, H.; Khalil, T.E.; Abubeah, R. The effect of dye chemical structure on adsorption on activated carbon: A comparative study. *Color. Technol.* **2014**, *130*, 205–214. [CrossRef]
90. Cheng, C.; Cheng, T.; Wang, Z.L.; Han, C.H. Removal of Zn^{2+} in aqueous solution by Linde F (K) zeolite prepared from recycled fly ash. *J. Indian Chem. Soc.* **2014**, *91*, 1–7.
91. Kumar, K.V.; Porkodi, K.; Rocha, F. Comparison of various error functions in predicting the optimum isotherm by linear and non-linear regression analysis for the sorption of basic red 9 by activated carbon. *J. Hazard. Mater.* **2008**, *150*, 158–165. [CrossRef] [PubMed]
92. Pan, M.; Lin, X.; Xie, J.; Huang, X. Kinetic, equilibrium and thermodynamic studies for phosphate adsorption on aluminum hydroxide modified palygorskite nano-composites. *RSC Adv.* **2017**, *7*, 4492–4500. [CrossRef]
93. Ho, Y.S. Citation review of Lagergren kinetic rate equation on adsorption reactions. *Scientometrics* **2004**, *59*, 171–177.
94. Ho, Y.S.; Mckay, G. Sorption of dye from aqueous solution by peat. *Chem. Eng. J.* **1998**, *70*, 115–124. [CrossRef]
95. Chen, C.; Cheng, T.; Shi, Y.; Tian, Y. Adsorption of Cu (II) from aqueous solution on fly ash based Linde F (K) zeolite. *Iran. J. Chem. Chem. Eng.* **2014**, *33*, 29–35.
96. Dogan, M.; Alkan, M. Adsorption kinetics of methyl violet onto perlite. *Chemosphere* **2003**, *50*, 517–528. [CrossRef]
97. Cheng, T.; Chen, C.; Tang, R.; Han, C.H.; Tian, Y. Competitive adsorption of Cu, Ni, Pb, and Cd from aqueous solution onto fly ash-based linde F(K) Zeolite. *Iran. J. Chem. Chem. Eng.* **2018**, *37*, 61–72.
98. Reffas, A.; Bouguettoucha, A.; Chebli, D.; Amrane, A. Adsorption of ethyl violet dye in aqueous solution by forest wastes, wild carob. *Desalin. Water Treat.* **2015**, *57*, 9859–9870. [CrossRef]

processes

MDPI

Article

Simultaneous Electrochemical Generation of Ferrate and Oxygen Radicals to Blue BR Dye Degradation

Mauricio Chiliquinga [1,2], Patricio J. Espinoza-Montero [3,*], Oscar Rodríguez [2], Alain Picos [2], Erick R. Bandala [4], S. Gutiérrez-Granados [2] and Juan M. Peralta-Hernández [2,*]

1 Departamento de Ingeniería Civil y Ambiental, Facultad de Ingeniería Civil y Ambiental, Escuela Politécnica Nacional. P.O. Box, Quito 17-01-2759, Ecuador; mauricioacade14@hotmail.com

2 Departamento de Química, División de Ciencias Naturales y Exactas, Universidad de Guanajuato, Cerro de la Venada s/n, Pueblito de Rocha, Guanajuato 36040, Mexico; om.rodrigueznarvaez@ugto.mx (O.R.); picosbrother@hotmail.com (A.P.); gutigs59@hotmail.com (S.G.-G.)

3 Escuela de Ciencias Químicas, Pontificia Universidad Católica del Ecuador, Avenida 12 de Octubre y Roca, Apartado, Quito 17-01-2184, Ecuador

4 Division of Hydrologic Sciences, Desert Research Institute, 755 E. Flamingo Road, Las Vegas, NV 89119-7363, USA; Erick.Bandala@dri.edu

* Correspondence: pespinoza646@puce.edu.ec (P.J.E.-M.); juan.peralta@ugto.mx (J.M.P.-H.); Tel.: +593-2299-1700 (ext. 1929) (P.J.E.-M.); +52-4737327555 (ext. 5416) (J.M.P.-H.)

Received: 10 March 2020; Accepted: 25 June 2020; Published: 28 June 2020

Abstract: In this study, electro-oxidation (EOx) and in situ generation of ferrate ions [Fe(VI)] were tested to treat water contaminated with Blue BR dye (BBR) using a boron-doped diamond (BDD) anode. Two electrolytic media (0.1 M $HClO_4$ and 0.05 M Na_2SO_4) were evaluated for the BDD, which simultaneously produced oxygen radicals ($^{\bullet}OH$) and [Fe(VI)]. The generation of [Fe(VI)] was characterized by cyclic voltammetry (CV) and the effect of different current intensity values (e.g., 7 mA cm^{-2}, 15 mA cm^{-2}, and 30 mA cm^{-2}) was assessed during BBR degradation tests. The discoloration of BBR was followed by UV-Vis spectrophotometry. When the EOx process was used alone, only 78% BBR discoloration was achieved. The best electrochemical discoloration conditions were found using 0.05 M Na_2SO_4 and 30 mA cm^{-2}. Using these conditions, overall BBR discoloration values up to 98%, 95%, and 87% with 12 mM, 6 mM, and 1 mM of $FeSO_4$, respectively, were achieved. In the case of chemical oxygen demand (COD) reduction, the EOx process showed only a 37% COD reduction, whereas combining [Fe(VI)] generation using 12 mM of $FeSO_4$ achieved an up to 61% COD reduction after 90 min. The evolution of reaction byproducts (oxalic acid) was performed using liquid chromatography analysis.

Keywords: advace oxitadion processes (AOP); electro-oxidation; ferrate ion; BBR dye

1. Introduction

Industrial effluents contain a wide variety of pollutants that are potentially harmful to humans, such as dyes and organic compounds, and most of these pollutants are difficult to remove using conventional water treatment systems (e.g., physicochemical, and/or biological processes) [1]. When released into the environment without the proper treatment, these effluents produce significant fluctuations in water quality parameters such as chemical oxygen demand (COD), biological oxygen demand (BOD), pH, color, and/or salinity [2], which increase public concerns when the proper treatments are not being implemented. Therefore, finding alternative treatment processes that are suitable to remove highly recalcitrant pollutants in industrial wastewater effluents is a pending science and technology need.

An alternative for water treatment known since the 1970s has been ferrate(VI) ion (FeO_4^{2-}), a very strong, chemically green oxidizer, highly stable in aqueous medium, especially at alkaline and neutral pH, where iron shows oxidation state +6 [3–6]. However, we must emphasize that ferrate's highest oxidizing power occurs preferably in an acidic medium, and decreases in alkaline pH (reaction 1 and 2) [4,7].

$$FeO_4^{2-} + 8H^+ + 3e^- \rightleftarrows Fe^{3+} + 4H_2O \qquad E^\circ = +2.20 \text{ V vs. SHE} \tag{1}$$

$$FeO_4^{2-} + 2H_2O + 3e^- \rightleftarrows Fe(OH)_3 + 5OH^- \qquad E^\circ = +0.72 \text{ V vs. SHE} \tag{2}$$

Ferrate anions can exist in various oxidation states: $[FeO_4]^{2-}$ with Fe(VI), $[FeO_4]^{3-}$ with Fe(V) or $[FeO_4]^{4-}$ with Fe(IV) [8]. However, Fe(VI) is the most stable form with broad industrial uses. The unique chemical properties of ferrate (VI) include:

(a) High redox potential, with the ability to oxidize harmful organic and inorganic species such as nitrosamines, phenol, nitrilotriacetic acid, hydrazine, thiourea, sulfides, cyanide, thioacetamide, ammonia, thiocyanate, soluble oils, as well as inactivate viruses and bacteria [5,8–11].
(b) A bifunctional ability to act as oxidizing and coagulating agent, due to the simultaneous release of ferric ions (Fe(III)), specifically under neutral or alkaline conditions [12–14].
(c) A higher redox potential than other oxidizers commonly used for water treatment such as chlorine (1.358 V vs. SHE), hydrogen peroxide (1.776 V vs. SHE), and ozone (2.076 V vs. SHE) [15].
(d) Ferrate use prevents the formation of carcinogenic disinfection byproducts (mainly trihalomethanes).
(e) It generates non-toxic reduction byproducts such as Fe(III) and Fe(II), which can be used for other purposes.
(f) Ferrate can also be used for biofouling control and for the removal of other contaminants such as metals, non-metals, and radionuclides [16–18].

Boron-doped diamond (BDD) electrodes have an inert surface, a wide electrochemical window, low adsorption, high corrosion resistance, and electrochemical stability. These electrodes are also considered a suitable surface upon which to carry out electro-oxidation (EOx) processes [19]. They have greater O_2 evolution overpotential than other electrodes (e.g., Pt and PbO_2) and low water oxidation reaction, combined with the capability to generate large amounts of oxygen radicals (e.g., $^\bullet OH$) [20]. Recent studies suggest that ferrate ions can be generated in situ from Fe(II) salts using a BDD anode surface under acidic conditions (0.1 M $HClO_4$) by applying a potential of 2.2 V vs. NHE [21].

Despite the existing information about the potential of ferrate for water treatment [22–27] and the attractive features of its generation in situ by EOx processes, more research is needed to identify the best reaction conditions to treat complex industrial effluents using this technology. The goal of this work was to investigate the best conditions to generate ferrate ions in situ using a BDD electrode by cyclic voltammetry as well as assess the feasibility of using in situ generated ferrate simultaneously with the EOx process to degrade a textile dye (e.g., Blue BR (BBR) dye), following water discoloration, COD abatement, as well as an assessment of the evolution of reaction byproducts (e.g., carboxylic acids).

As is well known, the interest in studying the electrochemical generation of the ferrate ion on boron-doped diamond (BDD) electrodes, has gained great relevance; this due to its potential applications in the environmental area, specifically in water treatment. This paper presents results on the simultaneous production of the ferrate ion and free radicals on a BDD surface; studies of the generation of both species in two different electrolytes and with the addition of different concentrations of iron sulfate are presented, we also study the effect of the current density on the degradation of an industrial dye; for this, the level of discoloration, reduction in COD and the evolution of a byproduct, namely carboxylic acid, are considered. We consider this research to be relevant because we present important results on how the action of both oxidants generated on the same electrode simultaneously will improve the degradation of the dye, which enhances its application in water treatment.

2. Materials and Methods

2.1. Materials

All reagents used in this investigation—perchloric acid ($HClO_4$), ferrous sulfate ($FeSO_4{\cdot}7H_2O$), sodium sulfate (Na_2SO_4), and sulfuric acid (H_2SO_4)—were obtained from Aldrich and J.T. Baker and used without any purification. The industrial grade BBR dye ($C_{32}H_{28}N_2Na_2O_8S_2$), widely used in the tanning industry in Mexico, was supplied by PCL (Guanajuato, Mexico). The BDD electrode was obtained from Metakem^{TM} (Germany).

2.2. Electrochemical Characterization

The redox behavior of $FeSO_4$ on BDD was evaluated by CV, employing a three-electrode cell: BDD as an anode (0.5 cm^2), Pt wire as a counter electrode, and Ag/AgCl as the reference electrode, using the BASi potentiostat/galvanostat and Epsilon-EC software at a scan rate ranging from 100 mV s^{-1} to 500 mV s^{-1}. As an electrolyte (blank), we used 0.1 M $HClO_4$ with 6 mM of $FeSO_4$ as analyte.

Additionally, the electrochemical response by cyclic voltammetry (CV) of three $FeSO_4$ concentrations (1 mM, 6 mM, and 12 mM), in two different electrolytes 0.1 M $HClO_4$ and 0.05 M Na_2SO_4 (adjusted to pH 3 with H_2SO_4) was evaluated. The experiments were performed in a three-electrode electrochemical cell (25 mL), using a BDD (0.5 cm^2) plate as a working electrode, platinum (Pt) wire as a counter electrode, and Ag/AgCl as a reference electrode. The CV tests were performed using a BASi Epsilon potentiostat/galvanostat. Specific concentrations of $FeSO_4$ were prepared from a stock solution in 100 mL of distilled water and added to the media in all the [Fe(VI)] production experiments.

2.3. Electrochemical Set Up

The discoloration assays were carried out in galvanostatic mode using three different current density values (e.g., 7 mA cm^2, 15 mA cm^2, and 30 mA cm^2) in a 100 mL tank reactor with constant magnetic stirring (200 rpm). One initial BBR dye concentration (50 mg/L) was used for all the experimental trials. The BBR dye (chemical formula = $C_{32}H_{28}N_2Na_2O_8S_2$, molecular weight = 678.68 g/mol, and λ_{max} = 695 nm) was used in this study. An accurately weighed quantity of the dye was dissolved in distilled water to prepare the stock solution (100 mg L^{-1}). Experimental solutions of desired concentration (50 mg/L) were obtained by successive dilution. The surface area of the BDD anode was 2.5 cm^2 and a Pt wire was used as the cathode. The study solution was monitored for color reduction using a UV-Vis spectrophotometer (Cintra 1010 device) at λ = 695 nm [27].

Water discoloration was carried out by testing two electrolytic media: 0.1 M $HClO_4$ and 0.05 M Na_2SO_4 adjusted to pH 3 with H_2SO_4. In both cases, three different $FeSO_4$ concentrations (1 mM, 6 mM, and 12 mM) were tested. To evaluate the amount of remaining oxidable material in solution, the chemical oxygen demand (COD) was quantified according to the Standard Methods (method 5220D); total nitrogen concentration was also estimated by the Standard Methods [28].

For the different parameters tested, the kinetics approach of the combined electrochemical degradation of BBR dye was determined by a pseudo-first order model, $Ln(A_0/A_t) = kt$, where A_0 is the initial absorbance, A_t is the absorbance after degradation time t, and k is the apparent rate constant.

Carboxylic acids were identified and quantified by ion-exclusion HPLC using an Agilent Technologies 1260 Infinity Series with a Bio-Rad Aminex HPX 87H (300 mm × 7.8 mm) column at 35 °C and by setting the photodiode detector at λ = 210 nm. For these determinations, 20 μL of the sample was injected into the HPLC, using 4 mM of H_2SO_4 as the mobile phase and flow rate 0.8 mL/min [29–32]. The quantification of the COD was carried out only for the best BBR degradation conditions (e.g., electrolytic solution of 0.05 M of Na_2SO_4 at a pH of 3, current density of 30 mA cm^{-2}, and 12 mM of $FeSO_4$) following the Standard Methods [29].

3. Results and Discussion

3.1. Cyclic Voltammetric (CV) Study

Figure 1 shows the redox behavior of 6 mM of $FeSO_4$ on BDD at different scan rates (100 mV s^{-1} to 500 mV s^{-1}) using 0.1 M $HClO_4$ as electrolyte. It is worth emphasizing that the linear behavior of the second peaks agree with the Randles–Sevcik equation (I_p vs. (scan rate)$^{1/2}$) (data not shown), indicating a diffusion-controlled process where the redox species are not adsorbed on the surface of the electrode [33]. In this study, oxidation peaks from Fe(II) to Fe(III) were observed at 0.54 V vs. Ag/AgCl, the overoxidation of Fe(III) to [Fe(VI)] was found at 0.90 V vs. Ag/AgCl, and reduction to Fe(II) was identified at 0.25 V vs. Ag/AgCl. These results support the theory of in situ formation of [Fe(VI)] on the BDD surface, as shown in Figure 1 [34]. In Figure 2 the electrogeneration of the ferrate ion can be confirmed as the concentration of $FeSO_4$ increases both the oxidation signal from Fe(II) to Fe(III) and the oxidation signal from Fe(III) to [Fe(VI)] increases proportionally. The number of oxidation and reduction signals matches with those described for Figure 1, even though a slight displacement in the potential axis was observed, probably related to the amount of $FeSO_4$ used for the media. However, ferrate ion production was found to increase as the concentration of $FeSO_4$ increased [35]. In a previous work, our research group suggested ferrate ion formation at potential values close to 2.5 V Ag/AgCl, using 0.1 M $HClO_4$ [21,35]. The difference in the oxidation and over-oxidation reaction behavior of Fe(II) to Fe(III) and Fe(VI), respectively, observed in this study could be related to the composition of the electrode surface (C-sp^2/C-sp^3 ratio), usage time, and the composition of BDD during its syntheses (e.g., doping level, or precursor used) [36].

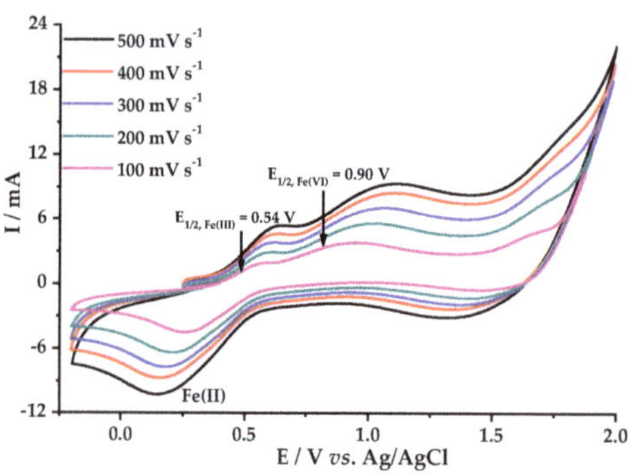

Figure 1. Behavior of 6 mM $FeSO_4$ at different voltammetric scan rate, in 0.1 M $HClO_4$, on 0.5 boron-doped diamond (BDD) cm^2 as working electrode; counter electrode—platinum wire.

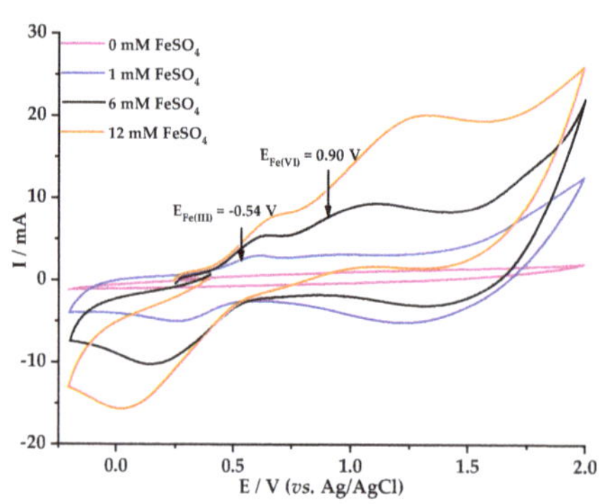

Figure 2. Voltammetric response of $FeSO_4$ at different concentrations in 0.1 M $HClO_4$, with BDD 0.5 cm^2 as working electrode, scanning speed 50 mV s^{-1}; counter electrode—platinum wire.

Figure 3 shows the redox behavior of 6 mM $FeSO_4$ on BDD, at different scan rates (50 mV s^{-1} to 100 mV s^{-1}) using 0.05 M Na_2SO_4 as electrolyte after adjusting pH to 3 with concentrated H_2SO_4. Two signals for oxidation from Fe(II) to Fe(III) at −0.30 V vs. Ag/AgCl were identified, plus another for overoxidation of Fe(II) to Fe(VI) at 0.78 V vs. Ag/AgCl. Figure 4 shows the effect of $FeSO_4$ concentration when a constant scan rate (50 mV s^{-1}) was used. It is worth noting that at increased concentrations of $FeSO_4$, the oxidation and reduction signals increase slightly and move significantly towards higher values of both oxidation and reduction potential. In addition, the second oxidation signal appeared at a lower potential value when Na_2SO_4 was used as the electrolyte (0.78 V, Figure 4), compared with the signal obtained when $HClO_4$ was used as the electrolyte (0.90 V, Figure 2). Therefore, better degradation of BBR would be expected when Na_2SO_4 is used as the electrolyte, as this reaction media would favor the formation of ferrate ion at lower potential values [37].

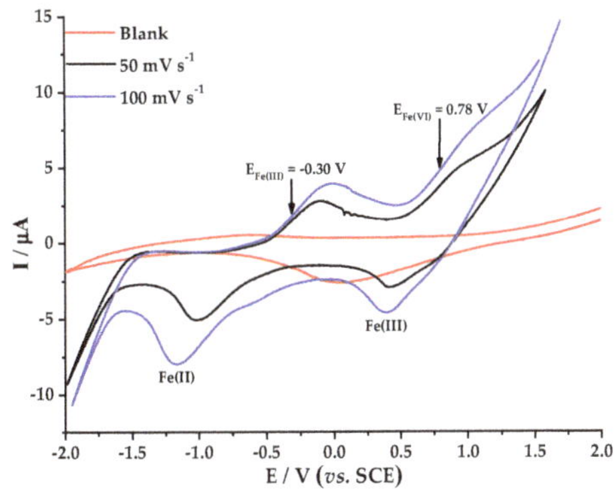

Figure 3. Behavior of 6 mM $FeSO_4$ at different voltammetric scan rate, 0.05 M Na_2SO_4, on BDD 0.5 cm^2 as working electrode; counter electrode—platinum wire.

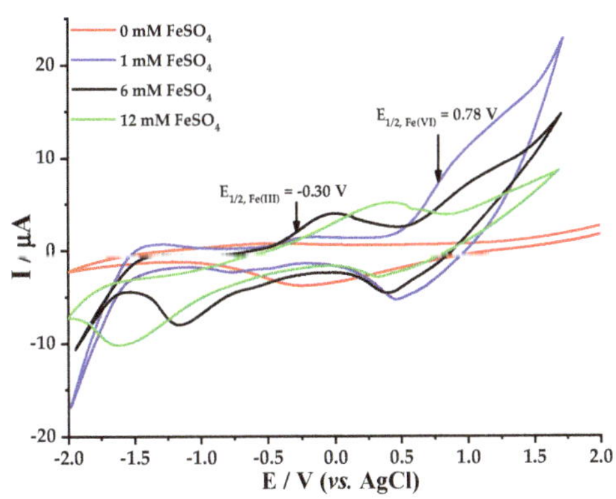

Figure 4. Voltammetric response of $FeSO_4$ at different concentrations, 0.05 M Na_2SO_4, on BDD 0.5 cm^2 as working electrode, scan rate 50 mV s^{-1}; counter electrode—platinum wire.

3.2. BBR Dye Degradation

3.2.1. Effect of the Electrolyte on the EOx Discoloration Process

The effect of the electrolyte type on the EOx discoloration was explored first. Figure 5 shows the discoloration curves as a function of reaction time for $HClO_4$ (Figure 5a) and Na_2SO_4 (Figure 5b) at the different current density values tested (e.g., 7 mA cm^{-2}, 15 mA cm^{-2}, and 30 mA cm^{-2}). As shown, no significant difference was found when the lower current density (7 mA cm^{-2}) was tested independently of the electrolyte type used. Nevertheless, a significant difference was found when higher current density values were tested. For example, an overall discoloration of 23% was achieved using 15 mA cm^{-2} and 0.1 M $HClO_4$, whereas up to 62% of overall discoloration (over twice the amount achieved reached for $HClO_4$) was achieved when 0.05 M Na_2SO_4 was used as the electrolyte.

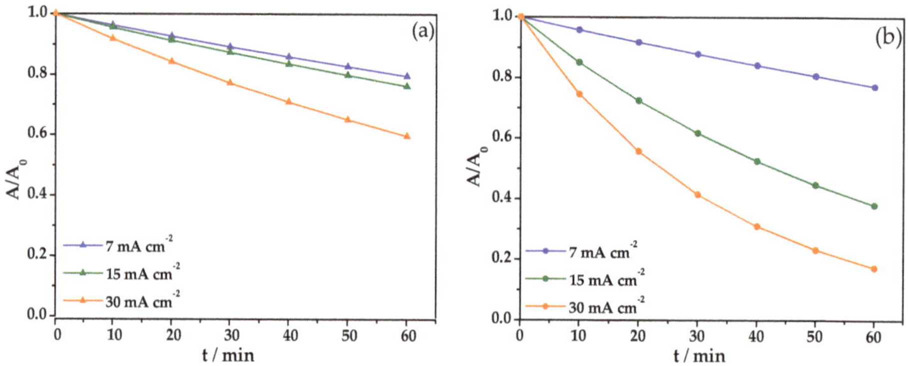

Figure 5. Removal of an aqueous solution of Blue BR (BBR) azo dye, C_0 = 50 mg/L, during the oxidation process in (**a**) 0.1 M of $HClO_4$ and (**b**) 0.05 M Na_2SO_4 with a solution volume of 0.1 L.

When the highest current density (30 mA cm^{-2}) was tested, the overall water discoloration achieved after 60 min was 49% for 0.1 M $HClO_4$, whereas the trials using 0.05 M Na_2SO_4 showed an overall water discoloration as high as 82%. It is worth noting that $HClO_4$ consistently produced lower overall discoloration than Na_2SO_4 under the same experimental conditions. These results could be due to halide electrolytes generating active halide species, which has been suggested in other studies [38]. In their report, Ambauen et al. [38] found that once hydroxyl radicals are formed, secondary reactions may occur when electrochemically active electrolytes are present, including the scavenging of the reactive oxygen species and the secondary production of other less reactive species that are capable of reacting with the organic pollutants under some conditions, resulting in partial chemical oxidation. However, based on the mechanisms for active chloride-mediated electrochemical oxidation, this type of side reaction is not favored because, in most cases, chlorinated byproducts possess a higher toxicity.

On the other hand, using Na_2SO_4 in the EOx-based discoloration of dye water has interesting potential. Previously, we used a BDD anode to remove different industrial sulfonated dye products from water [39]. In that study, we found BBR discoloration ranging from 25% to 95%, which was possibly related to the production of a low and steady concentration of oxygen radical on the BDD surface attacking the dye molecules. The best results were achieved using the highest current density value tested (e.g., 18 mA cm^{-2}), a pH of 3, a 0.5 L laboratory stirred tank reactor, and a 25 cm^2 BDD electrode. In this study, we achieved up to 82% of BBR discoloration for the same reaction time reported previously, but using a BDD electrode area one order of magnitude lower (e.g., 2.5 cm^2).

From this study, pseudo-first-order kinetics were calculated as described above. In the case of $HClO_4$ medium, the following results were obtained: (▲) 0.0038, (▲) 0.0045, (▲) 0.0086 min^{-1}, respectively. For the Na_2SO_4 supporting electrolyte, the following values were obtained: (♦) 0.0043, (♦) 0.0161, (♦) 0.0293 min^{-1}, respectively. For all the experimental fittings using the pseudo-first order kinetics model, the minimum R^2 value obtained was 0.99, suggesting that this kinetics model describes the chemical process fairly good. As shown, the *k*-values found for Na_2SO_4 were about 3.5 times higher than those observed when $HClO_4$ was used as the supporting electrolyte.

3.2.2. BBR Dye Degradation at Different Current Densities in 0.1 M $HClO_4$ and $FeSO_4$

Figure 6a shows water discoloration as a function of $FeSO_4$ concentration using a current density of 7 mA cm^{-2}. Slow discoloration was observed for the EOx process alone (♦), and 31% BBR degradation and *k*-value 0.0030 min^{-1} were achieved. By adding 1 mM of $FeSO_4$ to the system (▲), a 38% discoloration and *k*-value 0.0035 min^{-1} were achieved. Further increasing the $FeSO_4$ concentration (e.g., 6 mM (■), and 12 mM (•)) led to water discoloration values of 41% and 46%, respectively, and reaction rate constant values of 0.0038 min^{-1} and 0.0042 min^{-1}, respectively, were obtained after two hours of treatment [40,41].

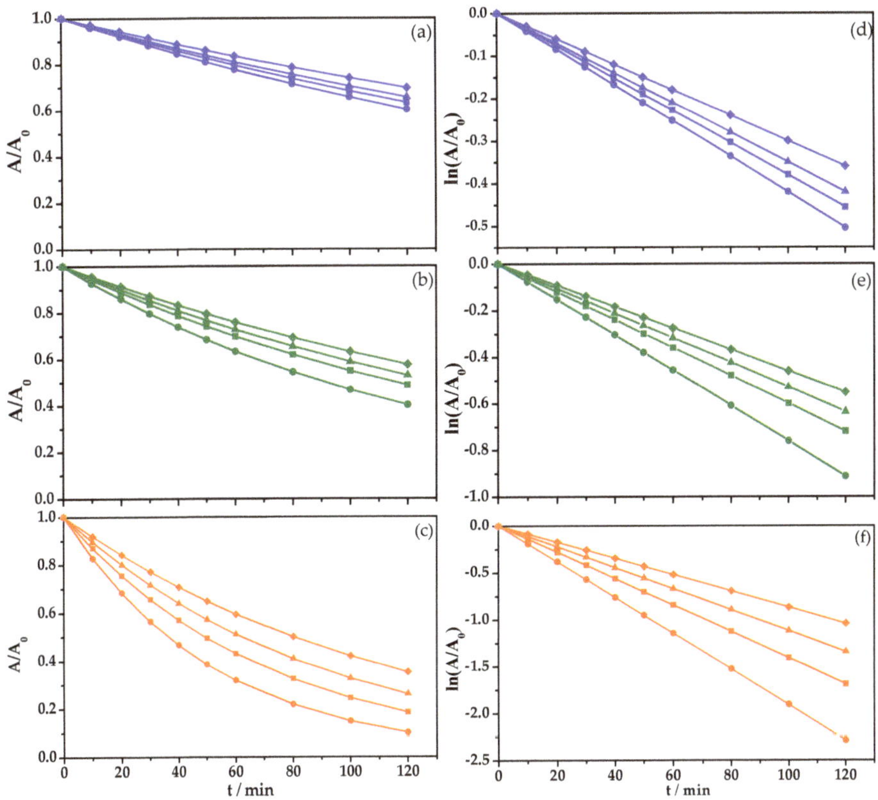

Figure 6. Removal of an aqueous solution of BBR azo dye, C_0 = 50 mg/L, during the oxidation process in 0.1 M of $HClO_4$ with a solution volume of 0.1 L, (**a**) j = 7 mA cm^{-2}, (**b**) j = 15 mA cm^{-2}, (**c**) j = 30 mA cm^{-2}, and fit to first order reaction kinetics (**d**), (**e**) y (**f**), respectively; (♦) BDD without $FeSO_4$, (▲) 1 mM of $FeSO_4$, (■) 6 mM of $FeSO_4$, and (•) 12 mM of $FeSO_4$.

When current density was increased to 15 mA cm^{-2} (Figure 6b), an improvement in discoloration tendency was observed, and the increase in $FeSO_4$ concentration produced a higher removal speed and discoloration percentage. For the EOx process alone (♦), a 46% discoloration and k_1-value of 0.0046 min^{-1} were observed. When 1 mM of $FeSO_4$ (▲) was used, a 50% discoloration and k_1-value of 0.0053 min^{-1} were achieved. Using 6 mM of $FeSO_4$ (■) produced a 54% discoloration and k_1-value of 0.0060 min^{-1}. Finally, the process with the greatest discoloration used 12 mM of $FeSO_4$ (•), generating a 63% discoloration after two hours of treatment and a k_1-value of 0.0076 min^{-1}.

Further increasing the current density to 30 mA cm^{-2} showed the best results, as shown in Figure 6c. Compared with the other conditions tested, the highest discoloration was achieved by adding 12 mM of $FeSO_4$ (•) into the solution, followed by adding 6 mM of $FeSO_4$ (■), 1 mM of $FeSO_4$ (▲), and finally using the EOx process alone (♦), which produced water discoloration values as high as 93%, 85%, 77%, and 67%, respectively, and k_1-values of 0.0191 min^{-1}, 0.0141 min^{-1}, 0.0112 min^{-1}, and 0.0087 min^{-1}, respectively.

3.2.3. BBR Dye Degradation at Different Current Densities in 0.05 M of Na_2SO_4 and $FeSO_4$

Figure 7a shows the degradation of BBR dye (50 mg L^{-1}) using 0.05 M Na_2SO_4 at pH 3 as the support electrolyte after 60 min. It is worth noting that continuous dye degradation was observed for the four different conditions tested: EOx alone and adding 1 mM, 6 mM, and 12 mM of $FeSO_4$ while operating at a constant current density (7 mA cm^{-2}). The EOx process alone (♦) showed 36% discoloration (k_1 = 0.0036 min^{-1}). When 1 mM of $FeSO_4$ (▲) was added, 91% discoloration was achieved (k_1 = 0.0052 min^{-1}).

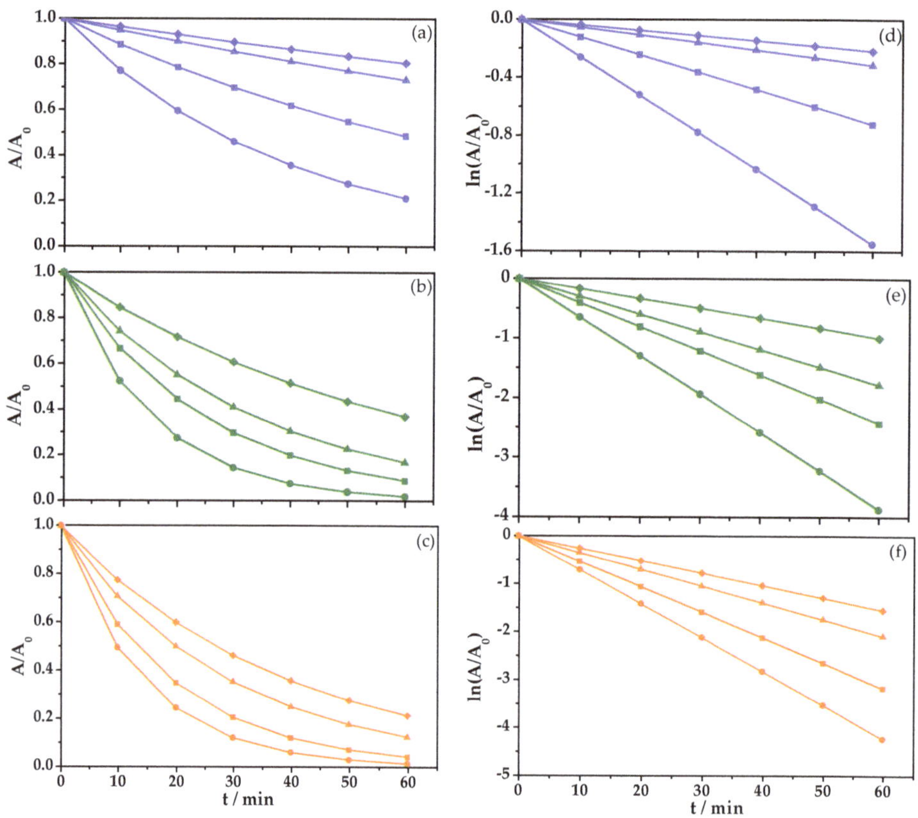

Figure 7. Removal of an aqueous solution of BBR azo dye, C_0 = 50 mg/L, during the oxidation process in 0.05 M of Na_2SO_4 and at a pH of 3 with a solution volume of 0.1 L, (**a**) j = 7 mA cm^{-2}, (**b**) j = 15 mA cm^{-2}, (**c**) j = 30 mA cm^{-2}, and fit to first order reaction kinetics (**d**), (**e**) y (**f**), respectively; (♦) BDD without $FeSO_4$, (▲) 1 mM of $FeSO_4$, (■) 6 mM of $FeSO_4$, and (•) 12 mM of $FeSO_4$.

Water discoloration as high as 94% was observed when 6 mM of $FeSO_4$ was added (■) (k_1 = 0.0120 min^{-1}), and finally the faster discoloration (e.g., 96%, k_1 = 0.0258 min^{-1}) was achieved after 60 min of treatment using 12 mM of $FeSO_4$ (•).

Increasing the current density to 15 mA cm^{-2} (Figure 7b) generated better results for BBR dye discoloration for shorter process times. When EOx was used alone (♦), 63% (k_1 = 0.0166 min^{-1}) was observed. When 1 mM of $FeSO_4$ (▲) was added, 80% dye removal (k_1 = 0.0297 min^{-1}) was achieved after 60 min. By increasing $FeSO_4$ to 6 mM (■), 89% (k_1 = 0.0404 min^{-1}) was reached and, when using 12 mM of $FeSO_4$ (•), 96% (k_1 = 0.0645 min^{-1}) was reached after 60 min.

The same tendency was observed when the highest current density (30 mA cm^{-2}) was applied, as shown in Figure 7c. In this case, using 12 mM of $FeSO_4$ (•) generated 98% discoloration (k_1 = 0.0703 min^{-1}), and using 6 mM of $FeSO_4$ (■) generated 95% discoloration (k_1 = 0.0528 min^{-1}). Using 1 mM of $FeSO_4$ (▲), 87% discoloration (k_1 = 0.0347 min^{-1}) was achieved, and finally EOx alone (♦) only reached 78% discoloration (k_1 = 0.0257 min^{-1}) after 60 min.

Based on these results, a combined BBR oxidation process is proposed. For the first case where we used $FeSO_4$ and $HClO_4$ as an electrolyte, the oxidation process would occur directly on the BDD electrode (EOx) and indirectly (chemically) through the hydroxyl radicals (resulting from water oxidation) and ferrate ions (resulting from $FeSO_4$ oxidation), as in Scheme 1, reactions 4, 5 and 8. For the second case, where $FeSO_4$ and Na_2SO_4 were used as electrolyte, the BBR oxidation process occurs directly on the BDD electrode (EOx, reaction 5) and indirectly (chemically) through the hydroxyl radicals (resulting from water oxidation, reactions 3 and 4), ferrate (VI) ions (resulting from $FeSO_4$ oxidation, reactions 7 and 8) and perosodisulfates (resulting from the oxidation of Na_2SO_4, reactions 9 and 10), as in the Scheme 2 [42]. The formation of peroxodisulfate causes the latter case to be more effective in the oxidation of the BBR.

Based on the above, the simultaneous processes of EOx and the generation of ferrate ions in situ can be summarized with the following reactions [21]:

1. Discharge of water into the system and generation of oxygen radicals ($^{\bullet}OH$).

$$BDD + H_2O \rightarrow BDD(^{\bullet}OH) + H^+ + e^- \tag{3}$$

2. Oxidation of contaminants by oxygen radicals ($^{\bullet}OH$).

$$BDD(^{\bullet}OH) + R \rightarrow BDD + nCO_2 + mH_2O \tag{4}$$

3. Oxidation of contaminants on BDD.

$$BDDpolarized(+) + R \rightarrow BDD(nCO_2) + mH_2O \tag{5}$$

4. Generation of [Fe(VI)] from $FeSO_4$ on BDD.

$$\text{oxidation}: Fe^{2+} + {}^{\bullet}OH \rightleftarrows Fe^{3+} + OH^- \tag{6}$$

$$\text{over} - \text{oxidation}: Fe^{3+} + 4H_2O \rightleftarrows FeO_4^{2-} + 8H^+ + 3e^- \tag{7}$$

5. Oxidation of organic compounds by [Fe(VI)] and $S_2O_8^{2-}$:

$$[Fe(VI)] + R \rightarrow Fe(III) + IS + nCO_2 + mH_2O \tag{8}$$

$$2SO_4^{2-} \rightarrow S_2O_8^{2-} + 2e^- \tag{9}$$

$$S_2O_8^{2-} + R \rightarrow 2SO_4^{2-} + IS + nCO_2 + mH_2O) \tag{10}$$

For this study, stages (e) and (f) were avoided by working at low $FeSO_4$ concentrations, which promoted a synergistic effect in stages (b) and (d).

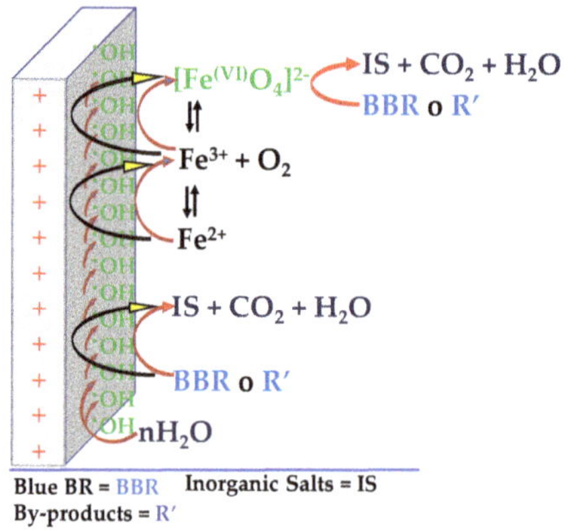

Scheme 1. Representation of the reactions involved on BDD substrate electrode held in $FeSO_4$ and 0.1 M of $HClO_4$.

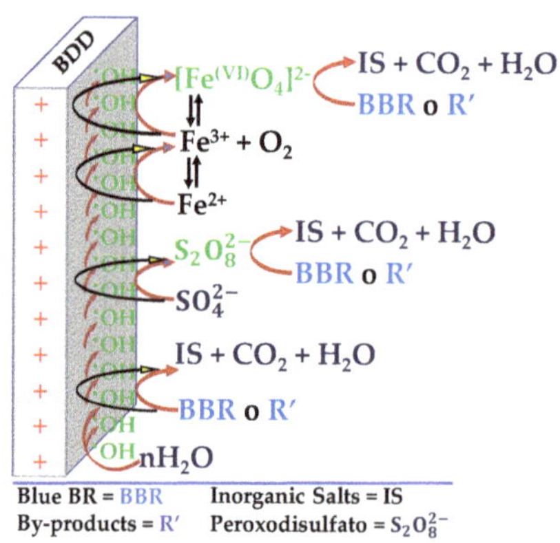

Scheme 2. Representation of the reactions involved on BDD substrate electrode held in $FeSO_4$ and 0.05 M Na_2SO_4.

3.2.4. COD Determinations and Byproduct Evolution

In Figure 8, COD reduction is shown by comparing EOx alone (♦) with the in situ generation of [Fe(VI)] (•). Electro-oxidation alone was able to achieve 37% COD removal after 90 min of treatment. In the case of ferrate generation, 61% COD removal was obtained, suggesting electrogenerated [Fe(VI)] ions directly influence a decrease in COD by carrying out pollutant degradation more efficiently and reducing the amount of oxidable matter in the solution [43,44].

Figure 8a shows the accumulation of a low amount of carboxylic acids generated during the EOx process alone (♦) and while testing the simultaneous EOx/[Fe(VI)] processes using 12 mM of Fe_2O_4 (•) at 30 mA cm^{-2}. The carboxylic acid quantified from BBR degradation was oxalic acid. Figure 8b shows that oxalic acid generation contributed only almost 20 mg/L using the EOx process alone, whereas adding 12 mM of $FeSO_4$ produced only 2 mg L^{-1} under the simultaneous EOx/[Fe(VI)] process. These results suggest that the degradative pathway of BBR produces carboxylic acid because of the combined action of $^{\cdot}OH$ and ferrate ions generated in the solution [45–49].

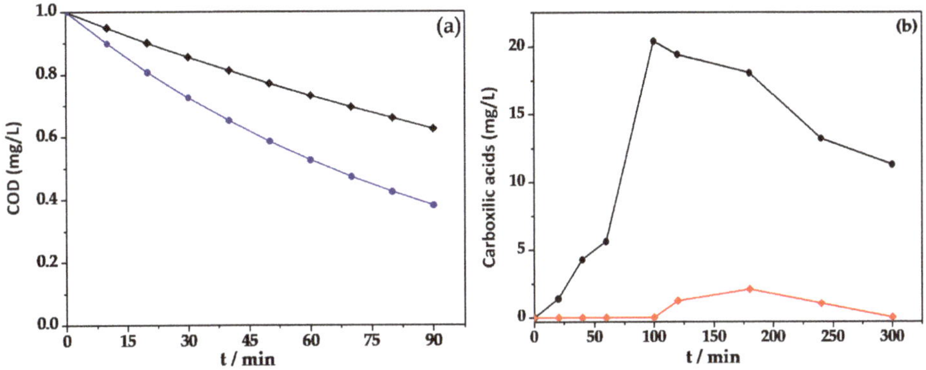

Figure 8. (**a**) Variation of COD abatement for 50 mg/L of BBR azo dye using BDD electrolysis without $FeSO_4$ (♦) and the simultaneous EOx/[Fe(VI)] process using 12 mM of $FeSO_4$ (•) in 0.05 M of Na_2SO_4 at a pH of 3 at 30 mA cm^{-2}. (**b**) Evolution of the concentration of oxalic acid produced during the degradation of 0.4 L of 50 mg/L BBR solution in 0.05 M of Na_2SO_4 at a pH of 3 and 35 °C using BDD without $FeSO_4$ (•) and the simultaneous EOx/[Fe(VI)] process using 12 mM of $FeSO_4$ (♦) at 30 mA cm^{-2}.

4. Conclusions

Electrochemical characterization using cyclic voltammetry confirms the in situ electrogeneration of ferrate ions on the BDD surface by adding $FeSO_4$ in both the sulfate and perchloric acid media. The use of $HClO_4$ produced discoloration efficiency close to 93% after 120 min of treatment when 12 mM of $FeSO_4$ was added. However, faster degradation rates were observed using the sulfate medium, which achieved up to 98% of color reduction in 60 min of treatment when 12 mM of $FeSO_4$ was added and a current density of 30 mA cm^{-2} was applied. A COD abatement as high as 63% was observed when the highest $FeSO_4$ concentration was added to the reaction mixture using the highest current density value. The byproduct analysis detected the presence of only one carboxylic acid (oxalic acid), which was rapidly oxidized by the simultaneous EOx/[Fe(VI)] process. These were identified as the best treatment conditions because they allowed some smaller amount reaction byproducts to accumulate.

Based on the obtained results, it was possible to demonstrate the viability of the simultaneous generation of both oxidant species, the free radical and the ferrate ion, which opens up an important opportunity to carry out not only the degradation of dyes, but also other organic compounds that are difficult to remove. This work provides an advance in the study of the simultaneous generation of oxidant species to the scientific community interested in this subject. In the future, the challenge will be understanding the reaction mechanisms.

Author Contributions: Conceptualization, P.J.E.-M. and J.M.P.-H.; methodology, P.J.E.-M. and J.M.P.-H.; validation, M.C.; formal analysis.; investigation, M.C., P.J.E.-M. and J.M.P.-H.; resources.; data curation, M.C. and P.J.E.-M.; writing—original draft preparation, M.C. and P.J.E.-M.; writing—review and editing, P.J.E.-M. and J.M.P.-H.; visualization.; supervision, J.M.P.-H.; project administration, J.M.P.-H.; funding acquisition J.M.P.-H., O.R., A.P., E.R.B. and S.G.-G. All authors have read and agreed to the published version of the manuscript.

Funding: The authors would like to acknowledge the economic support of the Universidad de Guanajuato.

Acknowledgments: O.M. Rodriguez-Narvaez and Alain Picos would like to thank CONACYT for a graduate fellowship. The authors are also grateful to Nicole Damon (DRI) for her editorial review.

Conflicts of Interest: The authors declare no conflict of interest.

References

1. Alcocer, S.; Picos, A.; Uribe, A.R.; Pérez, T.; Peralta-Hernández, J.M. Comparative study for degradation of industrial dyes by electrochemical advanced oxidation processes with BDD anode in a laboratory stirred tank reactor. *Chemosphere* **2018**, *205*, 682–689. [CrossRef]

2. Alsheyab, M.; Jiang, J.-Q.; Stanford, C. On-line production of ferrate with an electrochemical method and its potential application for wastewater treatment—A review. *J. Environ. Manag.* **2009**, *90*, 1350–1356. [CrossRef]
3. Ambauen, N.; Muff, J.; Mai, N.L.; Hallé, C.; Trinh, T.T.; Meyn, T. Insights into the kinetics of intermediate formation during electrochemical oxidation of the organic model pollutant salicylic acid in chloride electrolyte. *Water* **2019**, *11*, 1322. [CrossRef]
4. Bennett, J.A.; Wang, J.; Show, Y.; Swain, G.M. Effect of sp2-Bonded Nondiamond Carbon Impurity on the Response of Boron-Doped Polycrystalline Diamond Thin-Film Electrodes. *J. Electrochem. Soc.* **2004**, *151*, E306. [CrossRef]
5. Cataldo-Hernández, M.A.; Govindarajan, R.; Bonakdarpour, A.; Mohseni, M.; Wilkinson, D.P. Electrosynthesis of ferrate in a batch reactor at neutral conditions for drinking water applications. *Can. J. Chem. Eng.* **2018**, *96*, 1648–1655. [CrossRef]
6. Cataldo-Hernández, M.; Stewart, M.; Bonakdarpour, A.; Mohseni, M.; Wilkinson, D.P. Degradation of ferrate species produced electrochemically for use in drinking water treatment applications. *Can. J. Chem. Eng.* **2018**, *96*, 1045–1052. [CrossRef]
7. Cerreta, G.; Roccamante, M.A.; Oller, I.; Malato, S.; Rizzo, L. Contaminants of emerging concern removal from real wastewater by UV/free chlorine process: A comparison with solar/free chlorine and UV/H2O2 at pilot scale. *Chemosphere* **2019**, *236*, 124354. [CrossRef] [PubMed]
8. Cho, M.; Lee, Y.; Choi, W.; Chung, H.; Yoon, J. Study on Fe (VI) species as a disinfectant: Quantitative evaluation and modeling for inactivating *Escherichia coli*. *Water Res.* **2006**, *40*, 3580–3586. [CrossRef]
9. Coria, G.; Pérez, T.; Sirés, I.; Brillas, E.; Nava, J.L. Abatement of the antibiotic levofloxacin in a solar photoelectro-Fenton flow plant: Modeling the dissolved organic carbon concentration-time relationship. *Chemosphere* **2018**, *198*, 174–181. [CrossRef]
10. Dávila, O.O.; Bergeron, L.L.; Gutiérrez, P.R.; Jiménez, M.M.D.; Sirés, I.; Brillas, E.; Roig Navarro, A.F.; Arandes, J.B.; Llopis, J.V.S. Electrochemical oxidation of dibenzothiophene compounds on BDD electrode in acetonitrile–water medium. *J. Electroanal. Chem.* **2019**, *847*, 113172. [CrossRef]
11. Deng, Y.; Jung, C.; Liang, Y.; Goodey, N.; Waite, T.D. Ferrate (VI) decomposition in water in the absence and presence of natural organic matter (NOM). *Chem. Eng. J.* **2018**, *334*, 2335–2342. [CrossRef]
12. Denvir, A.; Pletcher, D. Electrochemical generation of ferrate Part I: Dissolution of an iron wool bed anode. *J. Appl. Electrochem.* **1996**, *26*, 815–822. [CrossRef]
13. Diaz, M.; Cataldo, M.; Ledezma, P.; Keller, J.; Doederer, K. Unravelling the mechanisms controlling the electro-generation of ferrate using four iron salts in boron-doped diamond electrodes. *J. Electroanal. Chem.* **2019**, *854*, 113501. [CrossRef]
14. Dos Santos, A.B.; Cervantes, F.J.; van Lier, J.B. Review paper on current technologies for decolourisation of textile wastewaters: Perspectives for anaerobic biotechnology. *Bioresour. Technol.* **2007**, *98*, 2369–2385. [CrossRef] [PubMed]
15. Elgrishi, N.; Rountree, K.J.; McCarthy, B.D.; Rountree, E.S.; Eisenhart, T.T.; Dempsey, J.L. A Practical Beginner's Guide to Cyclic Voltammetry. *J. Chem. Educ.* **2018**, *95*, 197–206. [CrossRef]
16. Eng, Y.Y.; Sharma, V.K.; Ray, A.K. Ferrate(VI): Green chemistry oxidant for degradation of cationic surfactant. *Chemosphere* **2006**, *63*, 1785–1790. [CrossRef]
17. Espinoza-Montero, P.J.; Vasquez-Medrano, R.; Ibanez, J.G.; Frontana-Uribe, B.A. Efficient anodic degradation of phenol paired to improved cathodic production of H2O2 at BDD electrodes. *J. Electrochem. Soc.* **2013**, *160*, G3171–G3177. [CrossRef]
18. Espinoza, C.; Romero, J.; Villegas, L.; Cornejo-Ponce, L.; Salazar, R. Mineralization of the textile dye acid yellow 42 by solar photoelectro-Fenton in a lab-pilot plant. *J. Hazard. Mater.* **2016**, *319*, 24–33. [CrossRef]
19. Wood, R.H. The Heat, Free Energy and Entropy of the Ferrate(VI) Ion. *J. Am. Chem. Soc.* **2002**, *80*, 2038–2041. [CrossRef]
20. Ibanez, J.G.; Tellez-Giron, M.; Alvarez, D.; Garcia-Pintor, E. Laboratory Experiments on the Electrochemical Remediation of the Environment. Part 6: Microscale Production of Ferrate. *J. Chem. Educ.* **2004**, *81*, 251. [CrossRef]
21. Jiang, J.-Q.; Lloyd, B. Progress in the development and use of ferrate(VI) salt as an oxidant and coagulant for water and wastewater treatment. *Water Res.* **2002**, *36*, 1397–1408. [CrossRef]
22. Jiang, J.Q. Advances in the development and application of ferrate(VI) for water and wastewater treatment. *J. Chem. Technol. Biotechnol.* **2014**, *89*, 165–177. [CrossRef]

23. Kanari, N.; Ostrosi, E.; Diliberto, C.; Filippova, I.; Shallari, S.; Allain, E.; Diot, F.; Patisson, F.; Yvon, J. Green Process for Industrial Waste Transformation into Super-Oxidizing Materials Named Alkali Metal Ferrates (VI). *Materials* **2019**, *12*, 1977. [CrossRef] [PubMed]
24. Lee, J.; Tryk, D.A.; Fujishima, A.; Park, S.-M. Electrochemical generation of ferrate in acidic media at boron-doped diamond electrodes. *Chem. Commun.* **2002**, 486–487. [CrossRef]
25. Lescuras-Darrou, V.; Lapicque, F.; Valentin, G. Electrochemical ferrate generation for waste water treatment using cast irons with high silicon contents. *J. Appl. Electrochem.* **2002**, *32*, 57–63. [CrossRef]
26. Liu, J.; Zhang, Z.; Liu, Z.; Zhang, X. Integration of ferrate (VI) pretreatment and ceramic membrane reactor for membrane fouling mitigation in reclaimed water treatment. *J. Membr. Sci.* **2018**, *552*, 315–325. [CrossRef]
27. Mácová, Z.; Bouzek, K.; Híveš, J.; Sharma, V.K.; Terryn, R.J.; Baum, J.C. Research progress in the electrochemical synthesis of ferrate(VI). *Electrochim. Acta* **2009**, *54*, 2673–2683. [CrossRef]
28. Martínez-huitle, C.A.; Brillas, E. Electrochemical Alternatives for Drinking Water Disinfection. *Angew. Int. Ed. Chem.* **2008**, *47*, 1998–2005. [CrossRef]
29. Morales, U.; Escudero, C.J.; Rivero, M.J.; Ortiz, I.; Rocha, J.M.; Peralta-Hernández, J.M. Coupling of the electrochemical oxidation (EO-BDD)/photocatalysis (TiO2-Fe-N) processes for degradation of acid blue BR dye. *J. Electroanal. Chem.* **2018**, *808*, 180–188. [CrossRef]
30. Oriol, R.; del Pilar Bernícola, M.; Brillas, E.; Cabot, P.L.; Sirés, I. Paired electro-oxidation of insecticide imidacloprid and electrodenitrification in simulated and real water matrices. *Electrochim. Acta* **2019**, *317*, 753–765. [CrossRef]
31. Pacheco-Álvarez, M.O.A.; Picos, A.; Pérez-Segura, T.; Peralta-Hernández, J.M. Proposal for highly efficient electrochemical discoloration and degradation of azo dyes with parallel arrangement electrodes. *J. Electroanal. Chem.* **2019**, *838*, 195–203. [CrossRef]
32. Panizza, M.; Cerisola, G. Application of diamond electrodes to electrochemical processes. *Electrochim. Acta* **2005**, *51*, 191–199. [CrossRef]
33. Villanueva-Rodríguez, M.; Hernández-Ramírez, A.; Peralta-Hernández, J.M.; Bandala, E.R.; Quiroz-Alfaro, M.A. Enhancing the electrochemical oxidation of acid-yellow 36 azo dye using boron-doped diamond electrodes by addition of ferrous ion. *J. Hazard. Mater.* **2009**, *167*, 1226–1230. [CrossRef] [PubMed]
34. Villanueva-Rodríguez, M.; Sánchez-Sánchez, C.M.; Montiel, V.; Brillas, E.; Peralta-Hernández, J.M.; Hernández-Ramírez, A. Characterization of ferrate ion electrogeneration in acidic media by voltammetry and scanning electrochemical microscopy. Assessment of its reactivity on 2, 4-dichlorophenoxyacetic acid degradation. *Electrochim. Acta* **2012**, *64*, 196–204.
35. Queiroz, N.M.P.; Sirés, I.; Zanta, C.L.P.S.; Tonholo, J.; Brillas, E. Removal of the drug procaine from acidic aqueous solutions using a flow reactor with a boron-doped diamond anode. *Sep. Purif. Technol.* **2019**, *216*, 65–73. [CrossRef]
36. Rice, E.W.; Baird, R.B.; Eaton, A.D.; Clesceri, L.S. *Standard Methods for the Examination of Water and Wastewater*; American Public Health Association: Washington, DC, USA, 2012; p. 541.
37. Ruiz, E.J.; Hernández-Ramírez, A.; Peralta-Hernández, J.M.; Arias, C.; Brillas, E. Application of solar photoelectro-Fenton technology to azo dyes mineralization: Effect of current density, Fe2+ and dye concentrations. *Chem. Eng. J.* **2011**, *171*, 385–392. [CrossRef]
38. Serrano, K.; Michaud, P.A.; Comninellis, C.; Savall, A. Electrochemical preparation of peroxodisulfuric acid using boron doped diamond thin film electrodes. *Electrochim. Acta* **2002**, *48*, 431–436. [CrossRef]
39. Sharma, V.K. Potassium ferrate(VI): Properties and applications. *ACS Div. Environ. Chem. Prepr.* **2000**, *40*, 131–132.
40. Sharma, V.K.; Burnett, C.R.; O'Connor, D.B. Ferrate(VI) and ferrate(V) oxidation of thiocyanate. *ACS Div. Environ. Chem. Prepr.* **2000**, *40*, 600–601.
41. Sharma, V.K. Potassium ferrate(VI): An environmentally friendly oxidant. *Adv. Environ. Res.* **2002**, *6*, 143–156. [CrossRef]
42. Sharma, V.K.; Kazama, F.; Jiangyong, H.; Ray, A.K. Ferrates (iron(VI) and iron(V)): Environmentally friendly oxidants and disinfectants. *J. Water Health* **2005**, *3*, 45–58. [CrossRef] [PubMed]
43. Sharma, V.K.; Rivera, W.; Joshi, V.N.; Millero, F.J.; O'Connor, D. Ferrate(VI) Oxidation of Thiourea. *Environ. Sci. Technol.* **1999**, *33*, 2645–2650. [CrossRef]

44. Talaiekhozani, A.; Talaei, M.R.; Rezania, S. An overview on production and application of ferrate (VI) for chemical oxidation, coagulation and disinfection of water and wastewater. *J. Environ. Chem. Eng.* **2017**, *5*, 1828–1842. [CrossRef]
45. Thiam, A.; Salazar, R. Fenton-based electrochemical degradation of metolachlor in aqueous solution by means of BDD and Pt electrodes: Influencing factors and reaction pathways. *Environ. Sci. Pollut. Res.* **2019**, *26*, 2580–2591. [CrossRef]
46. Pérez, T.; Sirés, I.; Brillas, E.; Nava, J.L. Solar photoelectro-Fenton flow plant modeling for the degradation of the antibiotic erythromycin in sulfate medium. *Electrochim. Acta* **2017**, *228*, 45–56. [CrossRef]
47. Potts, M.E.; Churchwell, D.R. Removal of Radionuclides in Wastewaters Utilizing Potassium ferrate(VI). *Water Environ. Res.* **1994**, *66*, 107–109. [CrossRef]
48. Ye, Z.; Brillas, E.; Centellas, F.; Cabot, P.L.; Sirés, I. Electro-Fenton process at mild pH using Fe(III)-EDDS as soluble catalyst and carbon felt as cathode. *Appl. Catal. B Environ.* **2019**, *257*, 117907. [CrossRef]
49. Lee, Y.; Cho, M.K.; Kim, J.Y.; Yoon, J. Chemistry of Ferrate (Fe(VI)) in Aqueous Solution and its Applications as a Green Chemical. *J. Ind. Eng. Chem.* **2004**, *10*, 161–171.

processes

MDPI

Article

Photocatalytic Treatment of Paracetamol Using TiO_2 Nanotubes: Effect of pH

S. Alejandro Lozano-Morales [1,*], Graciela Morales [1], Miguel Ángel López Zavala [2], Augusto Arce-Sarria [3] and Fiderman Machuca-Martínez [3,*]

1 CONACYT-Centro de Investigación en Química Aplicada (CIQA), Blvd. Enrique Reyna 140, Saltillo 25294, Coahuila, Mexico; graciela.morales@ciqa.edu.mx

2 Tecnologico de Monterrey, Escuela de Ingeniería y Ciencias, Av. Eugenio Garza Sada Sur No. 2501, Col. Tecnológico, Monterrey 64849, Nuevo León, Mexico; miganloza@itesm.mx

3 Escuela de Ingeniería Química, Universidad del Valle, Cali 760032, Columbia; augusto.arce@correounivalle.edu.co

* Correspondence: salozanom@conacyt.mx (S.A.L.-M.); fiderman.machuca@correounivalle.edu.co (F.M.-M.)

Received: 21 March 2019; Accepted: 30 April 2019; Published: 29 May 2019

Abstract: Pharmaceuticals are considered among the group of emerging contaminants. Paracetamol is a moderate painkiller, which has been detected in ground and surface water. Photodegradation of paracetamol at a wavelength of radiation of 254 nm with TiO_2 nanotubes was studied by UV-spectroscopy, HPLC and measurement of the potential zeta in dependence of the solution pH. The efficiency of the photodegradation of paracetamol (20 mg L^{-1}) was 99% after 100 min exposure. Application of the Langmuir-Hinshelwood equation allowed the evaluation of the rate constant. Non-organic by-products were detected under the conditions of the chromatographic analysis. The photoreaction was faster at pH 6.5, a value at which adsorption was favored, leading to higher efficiency.

Keywords: photocatalysis; TiO_2 nanotubes; emerging contaminants; paracetamol; pH

1. Introduction

The rapid development of modern industry and society occurs hand in hand with the increasing generation of residual liquids that contain highly toxic compounds with limited biodegradability. The high usage of chemical products such as dyes, pesticides and pharmaceuticals, which have recently been called "emerging contaminants", causes a growing problem because of their concentration and accumulation in the environment, particularly in aqueous effluents [1–3]. A type of water contaminants that is becoming more common are pharmaceuticals. They are mainly attributed to hospital and therapeutic waste, personal care products, and the pharmaceutical industry [4,5]. The presence of these products in drinking water is a public health problem because little is known about the potentially long-term toxic health effects that are associated with the ingestion of mixtures of these compounds in drinking water, even at very low concentrations [6–8].

Innovative treatments, such as Advanced Oxidation Processes (AOPs), are effective in drastically reducing the concentrations of pharmaceutical compounds in effluents. Generally, these processes are based on physicochemical methods that cause internal changes in the chemical structures of the contaminants to make them inert and/or decrease their toxicity. The concept of AOPs was initially established by Halling et al. [9], who defined them as processes that involve the generation and use of potential transitory species, especially the hydroxyl radical (HO•) because it is able to attack almost all organic compounds and can react up to 1012 times faster than other oxidants because of its very high oxidation potential of 2.80 eV.

Laboratory-scale reports and studies of treatment plants have found that paracetamol has no or very little biodegradability [10,11]. However, when it is subjected to a tertiary treatment, such as chloration, it produces highly toxic reaction by-products, such as *1,4-benzoquinone* and *N-acetyl-p-benzoquinone* imine [12,13]. Thus, there is a need to investigate highly efficient processes, such as AOPs, that have been recently used to degrade paracetamol, including ozonation, electrocatalysis, photolysis, electrolysis, sonolysis, photo-Fenton oxidation [14–18], and also heterogeneous photocatalysis using nanomaterials of titanium oxide (TiO_2) as photocatalysts [19,20].

It is particularly important to develop highly effective alternative methodologies that are harmless to the environment, such as heterogeneous photocatalysis, which employs nanocatalysts that are usually composed of TiO_2 because of its low toxicity, low cost and high performance [21,22]. The photocatalytic activity of nanocatalysts is governed by many properties, including its specific surface area, pore volume and distribution, crystal structure, phase composition, particle size and morpholog [9,23,24]. In particular, the nanotubes that are synthesized by hydrothermal processes have high surface- weight ratios of up to 120 m^2 per gram [2,25–30].

Many factors affect the reactions that have been proposed for the photodegradation of paracetamol in effluents, such as the catalyst and contaminant concentrations, the excitation wavelength of the catalyst, the oxygenation and the pH [31,32]. The latter has not been thoroughly investigated, which implies that the direct impact of pH on the reaction is efficiency is still uncertain. This investigation evaluates the relationship between the pH of paracetamol suspensions and the efficiency of the photocatalytic degradation using ultraviolet radiation with a wavelength of 254 nm in presence of nanostructured TiO_2 catalysts with a nanotubular morphology.

2. Materials and Methods

Reagents: All the chemical reagents used were analytical-grade. Paracetamol ($C_8H_9NO_2$), 99% purity, and titanium dioxide (P25), 99.5% were purchased from Aldrich. 99.1% Sodium hydroxide (NaOH) and 37% hydrochloric acid (HCl), from sigma Aldrich, were used. HPLC grade water was obtained using a Millipore Milli-Q system (Millipore Sigma Corporation, MA, USA).

Synthesis of TiO_2 nanotubes: The TiO_2 nanotubes were produced using a chemical process similar to that described by López et al. [30], TiO_2 powder (P25) with a crystal structure of 80% anatase and 20% rutile and an average particle size of 25 nm was used as a precursor for the synthesis of the nanotubes. To perform the synthesis, 0.3 g of TiO_2 powder was added to 30 mL of 10 M aqueous NaOH solution, and the suspension was vigorously agitated for 2 h at room temperature (30 °C). To perform the hydrothermal treatment, the mixture was transferred to an autoclave, and the temperature was increased to 110 °C for 72 h. The solid precipitate was separated by centrifugation at 10,000 rpm and re-suspended in 200 mL of a 0.1 M HCl solution with constant agitation for 3 h and then centrifuged again. The precipitated samples were washed with distilled water twice until the pH of the solution reached 6.6, which is approximately the pH value of the distilled water that was used. The washed samples were dried in a vacuum oven at 80 °C for 24 h

Preparation of paracetamol samples: An aqueous paracetamol stock solution (volume of 1.5 L) with a concentration of 20 mg L^{-1} was prepared. The solution was kept under constant magnetic agitation for 120 min on a stirring plate (LabTech model LMS-1003-Gemini BV Lab, Apeldoorn, The Netherlands) until complete solubilization of paracetamol. Six 250 mL aliquots were taken from the stock solution. The pH of each solution was modified using NaOH and HCl solutions to reach the desired pH values of 2.5, 4.5, 6.5, 8.5 and 10.5. To each aliquot 250 mL of the paracetamol solution, 100 mg of the TiO_2 nanotubes were added, and the suspension was then agitated for 30 min until it was completely dispersed.

Monitoring of the photodegradation of paracetamol by UV-Vis spectrophotometry: Batch cylindrical stainless steel photoreactor was used to perform the degradation of the samples. The photoreactor used a high-pressure Hg candle lamp as the energy source, which provided UV light with a wavelength of 254 nm and an intensity of 25 W cm^{-2}. The temperature was kept constant at 25 °C by means of a cooling system and was measured during each sampling during the 3 h of irradiation. The lamp was encapsulated in a quartz tube that was submerged in the center of the reactor along with the aqueous solution (Figure 1). The solution that surrounded the reaction system underwent constant agitation at air pressure and room temperature. The reaction kinetics inside the reactor was monitored by taking 1 mL samples every 10 min and analyzing them with a UV-Vis spectrometer (Hach model DR 5000-Loveland, CO, USA) within the range of 200 to 400 nm with 0.5 nm steps. The final paracetamol concentration Cf was determined with the aid of a calibration curve (m = 55.737, R^2 = 0.9995).

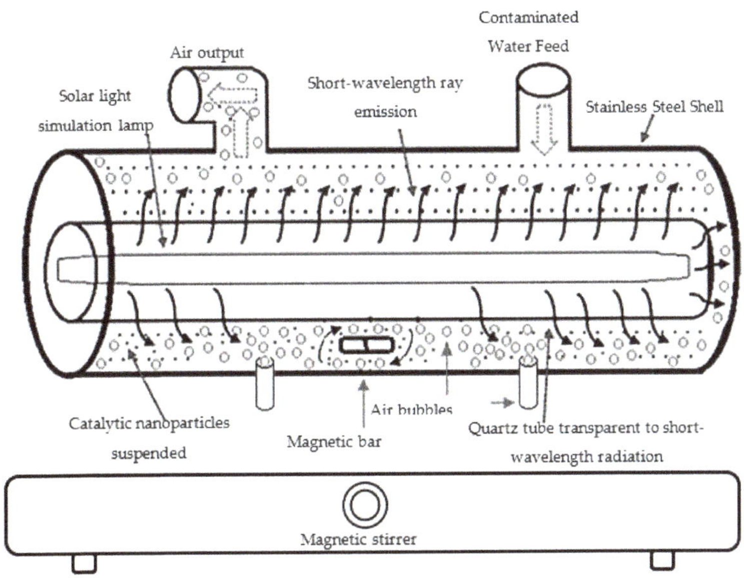

Figure 1. Experimental setup of the heterogeneous photocatalysis reactor.

Detection of photodegradation by-products by HPLC: The most efficient paracetamol photodegradation reaction was identified using a high-pressure liquid chromatograph (Agilent 1200- Santa Clara, CA, USA) with a C18 reversed-phase column. The chromatographic conditions were similar to those reported by López et al. [33,34]. However, they were optimized for paracetamol and its by-products by using a mobile phase of methanol and acetic acid in 1% water with a v/v ratio of 40/60. The equipment was operated at 25 °C with a flowrate of 1.5 mL min^{-1}. A diode array detector (DAD) was used with a principal wavelength of 254 nm, and intervals of 15 nm were used for the detection.

Measurement of the zeta potential: To measure the zeta potential of the TiO_2 nanotubes, an aqueous sample of paracetamol was prepared at a concentration of 0.5 mg L^{-1}. Adjustment of the pH was performed with 0.01 M HCl and 0.01 M NaOH solutions using a particle charge titration analyzer Stabino from Microtrac.

Data treatment: Generally, the photochemical oxidation reactions of organic compounds have a first-order kinetics behavior [35]. This is the case for the photocatalytic degradation of paracetamol [36]. The Langmuir-Hinshelwood model (Equation (1)) was applied because in heterogeneous catalysis processes, this model considers the decomposition at the catalyst surface to be the determining step. So, the first-order reaction rate is assumed to be the product of the kinetic constant that is associated with the decomposition and the adsorbate concentration [37,38].

$$r = \frac{dC}{dt} = -\frac{kKC}{1 + KC} \quad (1)$$

where k is the rate constant, which depends on the light intensity, K is the adsorption constant of the nanocatalyst, and C is the concentration of paracetamol. For low adsorption magnitudes and concentrations, ($KC < 1$). So, Equation (1) reduces to the first-order kinetics according to Equation (2).

$$\frac{dC}{dt} = -kKC \tag{2}$$

Separating variables and integrating between the initial conditions $t = 0$ and $C = C_0$ at time t and if $Kk = k'$, where k' is the rate constant for the photocatalytic degradation

$$\ln\left(\frac{C_0}{C}\right) = kKt = k't \tag{3}$$

When plotting Equation (3), the slope is the value of the reaction rate constant k' and has units of $\min^{-1}$. The half-life ($t_{1/2}$), which is an important parameter of photocatalytic degradation because it is the time that is required to reduce the paracetamol concentration by half (i.e., $C = C_{0/2}$ in Equation (3), is given by:

$$t_{1/2} = \frac{\ln 2}{k'} \tag{4}$$

The efficiency of the photocatalytic degradation of the paracetamol was quantified using Equation (5).

$$\text{Efficiency of paracetamol photodegradation } (\%) = \frac{\left(C_0 - C_f\right)}{C_0} \times 100 \tag{5}$$

where (C_f) and (C_0) are the paracetamol concentrations in the initial and final solutions, respectively.

3. Results and Discussion

3.1. Photodegradation of Paracetamol

Figure 2 shows the absorption spectrum of the paracetamol in the stock solution. The maximum absorption intensity was observed at a wavelength of approximately 243 nm. This change in absorbance at the wavelength of λmax is consistent with the maximum absorption band of paracetamol observed found in previous studies [39]. The absorbance at this wavelength was used to monitor the changes in concentration and to determine the photodegradation reaction rate constant k' from the experimental data.

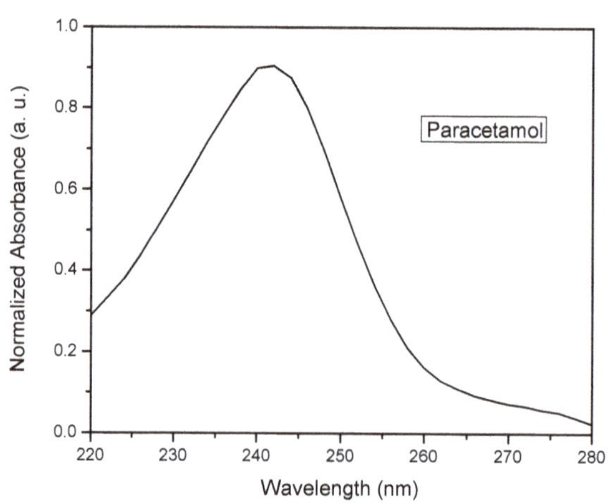

Figure 2. UV-vis absorption spectrum of normalized paracetamol.

Figure 3 shows the normalized decrease of the paracetamol concentration with the reaction time at different pH values. At acidic pH values between 2.5 and 4.5, the reaction rate constants are low because the •OH in the reagent medium interacts weakly with the paracetamol molecules because of the possible recombination of free H^+ protons and hydronium ions (H_3O^+) with the •OH that is generated in the photocatalytic process to form water molecules.

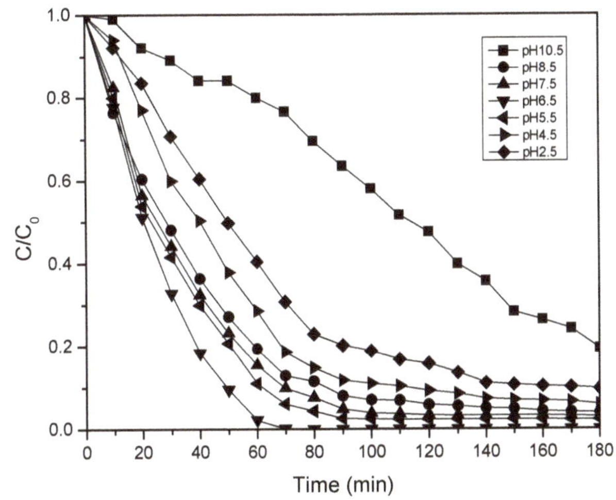

Figure 3. Photocatalytic activity of TiO_2 nanotubes in paracetamol solution at different pHs as function of irradiation time.

Table 1 shows that at a pH values of 6.5, the magnitudes of the reaction rate constant are higher, and correspondingly of the photodegradation efficiency. This can be attributed to the greater formation of hydroxyl radicals because at high pH values, more •OH groups are available at the nanocatalyst surface, and they can easily oxidize [40].

Table 1. Values of the photodegradation parameters calculated from Equations (3)–(5).

pH	K' (min^{-1})	R^2	$t_{1/2}$ (min)	Efficiency after 180 min (%)
2.5	0.016	0.9571	88.6	46
4.5	0.023	0.9305	60.1	67
5.5	0.029	0.9039	47.8	83
6.5	0.034	0.9641	27.4	99
7.5	0.027	0.9375	51.3	71
8.5	0.024	0.8422	49.9	68
10.5	0.008	0.8958	136.8	32

The graph of the zeta potential (Figure 4) shows the variation of the zeta potential of paracetamol and the catalyst, with the pH of the medium. The curves intersect at pH 6.5; a value at which the paracetamol molecule is not ionized and the surface charge of TiO_2 is close to zero [41].

Specific interactions of the polar groups and the aromatic moiety of paracetamol with TiO_2 lead to good adsorption, whereas the free generated and available •OH species have more interactions with the paracetamol molecules to degrade them. This explains why the reaction rate constant is higher at pH 6.5 (hollow points in Figure 4). The behavior of the rate constant depicted in Figure 4 also corresponds to the maximum value of the zeta potential. Also, the smaller the charge difference (intersect in Figure 4) between the substrate and the catalyst, the higher k.

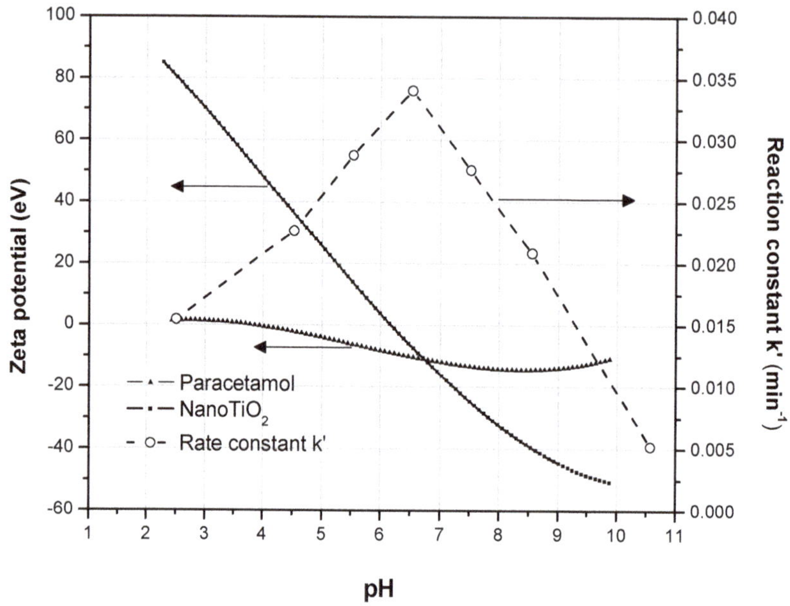

Figure 4. Graph of the zeta potential of the substrate and the photocatalyst with the variation of pH.

This result is consistent with the Langmuir-Hinshelwood kinetic model because it suggests that the adsorption-desorption equilibrium between the surface of the semiconductor and the fluid phase is indispensable for initiating the photodegradation of any substrate. Once this equilibrium is established, the photodegradation of the molecule starts through a mechanism that involves radicals. This adsorption-desorption equilibrium is produced precisely when the value of the zeta potential of paracetamol and the nanocatalyst are similar.

At the pH values of 8.5 to 10.5, the photodegradation rate decreases (Figure 4) because 10% paracetamol exists as anion already at pH 8.5 and 92% at pH 10.5 at higher pH values (pH > pKa). Thus, the increase in the pH gradually increases the electrostatic repulsion between the surface of the nanocatalyst and paracetamol, this considerably reduces the adsorption of paracetamol. Consequently, the reaction rate constant for the photodegradation of paracetamol is expected to decrease gradually at pH values higher than 8.5 [42].

3.2. Paracetamol Photodegradation at pH 6.5

To explore the intermediate products that result from the photodegradation of paracetamol, HPLC was used to analyze the treated solution after 100 min of sample treatment at pH 6.5. Only one peak is observed at a retention time of 0.43 min, both for the initial paracetamol sample and after of the photodegradation (Figure 5). This indicates that in our study, the molecule degrades into small molecules or non-detectable compounds under these measurement conditions.

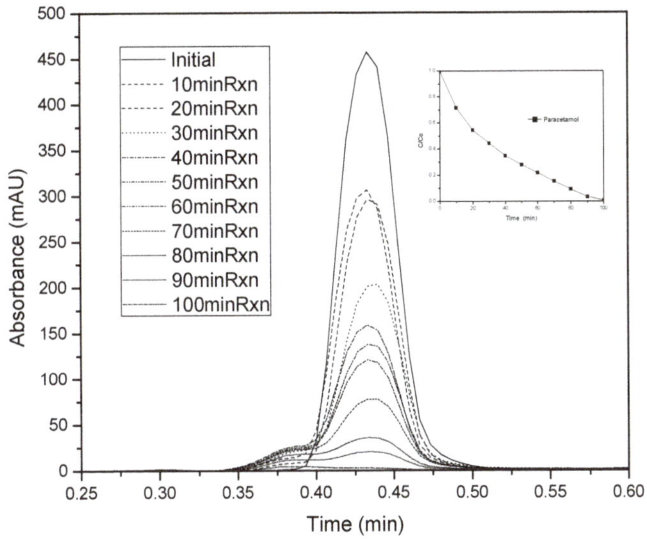

Figure 5. Paracetamol photodegradation kinetics with HPLC data for the samples at pH 6.5.

4. Conclusions

Paracetamol photodegradation in an aqueous solution using radiation with a wavelength of 254 nm is feasible. The best results were obtained at a pH of 6.5 and resulted in 99% photodegradation of paracetamol by catalytic oxidation under the studied conditions. The photodegradation of paracetamol can be performed at concentrations lower than 20 mg L^{-1} with relatively short exposure times (100 min) by using TiO_2 nanotubes. No traces of the metabolites were detected. With measurement of the zeta potential it was demonstrated that the most rapid photoreaction takes place at the pH, at which adsorption is favored. The pH of the photocatalytic process may influence the efficiency of the photodegradation.

Author Contributions: Designed the experiments, performed the experiments, analyzed and interpreted the data, wrote the paper, S.A.L.-M.; writing—review and editing, analyzed and interpreted the data, wrote the paper, G.M.; conceived and designed the experiments, contributed reagents, materials, analysis tools or data, M.Á.L.Z.; writing, methodology, formal analysis of the investigation, A.A.-S.; writing, methodology, formal analysis of the investigation and writing review and editing, F.M. M.

Funding: The authors are grateful to Universidad del Valle for the financial support to produce this work (GRANT 1004. Síntesis y evaluación de la actividad fotocatalítica TiO_2/WO_3 en el tratamiento de aguas residuales de la industria minera y la mineralización de contaminantes emergentes farmacéuticos).

Acknowledgments: The authors thank the support of the Consorcio del Agua FORDECYT No. 297116 of the National Council of Science and Technology of México (Consejo Nacional de Ciencia y Tecnología de México, CONACYT) and the Universidad del Valle of Cali Colombia for support during the internship. Arce-Sarria Thanks to Colciencias-Colombia for supporting his PhD Scolarship.

Conflicts of Interest: The authors declare no conflict of interest.

References

1. Wilkinson, J.; Hooda, P.S.; Barker, J.; Barton, S.; Swinden, J. Occurrence, fate and transformation of emerging contaminants in water: An overarching review of the field. *Environ. Pollut.* **2017**, *231*, 954–970. [CrossRef]
2. Awfa, D.; Ateia, M.; Fujii, M.; Johnson, M.S.; Yoshimura, C. Photodegradation of pharmaceuticals and personal care products in water treatment using carbonaceous-TiO_2composites: A critical review of recent literature. *Water Res.* **2018**, *142*, 26–45. [CrossRef] [PubMed]

3. Uheida, A.; Mohamed, A.; Belaqziz, M.; Nasser, W.S. Photocatalytic degradation of Ibuprofen, Naproxen, and Cetirizine using PAN-MWCNT nanofibers crosslinked TiO_2-NH_2 nanoparticles under visible light irradiation. *Sep. Purif. Technol.* **2019**, *212*, 110–118. [CrossRef]
4. Tobajas, M.; Belver, C.; Rodriguez, J.J. Degradation of emerging pollutants in water under solar irradiation using novel TiO_2-ZnO/clay nanoarchitectures. *Chem. Eng. J.* **2017**, *309*, 596–606. [CrossRef]
5. Zhao, L.; Deng, J.; Sun, P.; Liu, J.; Ji, Y.; Nakada, N.; Qiao, Z.; Tanaka, H.; Yang, Y. Nanomaterials for treating emerging contaminants in water by adsorption and photocatalysis: Systematic review and bibliometric analysis. *Sci. Total Environ.* **2018**, *627*, 1253–1263. [CrossRef] [PubMed]
6. Byrne, C.; Subramanian, G.; Pillai, S.C. Recent advances in photocatalysis for environmental applications. *J. Environ. Chem. Eng.* **2018**, *6*, 3531–3555. [CrossRef]
7. Richardson, S.D.; Kimura, S.Y. Emerging environmental contaminants: Challenges facing our next generation and potential engineering solutions. *Environ. Technol. Innov.* **2017**, *8*, 40–56. [CrossRef]
8. Sanganyado, E.; Gwenzi, W. Antibiotic resistance in drinking water systems: Occurrence, removal, and human health risks. *Sci. Total Environ.* **2019**, *669*, 785–797. [CrossRef] [PubMed]
9. Halling-Sørensen, B.; Nielsen, S.N.; Lanzky, P.F.; Ingerslev, F.; Lützhøft, H.C.H.; Jørgensen, S.E. Occurrence, fate and effects of pharmaceutical substances in the environment-A review. *Chemosphere* **1998**, *36*, 357–393. [CrossRef]
10. Barber, L.B.; Keefe, S.H.; Brown, G.K.; Furlong, E.T.; Gray, J.L.; Kolpin, D.W.; Meyer, M.T.; Sandstrom, M.W.; Zaugg, S.D. Persistence and potential effects of complex organic contaminant mixtures in wastewater-impacted streams. *Environ. Sci. Technol.* **2013**, *47*, 2177–2188. [CrossRef]
11. Shaheen, S.O.; Newson, R.B.; Ring, S.M.; Rose-Zerilli, M.J.; Holloway, J.W.; Henderson, A.J. Prenatal and infant acetaminophen exposure, antioxidant gene polymorphisms, and childhood asthma. *J. Allergy Clin. Immunol.* **2010**, *126*. [CrossRef] [PubMed]
12. Bedner, M.; MacCrehan, W.A. Transformation of acetaminophen by chlorination produces the toxicants 1,4-benzoquinone and N-acetyl-p-benzoquinone imine. *Environ. Sci. Technol.* **2006**, *40*, 516–522. [CrossRef]
13. Xagoraraki, I.; Hullman, R.; Song, W.; Li, H.; Voice, T. Effect of pH on degradation of acetaminophen and production of 1,4-benzoquinone in water chlorination. *J. Water Supply Res. Technol.-AQUA* **2008**, *57*, 381–390. [CrossRef]
14. Andreozzi, R.; Caprio, V.; Marotta, R.; Vogna, D. Paracetamol oxidation from aqueous solutions by means of ozonation and H_2O_2/UV system. *Water Res.* **2003**, *37*, 993–1004. [CrossRef]
15. Valdez, H.C.A.; Jiménez, G.G.; Granados, S.G.; de León, C.P. Degradation of paracetamol by advance oxidation processes using modified reticulated vitreous carbon electrodes with TiO_2 and $CuO/TiO_2/Al_2O_3$. *Chemosphere* **2012**, *89*, 1195–1201. [CrossRef]
16. Jagannathan, M.; Grieser, F.; Ashokkumar, M. Sonophotocatalytic degradation of paracetamol using TiO_2 and Fe^{3+}. *Sep. Purif. Technol.* **2013**, *103*, 114–118. [CrossRef]
17. Li, Y.; Song, W.; Fu, W.; Tsang, D.C.W.; Yang, X. The roles of halides in the acetaminophen degradation by UV/H_2O_2 treatment: Kinetics, mechanisms, and products analysis. *Chem. Eng. J.* **2015**, *271*, 214–222. [CrossRef]
18. Tan, C.; Gao, N.; Zhou, S.; Xiao, Y.; Zhuang, Z. Kinetic study of acetaminophen degradation by UV-based advanced oxidation processes. *Chem. Eng. J.* **2014**, *253*, 229–236. [CrossRef]
19. Su, C.C.; Cada, C.A.; Dalida, M.L.P.; Lu, M.C. Effect of UV light on acetaminophen degradation in the electro-Fenton process. *Sep. Purif. Technol.* **2013**, *120*, 43–51. [CrossRef]
20. Martinez-Haya, R.; Miranda, M.A.; Marin, M.L. Type I vs Type II photodegradation of pollutants. *Catal. Today* **2018**, *313*, 161–166. [CrossRef]
21. Scott, J.P.; Ollis, D.F. Integration of chemical and biological oxidation processes for water treatment: Review and recommendations. *Environ. Prog.* **1995**, *14*, 88–103. [CrossRef]
22. Verlicchi, P.; al Aukidy, M.; Zambello, E. Occurrence of pharmaceutical compounds in urban wastewater: Removal, mass load and environmental risk after a secondary treatment-A review. *Sci. Total Environ.* **2012**, *429*, 123–155. [CrossRef]
23. Baird, C.; Cann, M. *Enviromental Chemistry*, 5th ed.; W. H. Freeman and Company: New York, NY, USA, 2012.
24. Wen, Z.H.; Chen, L.; Meng, X.Z.; Duan, Y.P.; Zhang, Z.S.; Zeng, E.Y. Occurrence and human health risk of wastewater-derived pharmaceuticals in a drinking water source for Shanghai, East China. *Sci. Total Environ.* **2014**, *490*, 987–993. [CrossRef]

25. Kasuga, T.; Hiramatsu, M.; Hoson, A.; Sekino, T.; Niihara, K. Formation of Titanium Oxide Nanotube. *Langmuir* **1998**, *14*, 3160–3163. [CrossRef]
26. Idakiev, V.; Yuan, Z.-Y.; Tabakova, T.; Su, B.-L. Titanium oxide nanotubes as supports of nano-sized gold catalysts for low temperature water-gas shift reaction. *Appl. Catal. A Gen.* **2005**, *281*, 149–155. [CrossRef]
27. Ou, H.H.; Lo, S.L. Review of titania nanotubes synthesized via the hydrothermal treatment: Fabrication, modification, and application. *Sep. Purif. Technol.* **2007**, *58*, 179–191. [CrossRef]
28. Guo, Y.; Lee, N.H.; Oh, H.J.; Yoon, C.R.; Park, K.S.; Lee, W.H.; Li, Y.; Lee, H.G.; Lee, K.S.; Kim, S.J. Preparation of titanate nanotube thin film using hydrothermal method. *Thin Solid Films* **2008**, *516*, 8363–8371. [CrossRef]
29. Li, H.L.; Luo, W.L.; Tian, W.Y.; Chen, T.; Li, C.; Sun, M.; Zhu, D.; Liu, R.R.; Zhao, Y.L.; Liu, C.L. Fabrication and Photocatalytic Activity of Pt-Inserted Titania Nanotubes. *Spectrosc. Spectr. Anal.* **2009**, *29*, 1623–1626. [CrossRef]
30. Zavala, M.L.; Morales, S.L. Synthesis of stable TiO_2 nanotubes: Effect of hydrothermal treatment, acid washing and annealing temperature. *Heliyon* **2017**, *3*, e00456. [CrossRef] [PubMed]
31. O'Shea, K.E.; Pernas, E.; Saiers, J. Influence of mineralization products on the coagulation of TiO_2 photocatalyst. *Langmuir* **1999**, *15*, 2071–2076. [CrossRef]
32. Yang, L.; Yu, L.E.; Ray, M.B. Photocatalytic oxidation of paracetamol: Dominant reactants, intermediates, and reaction mechanisms. *Environ. Sci. Technol.* **2009**, *43*, 460–465. [CrossRef]
33. Zavala, M.Á.L.; Lara, C.R.J. Degradation of Paracetamol and Its Oxidation Products in Surface Water by Electrochemical Oxidation. *Environ. Eng. Sci.* **2018**, *35*, 1248–1254. [CrossRef]
34. Zavala, M.Á.L.; Estrada, E.E. Degradation of acetaminophen and its transformation products in aqueous solutions by using an electrochemical oxidation cell with stainless steel electrodes. *Water* **2016**, *8*, 383. [CrossRef]
35. Metcalf & Eddy, Inc.; Tchobanoglous, G.; Burton, F.L.; Stensel, H.D. Fundamentals of Biological Treatment. In *Wastewater Engineering Treatment Reuse*; McGraw-Hill: New York, NY, USA, 2003; pp. 611–615, 619–621.
36. Zhang, X.; Wu, F.; Wu, X.; Chen, P.; Deng, N. Photodegradation of acetaminophen in TiO_2 suspended solution. *J. Hazard. Mater.* **2008**, *157*, 300–307. [CrossRef]
37. Kumar, K.V.; Porkodi, K.; Rocha, F. Langmuir-Hinshelwood kinetics-A theoretical study. *Catal. Commun.* **2008**, *9*, 82–84. [CrossRef]
38. Ollis, D.F. Kinetics of liquid phase photocatalyzed reactions: An illuminating approach. *J. Phys. Chem. B* **2005**, *109*, 2439–2444. [CrossRef]
39. El-shahawy, A. DFT Cancer Energy Barrier and Spectral Studies of Aspirin, Paracetamol and Some Analogues. *Comput. Chem.* **2014**, *2*, 6–17. [CrossRef]
40. Shourong, Z.; Qingguo, H.; Jun, Z.; Bingkun, W. A study on dye photoremoval in TiO_2 suspension solution. *J. Photochem. Photobiol. A Chem.* **1997**, *108*, 235–238. [CrossRef]
41. Kulkarni, M.; Flašker, A.; Lokar, A.; Mrak-Poljšak, M.; Mazare, A.; Artenjak, A.; Čučnik, S.; Kralj, S.; Velikonja, A.; Schmuki, P.; et al. Binding of plasma proteins to titanium dioxide nanotubes with different diameters. *Int. J. Nanomed.* **2015**, *10*, 1359–1373. [CrossRef]
42. Borges, M.; García, D.; Hernández, T.; Ruiz-Morales, J.; Esparza, P. Supported Photocatalyst for Removal of Emerging Contaminants from Wastewater in a Continuous Packed-Bed Photoreactor Configuration. *Catalysts* **2015**, *5*, 77–87. [CrossRef]

processes

Article

Degradation of Aqueous Polycyclic Musk Tonalide by Ultraviolet-Activated Free Chlorine

Lili Wang [1,*] and Xiaowei Liu [2,3,*]

[1] Environmental Engineering, Jiyang College of Zhejiang A & F University, Zhuji 311800, China
[2] Institute of Water Resources & Ocean Engineering, Ocean College, Zhejiang University, Hangzhou 310058, China
[3] Institute of Municipal Engineering, College of Civil Engineering and Architecture, Zhejiang University, Hangzhou 310058, China
* Correspondence: liliwang@zafu.edu.cn (L.W.); liuxiaowei@zju.edu.cn (X.L.); Tel.: +86-571-88208721 (L.W.)

Received: 6 December 2018; Accepted: 7 February 2019; Published: 14 February 2019

Abstract: Chlorine-incorporating ultraviolet (UV) provides a multiple barrier for drinking water disinfection. Meanwhile, post-UV employment can promote the degradation of micropollutants by radical production from chlorine residual photolysis. This work studied the degradation of one such chemical, tonalide (AHTN), by low-pressure UV-activated free chlorine (FC) under typical UV disinfection dosage of <200 $mJ \cdot cm^{-2}$ and water matrix of filtered tank effluent. AHTN was rapidly degraded by UV/FC in accordance with pseudo-first-order kinetics. The reaction rate constants of AHTN with reactive chlorine species and hydroxyl radical (HO•) were estimated. Mechanistic exploration evidenced that under UV/FC, AHTN degradation was attributable to direct photolysis, ClO•, and HO•. The carbonyl side chain of AHTN served as an important attack site for radicals. Water matrices, such as natural organic matter (NOM), HCO_3^-, Cu^{2+}, PO_4^{3-}, and Fe^{2+}, showed noticeable influence on the UV/FC process with an order of NOM > HCO_3^- > Cu^{2+} > PO_4^{3-} > Fe^{2+}. Reaction product analysis showed ignorable formation of chlorinated intermediates and disinfection byproducts.

Keywords: polycyclic musks; degradation mechanism; UV/chlorine advanced oxidation process; water treatment

1. Introduction

Polycyclic musks (PCMs), as fragrance ingredients, have been extensively used in cosmetics, household cleaning products, and personal care products. Typical PCMs include tonalide (AHTN) and galaxolide (HHCB), which currently compose 85% of the total produced synthetic musk. These chemicals exist in various media, such as drinking water sources, owing to their hydrophobic characteristics, poor biodegradability, and frequent use. The adverse environmental effects of AHTN and HHCB on organisms have been reported [1–3]. Thus, AHTN and HHCB can pose a challenge to the health of consumers if they cannot be effectively intercepted by drinking water treatment processes (DWTPs). Such trepidation was confirmed by the survey of Stackelber et al. [4], who reported that typical DWTP through clarification, granular-activated-carbon filtration, and chlorine disinfection failed to comprehensively remove AHTN (~71.4%). Other studies have attempted to use chemical oxidation methods, such as ozonation and ferrate (VI) oxidation, to degrade AHTN; nonetheless, unsatisfied degradation was observed [5,6]. Similarly, ultrafiltration showed poor rejection of aqueous AHTN [7]. UV photolysis was proven efficient to remove AHTN and HHCB; however, considerable degradation intermediates with remarkable similarities to the structure of parent molecules were generated [8]. Therefore, additional efforts to control PCMs in potable water may be needed.

Advanced oxidation processes (AOPs) are effective methods for degrading and detoxifying aqueous contaminants by maximizing highly reactive radicals [9,10]. Combining medium-pressure UV with H_2O_2, a kind of AOP, was reported to show desirable degradation efficiency toward PCMs [11]. This hydroxyl radical (HO•)-dominated process showed a rate constant of 1.58 min^{-1} for HHCB degradation. The UV irradiation of free chlorine (UV/FC), as a novel AOP, has demonstrated its effectiveness in the attenuation of specific personal care products, taste and odor compounds, and antibiotics [9]. This process achieves contaminant degradation by three possible pathways: (1) direct reaction with FC, (2) direct photolysis, and (3) transformation mediated by radicals, such as HO• and/or reactive chlorine species (RCS, Cl•, $Cl_2^{\bullet -}$, and ClO•). Nonetheless, no other studies reported using UV/FC to degrade PCMs. Moreover, possible factors that may affect the degradation efficiency (i.e., pH, cations, and natural organic matter (NOM)) have not been investigated thus far.

In the current study, we aimed to (1) investigate the degradation efficiencies of AHTN by UV/FC; (2) to identify the primary contributor responsible for AHTN degradation and intermediate formation; (3) to obtain information on the toxicity profile accompanied by AHTN degradation; and (4) to evaluate the influence of relevant parameters of water-plant treated water, namely, NOM; common anions (Cl^-, NO_3^-, SO_4^{2-}, PO_4^{3-}, and HCO_3^-); and cations (Ca^{2+}, Fe^{3+}, Fe^{2+}, Mn^{2+}, Zn^{2+}, and Cu^{2+}). Distinguished from the high UV and chlorine doses delivered (>200 $mJ{\cdot}cm^{-2}$ for UV-C dosage; 5–10 $mg{\cdot}L^{-1}$ for chlorine dosage) in previous works [12–14], low chlorine concentration (<4.0 $mg{\cdot}L^{-1}$ as FC) and typical UV doses (product of fluence rate and exposure time, 40–100 $mJ{\cdot}cm^{-2}$ [15]) were adopted herein. AHTN was more retainable in potable water than HHCB, and thus, it was selected as the model compound to test the degradation characteristics of PCMs by UV/FC.

2. Materials and Methods

2.1. Materials

High purity AHTN, nitrobenzene (NB), 5,5-dimethyl-1-pyrroline N-oxide (DMPO), and benzoate (BA) with high purity were obtained from J&K Scientific (Beijing, China). XAD-4 and XAD-8 Amberlite resins were purchased from Sigma-Aldrich (San Francisco, CA, USA) and used to extract NOM in water plant-treated water. A total of 13 halogenated organic standards (chloroform, 1,1,1-trichloroethane, 1,1,2-trichloroethane, 1,1-dichloro-2-propanone, 1,1,1-trichloro-2-propanone, 1,2,3-trichloropropane, carbon tetrachloride, trichloroethylene, tetrachloroethylene, chloralhydrate, monochloroacetic acid, dichloroacetic acid, and trichloroacetic acid) were ordered from Center of National Standard Reference Material of China (Shanghai, China). High-performance liquid chromatography (HPLC)-grade acetonitrile and methyl tertiary-butyl ether (MTBE) were obtained from Merck (Darmstadt, Germany). H_3BO_3, $Na_2B_4O_7$, $NaNO_2$, NaCl, $CaCl_2$, $MnCl_2$, $ZnCl_2$, $FeCl_3$, $CuCl_2$, $NaHCO_3$, Na_3PO_4, and Na_2SO_4 were all analytical-reagent grade and traceable to Sinopharm Chemical Reagent Co., Ltd. (Shanghai, China). Sodium hypochlorite (NaOCl, 5%) was obtained from Aladdin (Shanghai, China) and calibrated by N,N-diethyl-p-phenylenediamine method. Ultrapure water (18.2 $M\Omega{\cdot}cm$) was used to prepare solutions. Three water plant-treated water samples were collected from drinking water plants of Hang Zhou (Zhejiang Province, China). The water samples were allowed to stand overnight to remove residual chlorine and stored at 4 °C before use.

NOM was extracted from the filtered tank effluent of Hangzhou Jiuxi drinking water plant. Amberlite resins were employed in NOM isolation, and NOM fractionation was achieved in accordance with previous works [16,17]. Hydrophobic fraction was collected by XAD-8 resin and XAD-4 resin. These two fractions were mixed for subsequent experiments.

2.2. Experimental Procedures

The experiment was performed at 25 °C in a 1.5-L glass reactor, which was wrapped with aluminum foil to block light. Figure S1 presents the set-up and construction of photoreactor. AHTN (1 $mg{\cdot}L^{-1}$) was initially prepared with ultrapure or real water, and pH was adjusted using

10 mM borate buffer. HCl and NaOH were used to adjust the pH for investigating the effects of cations. NaOCl stock solution (200 mg·L^{-1}) was then added to generate an initial concentration of 3.28 mg·L^{-1}. The mixture was vigorously stirred with a magnetic stirrer during the whole reaction process. The temperature of the reaction slurry was maintained at the set point with water bath. The UV lamp was lit for 5 min to obtain a stable UV output (Figure S2). Timing was started as soon as the UV light baffle was opened. Residual chlorine in samples was quenched by $NaNO_2$ before HPLC analysis. Each degradation experiment was performed in triplicate under identical conditions, and results were presented as the averages. Error bars represented the standard deviation of means (n = 3).

2.3. Determination of UV Fluence Rate and UV Dosage

The fluence rate in the photoreactor was determined using atrazine as actinometer [18]; a photon flux of 6.3×10^{-8} Einstein·s^{-1} was obtained. Average UV fluence rate (mW·L^{-1}) was calculated afterward. The photoreactor was non-standard collimated beam reactor. The UV light was casted to a cylindrical surface and not the horizontal one. The area of light projection varied with distance from light source. Thus, transforming the volume-averaged UV fluence rate into area-averaged UV fluence rate (mW·cm^{-2}), which was frequently used, presented difficulty. For comparisons with other UV-based processes, we assumed that the UV light projected on the cylindrical surface at a position of the effective light path length (L, cm), and a value of 0.067 mW·cm^{-2} was obtained. This value approximated the area-averaged UV fluence rate rather than the exact UV light distribution. UV fluence rates in presence of acetone, isopropanol, and NOM were calibrated in the same manner.

The effective path length of UV light (L = 7.03 cm) in the reactor was determined on the basis of H_2O_2 photolysis at a low concentration (0.1 mM, Figure S3). A detailed description is shown in Section S1 of the Supporting Information.

2.4. Analysis

AHTN was quantified by an Agilent 1200 HPLC (Agilent, Palo Alto, CA, USA). Separation was performed with an Agilent Eclipse XDB-C18 column (5 μm, 4.6 mm × 150 mm) at 30 °C. The mobile phase consisted of 90% acetonitrile and 10% H_2O, and had a flow rate of 1 mL·min^{-1}. Detection wavelength was set at 253 nm. A total of 10 μL of sample injection was employed.

To detect the possible presence of organic chlorinated products, we drew samples from the reactor (1 mg·L^{-1} AHTN initially in pH 7 borate buffer) after treatment by UV/FC (FC = 3.28 mg·L^{-1}) at different time points. We then added a dechlorination agent, i.e., sodium thiosulfate, to consume residual FC. Chlorinated products were analyzed by a Thermo Scientific TRACE 1300 gas chromatography (Thermo Fisher, Waltham, MA, USA). A total of 13 chlorinated byproducts (CBPs) were analyzed. The nonpolar chlorinated products were directly extracted with MTBE (50 mL samples with the addition of 2 mL MTBE). The polar chlorinated products were initially derivatized with methanol. A HP-5MS capillary column (30 m × 250 mm × 0.25 mm) was used for separating CBPs. The linearity (by R^2) of calibration data (0.1–40 μg·L^{-1}) was higher than 0.999. Table S1 provides the method detection limits (MDLs). Qualitative analysis was performed by matching retention times of samples with those of commercial standards. For comprehensive detection of degradation intermediates, samples were also concentrated on a Gilson GX-271 ASPEC apparatus (Gilson, Middleton, WI, USA) and qualified with an Agilent 6460 triple-quad HPLC-MS (Agilent, Palo Alto, CA, USA). The HPLC-mass spectrometer was equipped with an Agilent EclipseXDB-C18 column (5 μm, 4.6 mm × 150 mm), and column temperature was set at 40 °C. Parameter settings for the mass spectrometry (MS) were negative ion mode with a gas flow rate of 5 L·min^{-1} at 325 °C, a nebulizer pressure of 45 psi, sheath gas flow at 11 L·min^{-1} at 350 °C, a nozzle voltage of 0 or 500 V(+), a capillary voltage of 3000 V(+)/3500 V(−), and a fragmentor voltage of 135 V. The mobile phase for the HPLC-MS analysis was a 70/30 (v/v) mixture of 10 mM ammonium acetate with acetonitrile. The equipment was run at 1.0 mL·min^{-1}.

Inorganic chlorine species (Cl^-, ClO_2^-, ClO_3^-, and ClO_4^-), NO_3^-, and SO_4^{2-} were monitored using a Dionex ICS-2000 ion chromatograph (Chameleon 6.8, Sunnyvale, CA, USA) equipped with a Dionex IonPac AS19 analytical column (250 mm × 4 mm). An EluGen EGC-KOH cartridge and a continuously regenerated anion trap column (CR-ATC) were used. All analytes were detected by suppressed conductivity with an ASRS ULTRA II (4 mM) self-regenerating suppressor operating at 130 mA current.

Qualitative analysis of HO• was realized through a Bruker A200 electron paramagnetic resonance (EPR) 300E instrument (Bruker, Karlsruhe, Germany). The EPR spectrometer settings in the spin trapping experiments were as follows: center field, 351.194 mT; sweep width, 10.00 mT; modulation amplitude, 0.1 mT; sweep time, 41 s; microwave frequency, 9.858 GHz; microwave power, 2.25 mW; and receiver gain, 1.42×10^4.

Typical water-quality indexes were measured for the three collected effluent samples of waterworks filter. Total organic carbon (TOC) was determined via a Shimadzu TOC analyzer (Shimadzu, Kyoto, Japan). The concentrations of cations (Ca^{2+}, Mn^{2+}, Cu^{2+}, and total Fe) were determined using a PerkinElmer NexION 350Q ICP-MS Spectrometer (PerkinElmer, Shelton, CT, USA). UV absorbance at 254 nm (UV_{254}) was determined with a Shimadzu UV-250 spectrophotometer (Shimadzu, Kyoto, Japan). The pH was determined using an Orion 3-Star pH meter (Thermo Fisher, Shanghai, China). Bicarbonate was detected using chemical titration with a standard HCl solution. Phosphate concentration was determined using the molybdenum blue method.

2.5. Contributions of Different Radicals

Competitive kinetics methods were used to determine the second-order reaction rate constants of AHTN with HO• (k(HO• + AHTN)) and RCS (k(RCS + AHTN), RCS = Cl•, $Cl_2^{\bullet-}$, and ClO•). Table 1 summarizes the primary chemical reactions in the UV/FC system and rate constants. NB was selected as the probe compound to evaluate k(HO• + AHTN) and the steady-state concentration of HO• ($[HO\bullet]_{ss}$). The HO•-dominated system was generated by peroxymonosulfate activation using KOH while controlling the pH at 11, as most sulfate radicals were converted into HO• at such pH [19]. $[Cl\bullet]_{ss}$ was determined using both BA and NB as the probe compounds. 2,5-Dimethoxybenzoate (DMBA) was used as a probe compound for k(ClO• + AHTN) evaluation. Detailed information is provided in Supporting Information (Section S2, Figures S4–S6). k(HO• + AHTN) and k(ClO• + AHTN) measured 8.3×10^9 and 6.3×10^9 $M^{-1} \cdot s^{-1}$, respectively. $[HO\bullet]_{ss}$, $[Cl\bullet]_{ss}$, and $[ClO\bullet]_{ss}$ reached 2.6×10^{-14}, 2.8×10^{-15}, and 7.0×10^{-14} M, respectively. The pseudo first-order rate constants of AHTN with different reactive species (k'(R + AHTN), R = HO•, Cl•, $Cl_2^{\bullet-}$, and ClO•) can be calculated using Equations (1)–(5). Relative contributions of reactive species (G_R) can be calculated using Equations (6) and (7).

$$k'(\text{HO}\bullet + \text{AHTN}) = k(\text{HO}\bullet + \text{AHTN})[\text{HO}\bullet]_{ss} \tag{1}$$

$$k'(\text{Cl}\bullet + \text{AHTN}) = k(\text{Cl}\bullet + \text{AHTN})[\text{Cl}\bullet]_{ss} \tag{2}$$

$$k'\left(\text{Cl}_2^{\bullet-} + \text{AHTN}\right) = k\left(\text{Cl}_2^{\bullet-} + \text{AHTN}\right)\left[\text{Cl}_2^{\bullet-}\right]_{ss} \tag{3}$$

$$k'(\text{ClO}\bullet + \text{AHTN}) = k(\text{ClO}\bullet + \text{AHTN})[\text{ClO}\bullet]_{ss} \tag{4}$$

$$k'(\text{RCSs} + \text{AHTN}) = k'(\text{Cl}\bullet + \text{AHTN}) + k'\left(\text{Cl}_2^{\bullet-} + \text{AHTN}\right) + k'(\text{ClO}\bullet + \text{AHTN}) \tag{5}$$

$$k'_{\text{Total}} = k'(\text{HO}\bullet + \text{AHTN}) + k'(\text{RCSs} + \text{AHTN}) + k'_{\text{UV}} + k'_{\text{FC}} \tag{6}$$

$$G_R = k'(\text{R} + \text{AHTN})/k'_{\text{Total}} \tag{7}$$

Table 1. Principal reactions in the ultraviolet irradiation of free chlorine (UV/FC) system.

Equations	Reaction	Rate Constant	Reference
(I)	$HOCl \rightleftharpoons OCl^- + H^+$	$pK_{a1} = 7.5$	[20]
(II)	$HOCl + h\nu \rightarrow HO\bullet + Cl\bullet$	$\Phi_{HO\bullet} = 1.45$	[21]
(III)	$OCl^- + h\nu \rightarrow O^{\bullet-} + Cl\bullet$	$\Phi_{OCl-} = 0.97$	[20]
(IV)	$OCl^- + h\nu \rightarrow O(^1D) + Cl^-$	$\Phi_{O(1D)} = 0.133$	[10]
(V)	$OCl^- + h\nu \rightarrow O(^3P) + Cl^-$	$\Phi_{O(3P)\bullet} = 0.074$	[10]
(VI)	$HO\bullet \rightarrow O^{\bullet-} + H^+$	$pK_{a1} = 11.9$	[20]
(VII)	$HO\bullet + HOCl \rightarrow ClO\bullet + H_2O$	$k_7 = 2.0 \times 10^9$ $M^{-1}{\cdot}s^{-1}$	[20]
(VIII)	$HO\bullet + OCl^- \rightarrow ClO\bullet + OH^-$	$k_8 = 8.8 \times 10^9$ $M^{-1}{\cdot}s^{-1}$	[20]
(IX)	$HO\bullet + Cl^- \rightarrow ClOH^{\bullet-}$	$k_9 = 4.3 \times 10^9$ $M^{-1}{\cdot}s^{-1}$	[10]
(X)	$HO\bullet + OH^- \rightarrow O^{\bullet-} + H_2O$	$k_{10} = 1.3 \times 10^{10}$ $M^{-1}{\cdot}s^{-1}$	[20]
(XI)	$Cl\bullet + HOCl \rightarrow ClO\bullet + H^+ + Cl^-$	$k_{11} = 3.0 \times 10^9$ $M^{-1}{\cdot}s^{-1}$	[20]
(XII)	$Cl\bullet + OCl^- \rightarrow ClO\bullet + Cl^-$	$k_{12} = 8.2 \times 10^9$ $M^{-1}{\cdot}s^{-1}$	[22]
(XIII)	$Cl\bullet + Cl^- \rightarrow Cl_2^{\bullet-}$	$k_{13} = 6.5 \times 10^9$ $M^{-1}{\cdot}s^{-1}$	[23]
(XIV)	$Cl\bullet + H_2O \rightarrow ClOH^{\bullet-} + H^+$	$k_{14} = 4.5 \times 10^3$ $M^{-1}{\cdot}s^{-1}$	[24]
(XV)	$O^{\bullet-} + H_2O \rightarrow HO\bullet + OH^-$	$k_{15} = 1.8 \times 10^6$ $M^{-1}{\cdot}s^{-1}$	[25]
(XVI)	$O(^1D) + H_2O \rightarrow 2\,HO\bullet$	$k_{16} = 1.2 \times 10^{11}$ $M^{-1}{\cdot}s^{-1}$	[26]
(XVII)	$O(^3P) + O_2 \rightarrow O_3$	$k_{17} = 4.0 \times 10^9$ $M^{-1}{\cdot}s^{-1}$	[27]
(XVIII)	$O(^3P) + OCl^- \rightarrow ClO_2^-$	$k_{18} = 9.4 \times 10^9$ $M^{-1}{\cdot}s^{-1}$	[27]
(XIX)	$Cl_2^{\bullet-} \rightarrow Cl\bullet + Cl^-$	$K_{19} = 1.1 \times 10^5$ $M^{-1}{\cdot}s^{-1}$	[28]
(XX)	$ClOH^{\bullet-} \rightarrow HO\bullet + Cl^-$	$k_{20} = 6.1 \times 10^9$ $M^{-1}{\cdot}s^{-1}$	[23]
(XXI)	$ClOH^{\bullet-} + Cl^- \rightarrow Cl_2^{\bullet-} + OH^-$	$k_{21} = 1.0 \times 10^5$ $M^{-1}{\cdot}s^{-1}$	[29]
(XXII)	$ClOH^{\bullet-} + H^+ \rightarrow Cl\bullet + H_2O$	$K_{22} = 2.1 \times 10^{10}$ $M^{-1}{\cdot}s^{-1}$	[23]
(XXIII)	$2ClO\bullet + H_2O \rightarrow HOCl + H^+ + ClO_2^-$	$k_{23} = 2.5 \times 10^9 \cdot s^{-1}$	[23]
(XXIV)	$HO\bullet + CO_3^{2-} \rightarrow CO_3^{\bullet-} + OH^-$	$k_{24} = 3.9 \times 10^8$ $M^{-1}{\cdot}s^{-1}$	[30]
(XXV)	$HO\bullet + HCO_3^- \rightarrow CO_3^{\bullet-} + H_2O$	$k_{25} = 8.6 \times 10^6$ $M^{-1}{\cdot}s^{-1}$	[25]
(XXVI)	$HO\bullet + ClO_2^- \rightarrow ClO_2^{\bullet-} + OH^-$	$k_{26} = 6.3 \times 10^9$ $M^{-1}{\cdot}s^{-1}$	[31]
(XXVII)	$HO\bullet + ClO_2^{\bullet-} \rightarrow ClO_3^- + H^+$	$k_{27} = 4.0 \times 10^9$ $M^{-1}{\cdot}s^{-1}$	[31]

2.6. Toxicity Evaluation of Samples

Toxicity was evaluated using 1 L of samples of the reaction solutions. A Microtox Model 500 toxicity analyzer coupled with luminescent bacteria *Vibrio fisheri* was used. Samples were examined in quartz tube containing 2% sodium chloride in three dilutions. A toxic-free control experiment was conducted in three repeats using 2% sodium chloride, 3.5 mg·L^{-1} FC, 0.022 mM sodium thiosulfate, and 10 mM borate buffer. Luminescence was recorded after 15 min of incubation at 15 °C. The percentage of luminescence inhibition was recorded. Samples were concentrated by the freeze-drying method. Recovery for freeze-dried samples totaled from 95% to ~110%. The detoxification rate is defined in Equation (8).

$$\text{Detoxification rate } (\%) = (L_t - L_0)/L_0 \tag{8}$$

where L_0 refers to the initial loss rate of light emission and L_t denotes the loss rate of light emission at reaction time t (min).

2.7. Principal Factor Analysis

We used CANOCO for Windows package (version 4.5, Ter Braak & Smilauer, Wageningen, the Netherlands) to execute the principal component analysis. We initially performed detrended correspondence analysis (DCA) on the efficiency variance to determine the length of the ordination gradient. The length of the gradient along the first axis was 0.545 (<3.0); therefore, redundancy analysis (RDA) was performed. The degradation efficiency in UV/FC was expressed as response variables. Water quality parameters were set as environmental variables. The data were log (x + 1) transformed.

3. Results and Discussion

3.1. AHTN Degradation under UV/FC Conditions

AHTN decomposition in ultrapure water by UV, FC, and UV/FC were compared (Figure 1). The UV/FC rapidly degraded AHTN by 60–90% for 60 mJ·cm^{-2} at a FC dosage of 0.21–3.28 mg·L^{-1}. Direct UV photolysis showed a moderate degradation rate of 49.3% at 60 mJ·cm^{-2}. AHTN was resistant to FC alone. The measured molar absorbance of AHTN and quantum yield reached ε = 7911.7 M^{-1}·cm^{-1} and $\mathbf{\Phi}$ = 1.3 mol·Einstein^{-1}, respectively (Section S3 and Figures S8 and S9 in the Supporting Information). This finding illustrates the moderate degradation performance of UV photolysis. The reaction of AHTN with FC has been reported to initiate with a substitution reaction occurring at the acetyl side chain of AHTN (H substitution of α-carbon), successively followed by haloform reaction, decarboxylation, and methylation [32]. In the current study, the weak electron-withdrawing capability of carbonyl caused a slow reaction kinetics of FC with the methyl group of acetyl side chain. This result well explains the FC resistance of AHTN. The excellent degradation performance of UV/FC may be attributed to the formation of reactive species, such as HO•, Cl•, and $Cl_2^{\bullet-}$, which were generated from the UV photolysis of FC (Equations (I)–(XXII)).

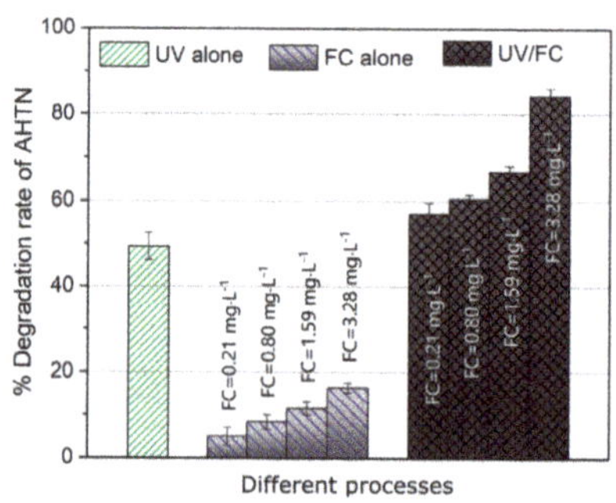

Figure 1. Degradation efficiency of tonalide (AHTN) by UV photolysis, chlorine, and UV/FC ([AHTN]$_0$ = 1.0 mg·L^{-1}, pH = 7.0, 25 ± 1 °C, and UV dose 60 mJ·cm^{-2}).

3.2. Identification of Reactive Species for AHTN Degradation by UV/FC

FC photolysis under irradiation at 253.7 nm mainly yielded HO•/$O^{\bullet-}$ and Cl• (Equations (I)–(VI)). $O^{\bullet-}$ was rapidly captured by H_2O to generate HO• under neutral pH condition (Equation (XV)) and exhibited low reactivity toward organic pollutants. Therefore, $O^{\bullet-}$ was usually considered to be less vital than HO• and Cl•. FC was added as NaOCl solution containing a small amount of Cl^-. Production of $Cl_2^{\bullet-}$ and $ClOH^{\bullet-}$ was expected via the reaction between Cl• or HO• with Cl^- (Equations (IX) and (XIII)). At neutral condition, $ClOH^{\bullet-}$ was unstable and rapidly decomposed into HO• (Equation (XX)). Both Cl• (2.4 V) and $Cl_2^{\bullet-}$ (2.0 V) are strong oxidants [20]. In the current study, another secondary radical, ClO•, was derived from the reaction of FC with HO•/Cl• (Equations (VII), (VIII), (XI), and (XII)). Sun et al. [24] observed that ClO• featured the same importance as HO• and Cl• in degrading caffeine by UV/FC. Therefore, the four radicals HO•, Cl•, $Cl_2^{\bullet-}$, and ClO• were suspected to contribute to AHTN degradation, in addition to direct photolysis and FC chlorination.

On the basis of the determined k(HO• + AHTN) and [HO•]$_{ss}$, k'(HO• + AHTN) was 2.1 × 10^{-4} s^{-1}. Given k'_{Total} (1.6 × 10^{-3} s^{-1}), k'_{UV} (7.0 × 10^{-4} s^{-1}), and k'_{FC} (1.0 × 10^{-4} s^{-1}) were known (Figure 2a), the relative contributions of HO• ($G_{HO\bullet}$) and RCSs (G_{RCSs}) could be initially determined. HO• and RCS, respectively, accounted for 13.1% and 36.9% of AHTN degradation compared with a G_{UV} value of 43.8% produced by direct UV photolysis. To ascertain the RCSs that played a major role, we calculated $G_{ClO\bullet}$ (30%) from the obtained k(ClO• + AHTN) (6.3 × 10^9 M^{-1}·s^{-1}) and [ClO•]$_{ss}$ (7.0 ×

10^{-14} M). Cl• and $Cl_2^{\bullet-}$ accounted for 6.9% of AHTN degradation. Figure 2b presents the contributions of relevant contributors.

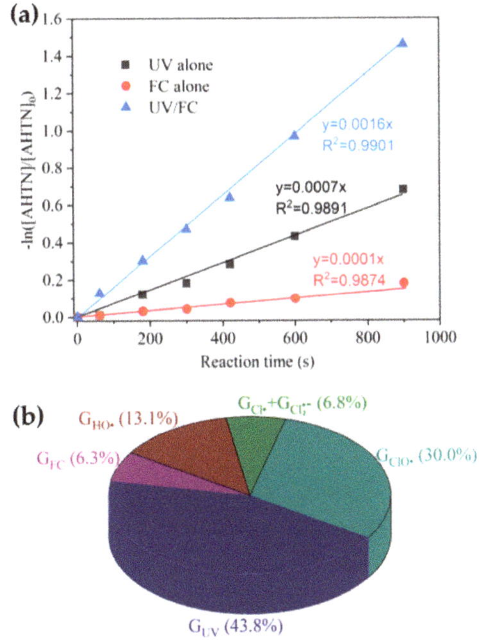

Figure 2. Degradation of AHTN by UV/FC—(**a**) first-order kinetic fitting and (**b**) contribution analysis of relevant contributors ($[AHTN]_0$ = 1.0 mg·L^{-1}; pH = 7.0; 25 ± 1 °C; $[FC]_0$ = 3.28 mg·L^{-1}; $[Cl^-]_0$ = 0.02 mM; and UV fluence rate 0.067 mW·cm^{-2}).

The results indicated that HO• and ClO• were the primary contributors among the four radicals, whereas Cl• and $Cl_2^{\bullet-}$ played minor roles. A direct support from EPR testing results confirmed the formation of HO• (Figure 3). When Cl• was generated from the photolysis of HOCl, Cl• was initially captured by H_2O molecules to form $ClOH^{\bullet-}$ at a rate of 2.5 × 10^5 s^{-1} (k(Cl• + H_2O) × [H_2O], Equation (XIV)). The initial concentration of Cl^- approximated 0.02 mM in the reaction system, corresponding to a Cl• scavenging rate of 1.3 × 10^5 s^{-1} (k(Cl• + Cl^-) × [Cl^-], Equation (XIII)). Furthermore, 3.28 mg·L^{-1} FC led to a Cl• scavenging rate of 2.0 × 10^5 s^{-1} (k(Cl•+ ClO^-) × [ClO^-] + k(Cl• + HClO) × [HClO], Equations (XI) and (XII)). Even if k(Cl• + AHTN) reached a level of ~10^{10} M^{-1}·s^{-1}, AHTN only led to a Cl• scavenging rate of ~10^4 s^{-1} (k(Cl• + AHTN) × [AHTN]). Therefore, under the conditions of the present work, Cl• was primarily captured by H_2O molecule, Cl^-, and FC, thereby leading to a low $[Cl\bullet]_{ss}$ concentration (2.8 × 10^{-15} M). Thus, $Cl_2^{\bullet-}$ may have contributed to 6.9% of AHTN degradation. An insignificant formation of organic chlorinated compounds in the investigation of chlorine balance (Figure 4) confirms the above speculation. In summary, the radical-induced AHTN elimination is primarily attributed to the attacking of HO• and ClO•. An identical conclusion was obtained in studies where caffeine was treated by UV/FC [24].

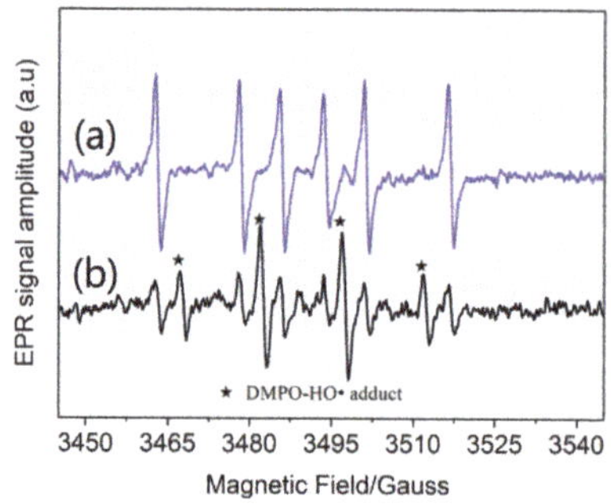

Figure 3. Derivative electron paramagnetic resonance (EPR) spectra of samples collected from (**a**) control blank (1 $g{\cdot}L^{-1}$ DMPO solution) and (**b**) UV/FC system ($[AHTN]_0$ = 1.0 $mg{\cdot}L^{-1}$; $[FC]_0$ = 3.28 $mg{\cdot}L^{-1}$; pH = 7.0; and 25 ± 1 °C).

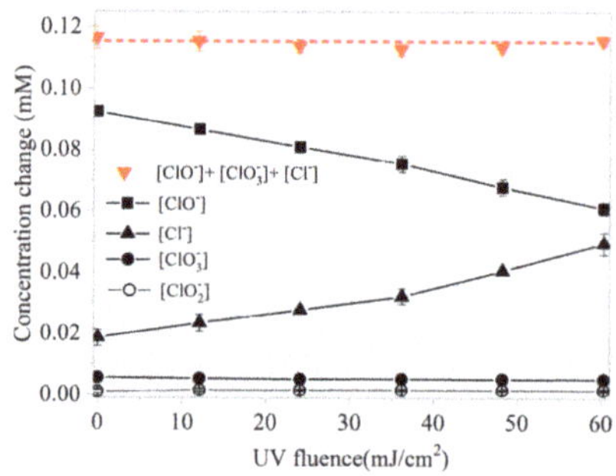

Figure 4. Mass balance of chlorine element during AHTN degradation by UV/FC ($[AHTN]_0$ = 1.0 $mg{\cdot}L^{-1}$; $[FC]_0$ = 3.28 $mg{\cdot}L^{-1}$; pH = 7.0; and 25 ± 1 °C).

3.3. *Toxicity Change, Intermediate Formation, and Degradation Pathway*

The toxicity change of treated water during target pollutant degradation must also be a concern due to the possible formation of intermediates with similar or even higher toxicity than their parent compounds [33,34]. Therefore, toxicity variation related to AHTN degradation by FC, UV, and UV/FC was examined, allowing the evaluation of the detoxification efficiency of UV/FC. As shown in Figure 5, almost no toxicity change was observed for AHTN solution when treated by FC and UV. By contrast, a weak increase in toxicity was detected for the solution treated by UV/FC. Such phenomena indicate the generation of products with higher toxicity than AHTN.

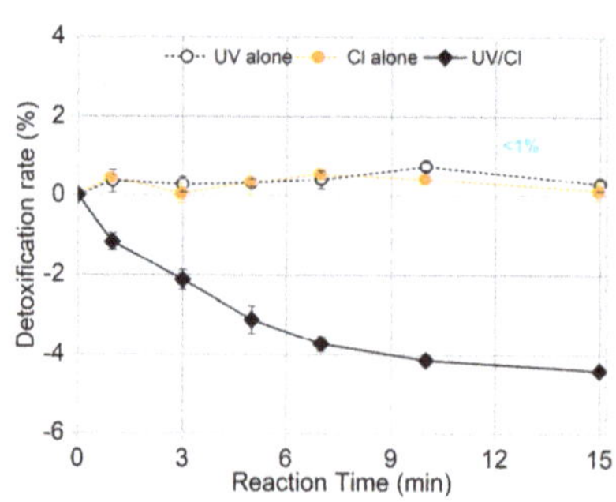

Figure 5. Toxicity evolution for AHTN degradation by UV, FC, and UV/FC ($[AHTN]_0$ = 1.0 $mg{\cdot}L^{-1}$; pH = 7.0; 25 ± 1 °C; and $[FC]_0$ = 3.28 $mg{\cdot}L^{-1}$).

To elucidate the observed toxicity elevation, we identified degradation intermediates. An intuitive image of intermediate formation was initially obtained from HPLC chromatogram (Figure S10). Three new peaks were detected with considerable abundance in the UV/FC system (Figure S10a). By contrast, a few intermediates were generated in UV (Figure S10b) and FC systems (Figure S10c). We performed LC-MS analysis to further qualify these intermediates. Figure 6 shows the mass spectrum and possible chemical structures of intermediates. Products (a), (c), (d), and (e) completely differed from those reported in AHTN chlorination [8,32] and UV photolysis [8]. The products were characterized by notable similarities to parent molecules. Such distinction in products confirms the difference in the major species that induced AHTN degradation. By combining the structural information of the products and the identification of contributors to AHTN degradation, a possible degradation pathway was proposed (Figure 6).

ClO• was reported as a powerful one-electron oxidant and non-reactive radical in hydrogen abstraction or addition reactions [35]. HO• can react with organic compounds in several different ways, such as C=C and C=N double bonds, H-atom abstraction, and electron transfer [36]. In the current study, the H-atom abstraction of AHTN resulted in the generation of carbon-centered radicals (Radical (I) and (II)). These carbon-centered species (Radical (I)) can react with dissolved oxygen to form peroxyl radicals, which can generate aldehydes (product (a)) through self-reaction and succeeding decomposition [37]. Another intermediate, with the same m/z value (244) as product (b), was observed during the UV photolysis of AHTN; this intermediate was proposed to be 6-ethyl-1,1,2,4,4,7-hexamethyltetralin [8]. The aldehyde functional group of product (a) was oxidized, whereas that of product (c) was subsequently decarboxylated. Radical (II) can undergo self-polymerization to form product (d). The appearance of product (e) would have been due to molecular branch trimming of product (b).

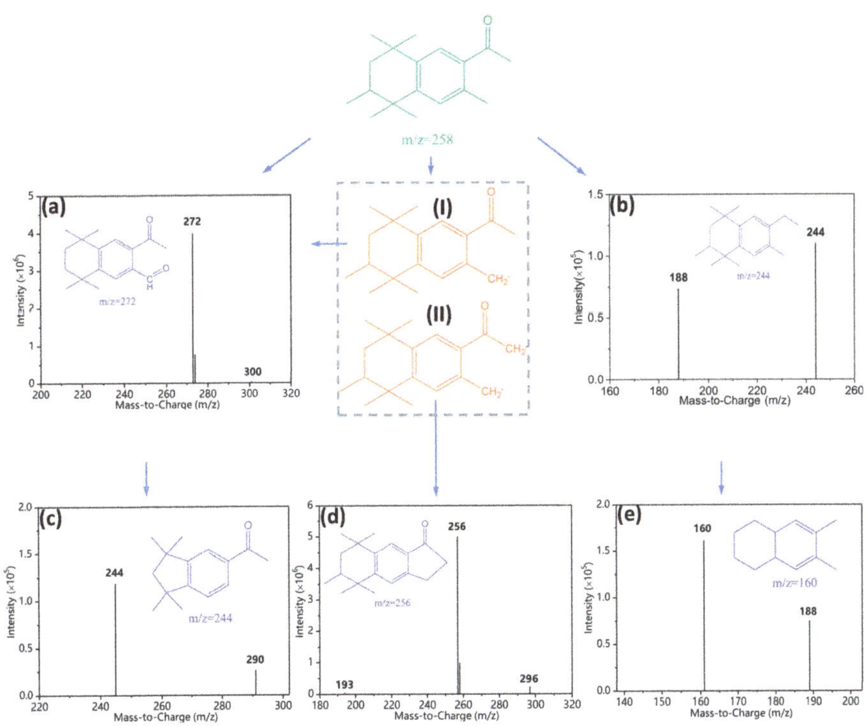

Figure 6. Degradation pathway of AHTN by UV/chlorine process ($[AHTN]_0$ = 1.0 mg·L^{-1}; pH = 7.0; 25 ± 1 °C; and $[FC]_0$ = 3.28 mg·L^{-1}).

To further confirm the formation of chlorinated intermediates, 13 CBPs were screened. The MDLs of the 13 CBPs ranged from 0.01 to 0.06 $\mu g{\cdot}L^{-1}$ (Table S1). Results revealed no targeted CBP at detectable concentrations in any of the samples. Notably, the decreased FC amount showed a linear correlation with an increased Cl^- level (k = 1.02, Figure 7). Thus, the final product of $HOCl/OCl^-$ was harmless Cl^-.

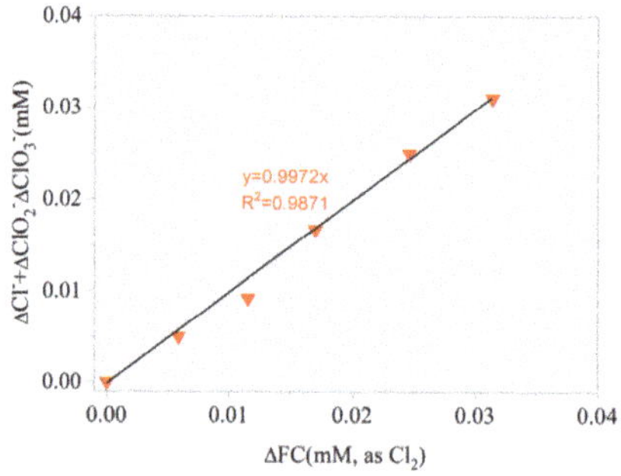

Figure 7. $\Delta[Cl^-] + \Delta[ClO_2^-] + \Delta[ClO_3^-]$ versus Δ[FC] during AHTN degradation by UV/FC ($[AHTN]_0$ = 1.0 $mg{\cdot}L^{-1}$; $[FC]_0$ = 3.28 $mg{\cdot}L^{-1}$; pH = 7.0; and 25 ± 1 °C).

3.4. Effect of Water Background on AHTN Degradation

The destruction of AHTN under actual water background may be more striking than the degradation performance in ultrapure water. Thus, AHTN degradation by UV/FC, conducted in FWs collected from local drinking water treatment plants, was investigated. Table 2 provides the water quality parameters of the three FWs. Figure 8 displays the degradation curves. From the obtained results, we can conclude that AHTN degradation in actual waters slowed down compared with the case of ultrapure water to a certain extent (~12%). Background components may have induced the degradation differentiation. To screen out water quality parameters that substantially influenced AHTN degradation, we individually studied the effects of common water quality parameters in ultrapure water, such as Ca^{2+}, Fe^{3+}, Fe^{2+}, Mn^{2+}, Zn^{2+}, Cu^{2+}, Cl^- NO_3^-, SO_4^{2-}, PO_4^{3-}, HCO_3^-, and NOM. The degradation of AHTN followed pseudo-first-order kinetics (Figure S11). Table 3 summarizes the rate constants.

Table 2. Water quality of the waters collected from filtered water (FW).

Water Matrices	Units	FW1#	FW2#	FW3#
pH	–	7.15	6.68	7.47
TOC	$mg{\cdot}L^{-1}$	1.93	4.00	8.77
HCO_3^-	$mg{\cdot}L^{-1}$	7.89	5.89	12.83
PO_4^{3-}	$mg{\cdot}L^{-1}$	ND	ND	ND
Cl^-	$mg{\cdot}L^{-1}$	15.71	8.91	34.69
NO_3^-	$mg{\cdot}L^{-1}$	9.39	8.49	3.90
UV_{254}	$cm^{-1}{\cdot}(mg{\cdot}L^{-1})^{-1}$	0.015	0.003	0.029
SO_4^{2-}	$mg{\cdot}L^{-1}$	18.33	11.35	22.41
Ca^{2+}	$mg{\cdot}L^{-1}$	7.22	9.52	10.36
Mn^{2+}	$mg{\cdot}L^{-1}$	0.012	0.007	0.006
Cu^{2+}	$mg{\cdot}L^{-1}$	0.076	0.001	ND
Zn^{2+}	$mg{\cdot}L^{-1}$	1.23	0.034	0.022
Total Fe	$mg{\cdot}L^{-1}$	0.16	0.14	0.061

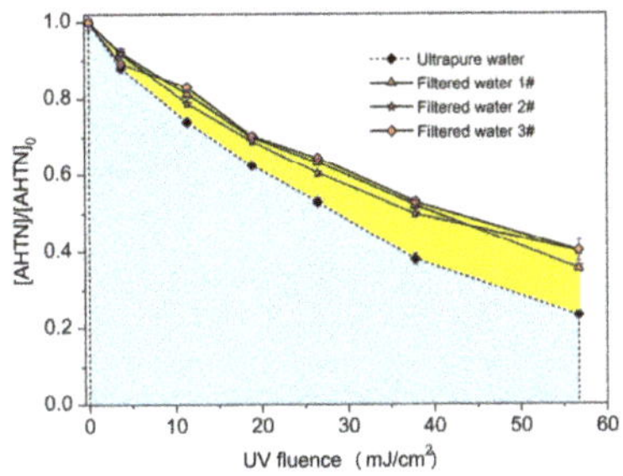

Figure 8. Degradation of AHTN by UV/FC under conditions relevant to tap water ($[AHTN]_0$ = 1.0 mg·L^{-1}; pH = 7.0; 25 ± 1 °C; and $[FC]_0$ = 3.28 mg·L^{-1}).

Table 3. Pseudo-first-order rate constants for AHTN degradation by UV/FC under various conditions.

Varying Parameters	Concentration	* Ionic Strength ($\times 10^{-5}$ M)	ΔpH	k (s^{-1})	R^2
Ca^{2+} (mM, as CaCl)	0.2	60	0.01	0.0014	0.997
	1.0	300	0.01	0.0013	0.997
	2.0	900	0.01	0.0014	0.999
Cl^- (mM, as NaCl)	0.5	50	0.01	0.0014	0.999
	1.0	100	0.01	0.0013	0.998
	4.0	400	0.01	0.0014	0.999
Cu^{2+} (μM, as $CuCl_2$)	0.6	0.18	0.01	0.0015	0.994
	1.8	0.54	0.01	0.0014	0.997
	3.0	0.9	0.01	0.0015	0.998
NOM (mg·L^{-1})	0.5	–	0	0.0012	0.999
	1.0	–	0	0.0011	0.998
	2.0	–	0	0.0010	0.999
	5.0	–	0	0.0007	0.997
Fe^{2+} (mM, as $FeCl_2$)	0.001	0.3	0.01	0.0015	0.998
	0.002	0.6	0.01	0.0014	0.998
	0.004	1.2	0.01	0.0014	0.999
Fe^{3+} (mM, as $FeCl_3$)	0.001	0.6	0.01	0.0015	0.999
	0.002	1.2	−0.04	0.0013	0.996
	0.004	2.4	0.01	0.0012	0.995
HCO_3^- (mM, as $NaHCO_3$)	0.1	10	0.01	0.0013	0.998
	0.5	50	0.01	0.0012	0.994
	1.0	100	0.01	0.0012	0.993
	2.0	200	0.01	0.0013	0.998
	4.0	400	0.01	0.0013	0.990
	10.0	1000	0.01	0.0009	0.996
Mn^{2+} (μM, as $MnCl_2$)	0.5	0.15	0.01	0.0015	0.999
	1.0	0.3	0.01	0.0014	0.998
	2.0	0.6	0.01	0.0014	0.999
NO_3^- (mM, as $NaNO_3$)	0.05	5	0.01	0.0015	0.996
	0.1	10	0.01	0.0012	0.998
	0.15	15	0.01	0.0013	0.999
	1.0	100	0.01	0.0013	0.999
PO_4^{3-} (μM, as Na_3PO_4)	0.1	0.06	0.01	0.0013	0.999
	0.2	0.12	0.01	0.0014	0.998
	0.4	0.24	0.01	0.0014	0.997
	1.0	0.6	0.01	0.0010	0.996
SO_4^{2-} (mM, as Na_2SO_4)	0.2	60	0.01	0.0013	0.996
	0.4	120	0.01	0.0014	0.998
	0.6	180	0.01	0.0012	0.997
	1.0	300	0.01	0.0010	0.994
Zn^{2+} (μM as $ZnCl_2$)	4.0	1.2	0.01	0.0013	0.998
	8.0	2.4	0.01	0.0014	0.998
	12.0	3.6	0.01	0.0014	0.999

* Ionic strength of borate buffer (54 mM) is not included.

As stated above, the direct photolysis and attack of HO• and ClO• caused AHTN degradation. Co-solutes, namely, Ca^{2+}, Fe^{3+}, Fe^{2+}, Mn^{2+}, Zn^{2+}, Cu^{2+}, Cl^-, NO_3^-, SO_4^{2-}, PO_4^{3-}, HCO_3^-, and NOM, may interfere with AHTN degradation via competition for photons and radicals. Table 4 lists the HO• quenching rates of these co-solutes. Information on the reactions between ClO• and these co-solutes is limited. Comparison of the redox potential of ClO• (1.5–1.8 V [35]) with those of co-solutes will allow speculation of ClO• scavenging by these compounds. Preliminarily, ClO• was assumed to be inert toward Ca^{2+}, Fe^{3+}, Zn^{2+}, Cu^{2+}, Cl^-, NO_3^-, SO_4^{2-}, PO_4^{3-}, and HCO_3^- [38,39]. Reaction kinetics between ClO• and Fe^{2+}/Mn^{2+} can be obtained from the rate constants of carbonate radical ($CO_3^{\bullet-}$) with Fe^{2+}/Mn^{2+} ($k(CO_3^{\bullet-} + Mn^{2+}) = 1.5 \times 10^7$ $M^{-1}{\cdot}s^{-1}$ and $k(CO_3^{\bullet-} + Fe^{2+}) = 3.6 \times 10^8$ $M^{-1}{\cdot}s^{-1}$) due to the close redox potentials of these two one-electron oxidants [35]. Regarding the scavenging of ClO• by NOM, a second-order rate constant of 4.5×10^4 $(mg{\cdot}L^{-1}\ C)^{-1}{\cdot}s^{-1}$ was reported [40]. Thus, for ClO•, quenching by Fe^{2+}/Mn^{2+} and NOM needs to be determined.

Table 4. Scavenging rate of hydroxyl radical (HO•) by AHTN and water matrices.

Water Matrices	Concentration (*C*)	Reaction Rate Constant with HO• (*k*)	Scavenging Rate (*C* × *k*)
AHTN	1.0 $mg{\cdot}L^{-1}$	8.3×10^9 $M^{-1}{\cdot}s^{-1}$ (this work)	3.2×10^4 s^{-1}
Cl^-	0.4–4 mM	1.4×10^9 $M^{-1}{\cdot}s^{-1}$ ([10])	$(0.6–5.6) \times 10^6$ s^{-1}
PO_4^{3-} *	0.1–1000 μM	1.5×10^5 $M^{-1}{\cdot}s^{-1}$ (HPO_4^{2-}, [25]) 2.0×10^4 $M^{-1}{\cdot}s^{-1}$ ($H_2PO_4^-$, [25])	(0.007–1.0) s^{-1}
NO_3^-	0.05–1 mM	$< 1.0 \times 10^5$ $M^{-1}{\cdot}s^{-1}$	$<(0.05–1.0) \times 10^2$ s^{-1}
HCO_3^- **	0.1–4 mM	8.6×10^6 $M^{-1}{\cdot}s^{-1}$ (HCO_3^-, [41]) $<1.0 \times 10^6$ $M^{-1}{\cdot}s^{-1}$ (H_2CO_3, [25])	$(0.07–2.9) \times 10^4$ s^{-1}
SO_4^{2-}	0.2–1 mM	6.9×10^5 $M^{-1}{\cdot}s^{-1}$ (HSO_4^{2-}, [25])	$\leq \sim 10^2$ s^{-1}
NOM	0.5–5 $mg{\cdot}L^{-1}$	2.5×10^4 $(mg{\cdot}L^{-1})^{-1}{\cdot}s^{-1}$ ([42])	$(1.3–13) \times 10^4$ s^{-1}
Cu^{2+}	0.6–3 μM	3.5×10^8 $M^{-1}{\cdot}s^{-1}$ ([25])	$(0.21–1.1) \times 10^3$ s^{-1}
Fe^{2+}	1–3 μM	3.2×10^8 $M^{-1}{\cdot}s^{-1}$ ([25])	$(3.2–9.6) \times 10^2$ s^{-1}
Fe^{3+}	1–4 μM	NA	
Mn^{2+}	0.5–2 μM	2.9×10^7 $M^{-1}{\cdot}s^{-1}$ ([25])	$(1.5–5.8) \times 10^1$ s^{-1}
Zn^{2+}	4–12 μM	$<5.0 \times 10^5$ $M^{-1}{\cdot}s^{-1}$ ([25])	<(2–6) s^{-1}
Ca^{2+}	0.2–3 mM	NA	
HOCl ***	46.2 μM	2.0×10^9 $M^{-1}{\cdot}s^{-1}$ (HOCl, [20]) 8.8×10^9 $M^{-1}{\cdot}s^{-1}$ (OCl^-, [20])	1.7×10^5 $M^{-1}{\cdot}s^{-1}$

* 61.7% HPO_4^{2-} and 38.3% $H_2PO_4^-$ at pH = 7; ** 81.7% HCO_3^- and 18.3% H_2CO_3 at pH = 7; and *** 76.0% HOCl and 24.0% OCl^- at pH = 7.

Addition of 0.4–4 mM Cl^- caused almost no influence on AHTN degradation (Table 2). Although Cl^- can scavenge HO• (a primary contributor for AHTN degradation) at a rate of $(0.6–5.6) \times 10^6$ s^{-1} (Table 3) to form $ClOH^{\bullet-}$ (Equation (IX)), $ClOH^{\bullet-}$ was unstable and rapidly decomposed into HO• and Cl^- (Equation (XX)). HO•-based AOPs were insensitive to low Cl^- concentration [43]. Thus, 0.4–4 mM Cl^- was not expected to exert considerable effect on AHTN degradation.

Similarly, Cl^-, PO_4^{3-}, SO_4^{2-}, and NO_3^- at an environmental concentration level almost caused no remarkable influence, as presented in Table 2. Approximately, 16% inhibition was obtained until the concentration of PO_4^{3-} increased to as high as 1 mM. The weak scavenging of HO• by PO_4^{3-}, SO_4^{2-}, and NO_3^- ($\leq \sim 10^2$ s^{-1}, Table 3) compared with that by AHTN (3.2×10^4 s^{-1}, Table 3) can reasonably explain this phenomenon.

HCO_3^- can capture HO• ($(0.09–8.6) \times 10^4$ s^{-1} for 0.1–10 mM HCO_3^-, Table 3) to yield $CO_3^{\bullet-}$, which is a weak oxidant compared with the former. AHTN captured HO• at a rate of 3.2×10^4 s^{-1}. Thus, a considerable proportion of HO• would transform into $CO_3^{\bullet-}$, once HCO_3^- concentration increased above a certain threshold. This partial HO• conversion may illustrate the deterioration of degradation efficiency at a higher concentration of HCO_3^- (Table 2).

As an important component of drinking water, the dissolved NOM often exerts noticeable effect on the chemical oxidation of pollutants. Table 2 shows the results on AHTN degradation in the presence of NOM. Overall, the degradation process is sensitive toward NOM. The presence of NOM at 0.5 mg·L^{-1} decreased the degradation efficiency by approximately 10%. NOM reacted with HO• and ClO• at rates of 2.5 × 10^4 and 4.5 × 10^4 (mg·L^{-1} C)$^{-1}$·s^{-1}, respectively. Thus, 0.5 mg·L^{-1} NOM showed a scavenging rate of 1.3 × 10^4 s^{-1} to HO• and 2.3 × 10^4 s^{-1} to ClO•. The results were comparable to the scavenging rate of HO• and ClO• by AHTN (Table 3). By analogy, 50% decrease in reaction rate constant occurred in the presence of 5.0 mg·L^{-1} NOM.

Regarding the presence of common cations, such as Ca^{2+}, Fe^{3+}, Fe^{2+}, Mn^{2+}, Zn^{2+}, and Cu^{2+}, at concentrations relevant to environmental levels, a subtle inhibition influence on the degradation process was observed (k = 0.0012–0.0015 s^{-1} and k_{CB} = 0.0016 s^{-1}, Table 2). The influence of solution pH change and ion strength change induced by the introduction of cations was excluded (Table 2). As shown in Figure 6, numerous carbonyl-containing intermediates were produced during the degradation process. These carbonyl-containing intermediates can complex with metal ions. Studies have reported the degradation enhancement of nitrophenolic compounds by HO•-based AOP due to Fe^{3+} complexation [44]. Variations in oxidation characteristic of pollutants evoked by complexation with cations have been proposed [45]. In other words, metal ion complexation with transformation intermediates may increase the competition with parent molecule toward reactive species, such as HO•. A similar influencing mechanism was speculated to operate herein.

Finally, RDA was performed to screen out principal factors from the above water quality parameters. RDA results indicated that 98.1% of the response variables (gray circles in Figure 9) were explained by the environmental variables. The length of the arrows in the ordination biplot (Figure 9) indicates the relationship strength of the environmental variable and the response variables. The correlation among the variables is positively related to the cosine value of the intersection angles between two arrows. In accordance with these two rules, the five NOM, HCO_3^-, Cu^{2+}, PO_4^{3-}, and Fe^{2+} considerably influenced the degradation efficiency in the order of NOM > HCO_3^- > Cu^{2+} > PO_4^{3-} > Fe^{2+}.

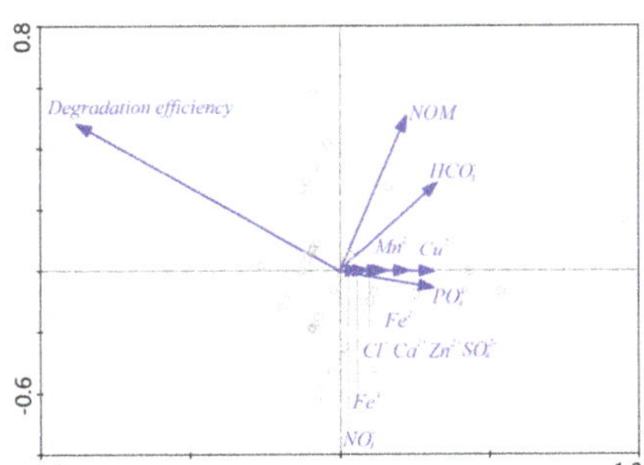

Figure 9. Redundancy analysis (RDA) analysis of influence of investigated water matrices.

4. Conclusions

UV/chlorine AOP degraded AHTN efficiently in accordance with pseudo first-order kinetics. The first-order rate constant in the UV/chlorine AOP were, respectively, 16 and 2.3 times higher than those in chlorination alone and direct UV photolysis under typical chlorine (3.28 mg·L^{-1}) and UV (60 mJ·cm^{-2}) dosage at pH 7. Among the common water matrix components, NOM, HCO_3^-, Cu^{2+}, PO_4^{3-}, and Fe^{2+} showed a noticeable influence on AHTN degradation by the UV/chlorine AOP. However, other co-solutes, namely, Ca^{2+}, Fe^{3+}, Mn^{2+}, Zn^{2+}, Cl^-, NO_3^-, and SO_4^{2-}, failed to show the same result. AHTN degradation in UV/chlorine AOP was induced by UV photolysis and attack of ClO• and HO•. Formation of chlorinated intermediates was irrelevant under current experimental conditions. Five chlorine-free intermediates were identified. After treatment with UV/chlorine AOP, the toxicity of AHTN mixture weakly increased.

Supplementary Materials: The following are available online at http://www.mdpi.com/2227-9717/7/2/95/s1, Section S1: Determination of effective optical path length. Section S2: Determination of k(HO• + AHTN) and k(RCS + AHTN). Section S3: Determination of $\varepsilon_{AHTN,254}$ and $\Phi_{AHTN,254}$. Figure S1: Set-up and construction of photoreactor; Figure S2: Energy output profile of UV lamp. Figure S3: UV Photolysis of H_2O_2. Figure S4: Degradation of AHTN and NB by (a) PMS oxidation in presence of 10 μM isopropanol and (b) direct UV photolysis. Figure S5: -ln ($[AHTN]/[AHTN]_0$) Vs -ln ($[NB]/[NB]_0$). Figure S6: $\frac{k(HO\bullet+BA)}{k(HO\bullet+BA)} \ln\left(\frac{[NB]}{[NB]_0}\right) - \ln\left(\frac{[BA]}{[BA]_0}\right)$ Vs t. Figure S7: Degradation of AHTN and DMOB in the ClO• system. Figure S8: UV absorbance of AHTN at different concentration. Figure S9: Degradation of AHTN corresponding to different absorbed photon. Figure S10: Chromatogram for HPLC of AHTN degradation by (a) FC, (b) UV, and (c) UV/FC. Figure S11: AHTN degradation by UV/FC under various conditions. Table S1: Method detection limits (MDL) of halogenated DBPs.

Author Contributions: Conceptualization, L.W. and X.L.; methodology, X.L.; validation, L.W.; formal analysis, L.W.; investigation, L.W.; resources, X.L.; data curation, L.W.; writing—original draft preparation, L.W.; writing—review and editing, X.L.; visualization, L.W.; supervision, X.L.; project administration, L.W. and X.L.; and funding acquisition, L.W. and X.L.

Funding: This work was financially supported by the Natural Science Foundation of Zhejiang Province (Grant No. LQ19E080023), Jiyang College of Zhejiang A & F University (Grant No. JY2015RC001), and the special S & T project on the treatment and control of water pollution (Grant No. 2017ZX07201-003).

Conflicts of Interest: The authors declare no conflict of interest.

References

1. Parolini, M.; Magni, S.; Traversi, I.; Villa, S.; Finizio, A.; Binelli, A. Environmentally relevant concentrations of galaxolide (HHCB) and tonalide (AHTN) induced oxidative and genetic damage in Dreissena polymorpha. *J. Hazard. Mater.* **2015**, *285*, 1–10. [CrossRef] [PubMed]
2. Bitsch, N.; Dudas, C.; Korner, W.; Failing, K.; Biselli, S.; Rimkus, G.; Brunn, H. Estrogenic activity of musk fragrances detected by the E-screen assay using human MCF-7 cells. *Arch. Environ. Contam. Toxicol.* **2002**, *43*, 257–264. [CrossRef] [PubMed]
3. Yamauchi, R.; Ishibashi, H.; Hirano, M.; Mori, T.; Kim, J.W.; Arizono, K. Effects of synthetic polycyclic musks on estrogen receptor, vitellogenin, pregnane X receptor, and cytochrome P450 3A gene expression in the livers of male medaka (*Oryzias latipes*). *Aquat. Toxicol.* **2008**, *90*, 261–268. [CrossRef] [PubMed]
4. Stackelberg, P.E.; Gibs, J.; Furlong, E.T.; Meyer, M.T.; Zaugg, S.D.; Lippincott, R.L. Efficiency of conventional drinking-water-treatment processes in removal of pharmaceuticals and other organic compounds. *Sci. Total. Environ.* **2007**, *377*, 255–272. [CrossRef] [PubMed]
5. Li, W.; Nanaboina, V.; Chen, F.; Korshin, G.V. Removal of polycyclic synthetic musks and antineoplastic drugs in ozonated wastewater: Quantitation based on the data of differential spectroscopy. *J. Hazard. Mater.* **2016**, *304*, 242–250. [CrossRef] [PubMed]
6. Barışçı, S.; Dimoglo, A. Review on the stability of ferrate (vi) species in aqueous medium and oxidation of pharmaceuticals and personal care products (PPCPs) by ferrate (VI): Identification of transformation by-products. In *Ferrites and Ferrates: Chemistry and Applications in Sustainable Energy and Environmental Remediation*, 1st ed.; Sharma, V.K., Doong, R.A., Kim, H., Varma, R.S., Dionysiou, D.D., Eds.; American Chemical Society: Washington, DC, USA, 2016; Volume 1238, pp. 287–335, ISBN 9780841231870.
7. Snyder, S.A.; Adham, S.; Redding, A.M.; Cannon, F.S.; DeCarolis, J.; Oppenheimer, J.; Wert, E.C.; Yoon, Y. Role of membranes and activated carbon in the removal of endocrine disruptors and pharmaceuticals. *Desalination* **2007**, *202*, 156–181. [CrossRef]
8. Godayol, A.; Gonzalez-Olmos, R.; Sanchez, J.M.; Anticó, E. Assessment of the effect of UV and chlorination in the transformation of fragrances in aqueous samples. *Chemosphere* **2015**, *125*, 25–32. [CrossRef] [PubMed]
9. Kong, X.; Jiang, J.; Ma, J.; Yang, Y.; Liu, W.; Liu, Y. Degradation of atrazine by UV/chlorine: Efficiency, influencing factors, and products. *Water Res.* **2016**, *90*, 15–23. [CrossRef] [PubMed]
10. Remucal, C.K.; Manley, D. Emerging investigators series: The efficacy of chlorine photolysis as an advanced oxidation process for drinking water treatment. *Environ. Sci. Water Res. Technol.* **2016**, *2*, 565–579. [CrossRef]
11. Felis, E.; Alder, A.C.; Surmacz-Gorska, J.; Miksch, K. Advanced oxidation of the polycyclic musk fragrances with using UV and UV/H_2O_2 processes. *Arch. Environ. Prot.* **2008**, *34*, 13–23. [CrossRef]
12. Wang, D.; Bolton, J.R.; Andrews, S.A.; Hofmann, R. Formation of disinfection by-products in the ultraviolet/chlorine advanced oxidation process. *Sci. Total. Environ.* **2015**, *518–519*, 49–57. [CrossRef] [PubMed]

13. Zhang, T.; Lin, Y.; Xu, B.; Xia, S.; Tian, F.; Gao, N. Effect of UV irradiation on the proportion of organic chloramines in total chlorine in subsequent chlorination. *Chemosphere* **2016**, *144*, 940–947. [CrossRef] [PubMed]
14. Deng, L.; Huang, C.; Wang, Y. Effects of combined UV and chlorine treatment on the formation of trichloronitromethane from amine precursors. *Environ. Sci. Technol.* **2014**, *48*, 2697–2705. [CrossRef] [PubMed]
15. Liu, C.; Qiang, Z.; Tian, F.; Zhang, T. Photodegradation of etridiazole by UV radiation during drinking water treatment. *Chemosphere* **2009**, *76*, 609–615. [CrossRef] [PubMed]
16. Leenheer, J.A. Comprehensive approach to preparative isolation and fractionation of dissolved organic carbon from natural waters and wastewaters. *Environ. Sci. Technol.* **1981**, *15*, 578–587. [CrossRef] [PubMed]
17. Swietlik, J.; Dabrowska, A.; Raczyk-Stanislawiak, U.; Nawrocki, J. Reactivity of natural organic matter fractions with chlorine dioxide and ozone. *Water Res.* **2004**, *38*, 547–558. [CrossRef] [PubMed]
18. Rosenfeldt, E.; Linden, K.; Canonica, S.; von Gunten, U. Comparison of the efficiency of OH radical formation during ozonation and the advanced oxidation processes O_3/H_2O_2 and UV/H_2O_2. *Water Res.* **2006**, *40*, 3695–3704. [CrossRef] [PubMed]
19. Guan, Y.; Ma, J.; Li, X.; Fang, J.; Chen, L. Influence of pH on the formation of sulfate and hydroxyl radicals in the UV/peroxymonosulfate system. *Environ. Sci. Technol.* **2011**, *45*, 9308–9314. [CrossRef] [PubMed]
20. Fang, J.; Fu, Y.; Shang, C. The roles of reactive species in micropollutant degradation in the UV/free chlorine system. *Environ. Sci. Technol.* **2014**, *48*, 1859–1868. [CrossRef] [PubMed]
21. Watts, M.J.; Linden, K.G. Chlorine photolysis and subsequent OH radical production during UV treatment of chlorinated water. *Water Res.* **2007**, *41*, 2871–2878. [CrossRef] [PubMed]
22. Zehavi, D.; Rabani, J. Oxidation of aqueous bromide ions by hydroxyl radicals. Pulse radiolytic investigation. *J. Phys. Chem.* **1972**, *76*, 312–319. [CrossRef]
23. Kläning, U.K.; Wolff, T. Laser flash photolysis of HClO, ClO^-, HBrO, and BrO^- in aqueous solution. Reactions of Cl^- and Br^- Atoms. *Berichte der Bunsengesellschaft für Physikalische Chemie* **1985**, *89*, 243–245. [CrossRef]
24. Sun, P.; Lee, W.; Zhang, R.; Huang, C. Degradation of DEET and caffeine under UV/Chlorine and simulated sunlight/chlorine conditions. *Environ. Sci. Technol.* **2016**, *50*, 13265–13273. [CrossRef] [PubMed]
25. Buxton, G.V.; Greenstock, C.L.; Helman, W.P.; Ross, A.B. Critical review of rate constants for reactions of hydrated electrons, hydrogen atoms and hydroxyl radicals (•OH/•O^-) in aqueous solution. *J. Phys. Chem. Ref. Data* **1988**, *17*, 513–886. [CrossRef]
26. Molina, M.J.; Ishiwata, T.; Molina, L.T. Production of hydroxyl from photolysis of hypochlorous acid at 307–309 nm. *J. Phys. Chem.* **1980**, *84*, 821–826. [CrossRef]
27. Kläning, U.K.; Sehested, K.; Wolff, T. Ozone formation in laser flash photolysis of oxoacids and oxoanions of chlorine and bromine. *J. Chem. Soc. Faraday Trans.* **1984**, *80*, 2969–2979. [CrossRef]
28. Jayson, G.; Parsons, B.; Swallow, A.J. Some simple, highly reactive, inorganic chlorine derivatives in aqueous solution. Their formation using pulses of radiation and their role in the mechanism of the Fricke dosimeter. *J. Chem. Soc. Faraday Trans.* **1973**, *69*, 1597–1607. [CrossRef]
29. Grigor'ev, A.; Makarov, I.; Pikaev, A. Formation of Cl_2^- in the bulk of solution during radiolysis of concentrated aqueous solutions of chlorides. *Khimiya Vysokikh Ehnergij* **1987**, *21*, 99–102.
30. Grebel, J.E.; Pignatello, J.J.; Mitch, W.A. Effect of halide ions and carbonates on organic contaminant degradation by hydroxyl radical-based advanced oxidation processes in saline Waters. *Environ. Sci. Technol.* **2010**, *44*, 6822–6828. [CrossRef] [PubMed]
31. Zuo, Z.; Katsumura, Y.; Ueda, K.; Ishigure, K. Reactions between some inorganic radicals and oxychlorides studied by pulse radiolysis and laser photolysis. *J. Chem. Soc. Faraday Trans.* **1997**, *93*, 1885–1891. [CrossRef]
32. Kuhlich, P.; Göstl, R.; Teichert, P.; Piechotta, C.; Nehls, I. Transformations of polycyclic musks AHTN and HHCB upon disinfection with hypochlorite: Two new chlorinated disinfection by-products (CDBP) of AHTN and a possible source for HHCB-lactone. *Anal. Bioanal. Chem.* **2011**, *399*, 3579–3588. [CrossRef] [PubMed]
33. Radjenovic, J.; Godehardt, M.; Hein, A.; Farre, M.; Jekel, M.; Barcelo, D. Evidencing generation of persistent ozonation products of antibiotics roxithromycin and trimethoprim. *Environ. Sci. Technol.* **2009**, *43*, 6808–6815. [CrossRef] [PubMed]

34. Chen, P.; Linden, K.G.; Hinton, D.E.; Kashiwada, S.; Rosenfeldt, E.J.; Kullman, S.W. Biological assessment of bisphenol A degradation in water following direct photolysis and UV advanced oxidation. *Chemosphere* **2006**, *65*, 1094–1102. [CrossRef] [PubMed]
35. Alfassi, Z.B.; Huie, R.E.; Mosseri, S.; Neta, P. Kinetics of one-electron oxidation by the ClO radical. *Int. J. Radiat. Appl. Instrum. C* **1988**, *32*, 85–88. [CrossRef]
36. von Sonntag, C. Advanced oxidation processes: Mechanistic aspects. *Water Sci. Technol.* **2008**, *58*, 1015–1021. [CrossRef] [PubMed]
37. Cooper, W.J.; Cramer, C.J.; Martin, N.H.; Mezyk, S.P.; O'Shea, K.E.; Sonntag, C.V. Free radical mechanisms for the treatment of methyl tert-butyl ether (MTBE) via advanced oxidation/reductive processes in aqueous solutions. *Chem. Rev.* **2009**, *109*, 1302–1345. [CrossRef] [PubMed]
38. Yang, Y.; Pignatello, J.J.; Ma, J.; Mitch, W.A. Comparison of halide impacts on the efficiency of contaminant degradation by sulfate and hydroxyl radical-based advanced oxidation processes (AOPs). *Environ. Sci. Technol.* **2014**, *48*, 2344–2351. [CrossRef] [PubMed]
39. Wu, Z.; Fang, J.; Xiang, Y.; Shang, C.; Li, X.; Meng, F.; Yang, X. Roles of reactive chlorine species in trimethoprim degradation in the UV/chlorine process: Kinetics and transformation pathways. *Water Res.* **2016**, *104*, 272–282. [CrossRef] [PubMed]
40. Bard, A.J.; Parsons, R.; Jordan, J. (Eds.) *Standard Potentials in Aqueous Solution*, 1st ed.; Marcel Dekker: New York, NY, USA, 1985; pp. 127–413. ISBN 0-8247-7291-1.
41. Neta, P.; Huie, R.E.; Ross, A.B. Rate constants for reactions of inorganic radicals in aqueous solution. *J. Phys. Chem. Ref. Data* **1988**, *17*, 1027–1284. [CrossRef]
42. Guo, K.; Wu, Z.; Shang, C.; Yao, B.; Hou, S.; Yang, X.; Song, W.; Fang, J. Radical chemistry and structural relationships of PPCP degradation by UV/Chlorine treatment in simulated drinking water. *Environ. Sci. Technol.* **2017**, *51*, 10431–10439. [CrossRef] [PubMed]
43. Piscopo, A.; Robert, D.; Weber, J.V. Influence of pH and chloride anion on the photocatalytic degradation of organic compounds. *Appl. Catal. B Environ.* **2001**, *35*, 117–124. [CrossRef]
44. Abe, K.; Tanaka, K. Effect of Fe^{3+} on UV-illuminated ozonation of nitrophenolic compounds. *Chemosphere* **1999**, *38*, 2747–2752. [CrossRef]
45. Wang, H.; Yao, H.; Sun, P.; Li, D.; Huang, C. Transformation of tetracycline antibiotics and Fe (II) and Fe (III) species induced by their complexation. *Environ. Sci. Technol.* **2015**, *50*, 145–153. [CrossRef] [PubMed]

processes

MDPI

Article

Investigations on Ozone-Based and UV/US-Assisted Synergistic Digestion Methods for the Determination of Total Dissolved Nitrogen in Waters

Xiaofang Sun, Huixuan Chen, Zhengyu Liu, Mengfei Zhou, Yijun Cai, Haitian Pan and Luyue Xia *

College of Chemical Engineering, Zhejiang University of Technology, Hangzhou 310014, China; zgdsxf@zjut.edu.cn (X.S.); ZjutChx@126.com (H.C.); zjutlzy0811@126.com (Z.L.); mfzhou@zjut.edu.cn (M.Z.); hgybcyj@zjut.edu.cn (Y.C.); htpan@zjut.edu.cn (H.P.)

* Correspondence: lyxia@zjut.edu.cn; Tel.: +86-0571-88320329

Received: 3 April 2020; Accepted: 21 April 2020; Published: 23 April 2020

Abstract: Over the past two decades, the alkaline persulfate oxidation (PO) with thermal and/or ultraviolet (UV) assisted digestion method has been widely used for digestion of nitrogen containing compounds (N-compounds) in water quality routine analysis in laboratory or on-line analysis, due to its simple principle, high conversion rate, high percent recovery, low-cost. However, this digestion method still has some inevitable problems such as complex operations, high contamination potential, batch N blanks, higher reaction temperature (120–124 °C) and time-consuming (30–60 min). In this study, ozone (O_3) was selected as the oxidant for digestion of N-compounds through analysis and comparison firstly. Secondly, we proposed and compared the UV and/or ultrasound (US) combined with ozone (UV/O_3, US/O_3 and UV/US/O_3) synergistic digestion methods based on O_3 with sole O_3 oxidation method on digestion efficiency (digestion time and conversion rate) of standard N-compounds. Simultaneously, the influence of reaction temperature, pH of water sample, concentration of O_3 and mass flow rate, UV intensity, US frequency and power on digestion efficiency were investigated, and then the optimum parameters for digestion system were obtained. Experimental results indicated that UV radiation can effectively induce and promote the decomposition and photolysis of O_3 in water to generate hydroxyl radicals (•OH), while US can promote the diffusion and dissolution of O_3 in water and intensify the gas-liquid mass transfer process for the reaction system. Meanwhile, results showed that the UV/US/O_3 synergistic digestion method had the best digestion efficiency under the optimum conditions: water sample volume, 10 mL; pH of water sample, 11; O_3 mass flow rate, 3200 mg/h; reaction temperature, 30 °C; digestion time, 25 min; UV lamp power, 18 W; distance between UV lamp and reactor, 2 cm; US frequency, 20 kHz; US power, 75 W. The conversion rate (CR) of synthetic wastewater samples varied from 99.6% to 101.4% for total dissolved nitrogen (TDN) in the range of 1.0~4.0 mg/L. The UV/US/O_3 synergistic digestion method would be an effective and potential alternative for digestion of N-compounds in water quality routine analysis in laboratory or on-line analysis.

Keywords: total dissolved nitrogen; digestion method; digestion efficiency; intensification; ozone

1. Introduction

Total nitrogen (TN) is an important indicator for water quality determination and on-line monitoring [1]. The excess of nitrogen in natural waters may cause eutrophication, which is one of the most serious threats to the aquatic ecosystems, and even do harm to human healthy [2].

Total nitrogen refers to the sum of total dissolved nitrogen (TDN) and suspended sediment nitrogen (SSN) in water sample [3]. In which, the TDN defined as the dissolved species that can passes through a 0.2 or 0.45 μm membrane filter, and the SSN defined as the particulate fraction is retained by the filter [4]. However, determination of TDN has always been a challenging task and contemporary strategies still rely on methods which have some limitations.

Generally, TDN consists of two fractions: an inorganic fraction (i.e., dissolved inorganic nitrogen-DIN), composed of dissolved free ammonia (NH_3), ammonium (NH_4^+), nitrite (NO_2^-), nitrate (NO_3^-); an organic fraction (i.e., dissolved organic nitrogen-DON), the composition of which is unknown, but which usually include urea, proteins, amino acid, acylamide, fulvic acid and humic acid, etc. In order to determine and quantify TDN in waters, all N-compounds should be converted to a single species, such as ammonia, ammonium, nitrite and nitrate. In general, ammonia and ammonium can be distilled off in the form of ammonia and determined by titration with standardized mineral acid, nitrate and nitrite (need reduction reaction, diazo-coupling reaction and Griess reaction) can be measured by various techniques [5], but the most widely used and convenient methods are spectroscopic methods [6], such as the ultraviolet (UV)/Vis spectrophotometric method. Therefore, the first step for the determination of TDN is the digestion of the water sample in order to convert DON into an inorganic form of nitrogen, it is the most tedious and time-consuming step and the greatest source of errors in the analytical determination of TDN.

Acid Kjeldahl digestion is a traditional digestion method, which simply involves heating and boiling the sample in the presence of concentrated sulfuric acid and a mercury or copper catalyst at high temperature for several hours causing the mineralization of organic nitrogenous species to ammonium [7]. However, this digestion method involves issues of use strong acid and toxic metal catalysts, time-consuming, expensive, and only measures organic nitrogen, free ammonia and ammonium, therefore the separate determination of nitrite and nitrate by another method was required in order to determine TDN. UV photo-oxidation, first reported in 1966 by Armstrong et al. [8], which utilizes UV radiation as a source to generate strongly oxidizing hydroxyl radicals with 30% hydrogen peroxide (H_2O_2) as the oxidant. The main limitations of this method are that it is highly variable, has low oxidation efficiency and poor percent recovery. A wet digestion method was introduced by Koroleff who converted N-compounds to nitrate in aqueous alkaline persulfate (PS, $S_2O_8^{2-}$) [9]. Persulfate has been used and favored in flow analysis applications for it produces fewer bubbles upon decomposition compared with hydrogen peroxide. If the digestion is performed under alkaline conditions, then the sole mineral species generated is nitrate. The United States Geological Survey (USGS) states that alkaline persulfate digestion is less toxic, more accurate and sensitive to Kjeldahl digestion for nitrogen determination in environmental samples [10]. Another digestion method for TDN was later introduced with high-temperature oxidation (HTO) or named high temperature combustion (HTC) [11–13]. In this method, N-compounds were converted into •NO under high temperature more than 950 °C, and then reacted with O_3 to form NO_2^* and emitted the photon which could be detected with a chemiluminescent detector, which needs a small amount of sample, but requires expensive equipment and is not suitable for field analysis and on-line monitoring.

Within these four methodologies, alkaline persulfate digestion remains the most mature and popular method for determination of TDN in natural waters [14–17]. This approach does not need expensive instruments, is easy to implement, and can be employed for simultaneous determination of total nitrogen and total phosphorus [18–20]. During the last 25 years, some published procedures using alkaline persulfate include batch mode [21] and flow analysis methods [22–25] with both thermal and UV radiation assisted digestion [26,27]. Flow analysis methods with UV mineralization allow high sample throughput, but the limited digestion time may result in partial conversion with interference of unreacted persulfate on the Cd/Cu column which is commonly employed in nitrate analyzers. Nevertheless, thermal digestion in batch mode is more reliable and easier controlled. In which, reaction is performed at 121 °C for 30–60 min in H_3BO_3/NaOH buffer at pH of 9.7 [28]. In the past two decades, the alkaline persulfate oxidation (PO) with thermal or UV (UV/PO) has been widely used in total nitrogen routine analysis in laboratory or on-line analysis due to its simple principle, high conversion rate, high percent recovery, and low-cost. Nowadays, total nitrogen (TN) on-line monitors are widely sold in the international market and used in environmental monitoring, such as famous NPW-160H (HACH), TNP-4200 (SHIMADZU) and WPA-58 (KEM), which are basically used for the alkaline persulfate oxidation with thermal or UV assisted digestion method.

However, this method still has some inevitable problems such as complex operations, high contamination potential, batch N blanks [29], higher reaction temperature (120–124 °C) and time-consuming (30–60 min). As we know, UV/PO and UV/H_2O_2 are the advanced oxidation processes (AOPs) [30,31] based on digestion methods, in which the oxidant is the key of digestion, under UV radiation, PS can generate strongly reactive yet selective sulfate radicals ($\bullet SO_4^-$), and H_2O_2 can generate hydroxyl radicals (•OH) with super oxidation capacity, which can react with the majority of organic compounds and degrade the contaminants in waters non-selectively [32]. According to the commonly used oxidants in water treatment [33], besides the PS and H_2O_2, ozone (O_3) may be a compromised alternative for N-compounds digestion, and which can generate •OH by self-decomposition in basic medium or under certain conditions such as UV, ultrasound (US) and microwave (MW) [34,35]. Although ozone has been widely used in water treatment as disinfectant and oxidant, in which transformation of organic micropollutants occurs via direct reaction with O_3 or indirect reaction with hydroxyl radicals [36–40], fewer studies have reported its use for digestion and determination of TDN in waters. Considering the oxidation capacity, safety, self-degradability, removal ability and convenience of oxidant, O_3 was selected as the oxidant for digestion of N-compounds, and then will use UV and/or US to intensify its digestion ability in this study.

With this background, the main objectives of this study were: (a) to compare the UV/O_3, US/O_3 and UV/US/O_3 synergistic digestion methods based on O_3 with sole O_3 oxidation method on digestion efficiency (digestion time and conversion rate) of standard N-compounds; (b) to investigate the influence of reaction temperature, pH of water sample, concentration of O_3 and mass flow rate, UV intensity, US frequency and power on digestion efficiency, and then to obtain the optimum parameters for digestion system; (c) to seek and develop a novel and reliable digestion method for safe, environmentally-friendly, convenient and efficient digestion of nitrogen containing organic and inorganic compounds for improving the precision and stability of determination of total dissolved nitrogen in natural waters and waste waters. Section 2 presented the novel methodology for determination of TDN, including the reagents, solutions, apparatus, experimental devices, and detailed procedure. Section 3 consisted of experimental results and discussion. Section 4 addressed the conclusions of this research and directions for future work.

2. Materials and Methods

2.1. Reagents and Solutions

All chemicals used in experiments were of analytical grade and superior grade, all solutions were prepared with nitrogen-free water. The N-compounds used in this study included representative dissolved inorganic nitrogen (DIN) and dissolved organic nitrogen (DON), such as potassium nitrate, sodium nitrate, ammonia chloride and carbamide, which are frequently occurring components in natural waters or waste waters from factories. All stock solutions of them were prepared with nitrogen concentration of 100 mg/L, then all working standard solutions were prepared by diluting from the stock solutions, such as 10 mg/L, 4 mg/L and 1 mg/L. In addition, the concentration of sodium hydroxide and hydrochloric acid solutions used to regulate the pH of water sample were 0.01 mol/L.

2.2. Apparatus and Devices

In order to establish a suitable digestion method for N-compounds in waters, different process intensification methods (UV, US, and combination of UV & US) were proposed and investigated in this study.

In the first process intensification scheme, two UV lamps were used to assist O_3 for oxidation of N-compounds, and the schematic diagram of UV/O_3 synergistic digestion experimental device is presented in Figure 1.

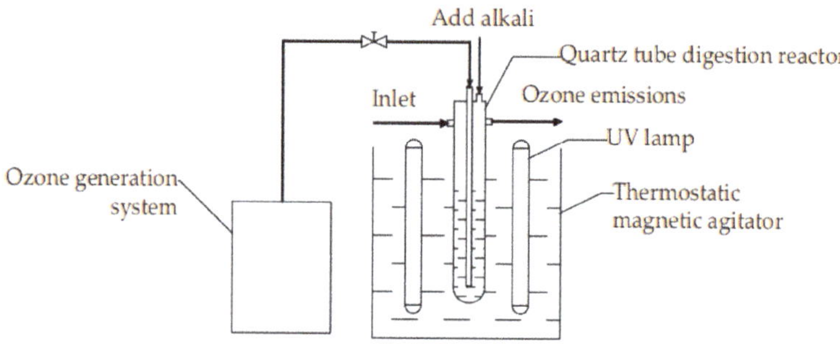

Figure 1. Schematic diagram of UV/O_3 synergistic digestion experimental device.

As shown in Figure 1, two UV lamps hanged around the quartz tube digestion reactor are 18 W low-press mercury lamps with maximum emission at 254 nm (Cnlight Co., Guangdong, China), and the jacket of the lamp was made up of quartz for strong and sharp emission. In this device, the quartz tube digestion reactor is a straight and cylindrical reactor having 12 mm inner diameter (i.d.) and 180 mm height. A 1–10 mL adjustable micropipette (Lichen Instruments Co., Shanghai, China) was used to transfer the working standard solutions and reagents. The self-made electrolytic ozone generation system was consisted of four DJ800 polymer electrolysis membrane (PEM) electrode modules, two 3–5 V/12 A (w) power modules and four water tanks, in which the mixed gas of ozone/oxygen with a mass concentration approximately of 18–20 wt/wt% (250–280 mg/L, 3 L/h) were prepared and gathered in two anode water tanks by electrolyzing deionized water, two cathode water tanks gathered the by-products H_2. In this system, two PEMs, two water tanks and one power module form one ozone generator, and two generators run in parallel. In addition, in order to keep reaction temperature in constant, the DF-101S thermostatic magnetic agitator (Lichen Instruments Co., Shanghai, China) was used to control the digestion reaction temperature (±0.1 °C).

The second scheme used US to assist O_3 for oxidation of N-compounds, and the schematic diagram of US/O_3 synergistic digestion experimental device is presented in Figure 2.

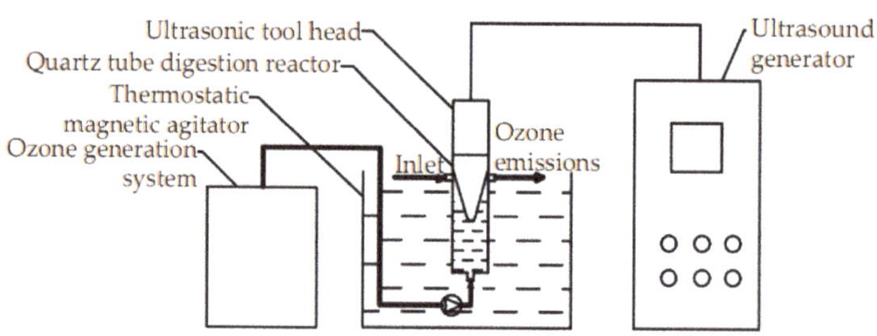

Figure 2. Schematic diagram of US/O_3 synergistic digestion experimental device.

As shown in Figure 2, the ultrasound generator with 10 mm ultrasonic probe was used and the optimum working power range of 50–100 W. Considering the convenience of installation for ultrasonic probe and keep it under the surface of digested solution, the quartz tube digestion reactor was designed with 20 mm inner diameter (i.d.) and 80 mm height. In addition, the ozone generation system and thermostatic magnetic agitator were used same as scheme one in this device.

In the third scheme, UV in combination with US digestion method was used to assist and intensify the digestion ability of O_3, in which the UV/US/O_3 synergistic digestion experimental device was applied.

Accordingly, an 754PC ultraviolet-visible spectrophotometer (Jinghua Instruments Co., Shanghai, China) and two quartz colorimetric cells with light path of 10 mm were used for TDN determination in detection system.

2.3. *Procedure and Methods*

The detailed procedure was as follows:

(1) Turn on the ozone generation system and close the outlet valve of O_3 synchronously;
(2) Turn on the thermostatic magnetic agitator and control the digestion reaction temperature at 30 °C;
(3) Transfer 9 mL of standard solution or water sample of carbamide and 1 ml of sodium hydroxide solution (0.01 mol/L) to the quartz tube digestion reactor by adjustable micropipette;
(4) Turn on the UV lamps (scheme 1 and scheme 3), US generator (scheme 2 and scheme 3) and open the outlet valve of O_3, high concentration of ozone gas bubbled through an ozone bubble tube into the bottom of digestion reactor for digestion of N-compounds;
(5) Several minutes later (5, 10, 15, 20 and 25 min), turn off the UV lamps, US generator and ozone generation system;
(6) Heat the digested standard solution or water sample to 50 °C and drive out the dissolved O_3 for 5 min;
(7) Transfer 5 mL of digested standard solution or water sample to the quartz colorimetric cell for the detection of absorbance at 220 nm and 275 nm, respectively;
(8) Calculate the concentrations of water samples using the prepared calibration curve, according to the net absorbances calculated by subtracting the corrected absorbances of digested blank solutions from the corrected absorbances of digested standard solutions.

In this procedure, the calibration curve was prepared by nitrate standard solutions and went through the whole digestion and detection procedure, in which the concentration of TDN was in the range of 0.5~4.0 mg/L and the mean N blank is 0.020 in 14 blank tests. The calibration curve of TN is shown in Figure 3.

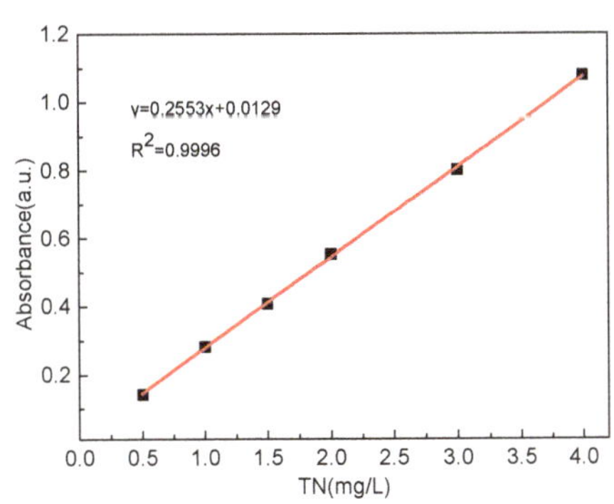

Figure 3. Calibration curve of TN.

Moreover, the percent conversion of N-compounds to their nitrate forms were calculated with respect to the signal generated using standards of potassium nitrate at an equivalent nitrogen concentration.

3. Results and Discussion

3.1. UV/O_3 Digestion

According to the basic principles of chemical reaction kinetics, the concentration of oxidant dissolved in water is a very important factor for the gas-liquid heterogeneous reaction system of O_3 with N-compounds. By Henry's law, under certain temperature conditions, the higher the concentration of O_3 in gas phase, the higher those in aqueous solution. Consequently, improving the concentration of O_3 in gas phase and the quantity of ozone gas are important to intensify the ozone mass transfer process. On the other hand, ozone dissolved in aqueous solution can self-decompose and form the intermediate products of hydrogen peroxide, which can be motivated by the energy of photon to generate •OH under the condition of ultraviolet radiation [41]. Meanwhile, the photolysis of ozone can generate highly active oxygen radicals (•O), and then react with water to form •OH [42]. So the digestion efficiency of UV/O_3 is mainly affected by UV radiation and O_3 concentration dissolved in water, for which the electrolytic ozone generation system with 4 PEM modules was constructed and the ozone gas with a mass concentration approximately of 18–20 wt/wt% was prepared in this study.

In order to investigate the effect of reaction temperature on digestion efficiency, the univariate experiments were conducted by increasing temperature, in which the digested nitrogen containing compound was the carbamide solution with concentration of 10 mg/L and without the addition of sodium hydroxide, and the reaction time was 10 min. Results showed that the digested solution had the highest absorbance and the best conversation rate when the reaction temperature at 30 °C. As we known, although the saturation concentration of ozone in liquid phase is higher at lower temperature, but the reaction rate constant and the rate of ozone self-decomposition are lower, so the efficiency of ozone conversion to •OH is lower too. However, when the temperature is higher than 35 °C, although the reaction rate is faster, the low concentration of ozone in solution leads to the decline of the probability of contact between ozone and reactant, and the digestion efficiency declines spontaneously. In the actual digestion process, the reaction temperature may be higher than 30 °C due to the long time radiation of UV lamps. Therefore, it is very important to avoid the decline of digestion efficiency by dissipating the heat of digestion system.

For the UV intensity, according to the Lambert law, UV lamp power and distance between UV lamp and reactor are the main two factors affected the UV intensity and photolysis rate. For the convenience of investigation, the distance between UV lamp and reactor (2 cm) was fixed at first, and then three types of UV lamps with different power (15 W, 18 W and 23 W) were used and investigated under the conditions of carbamide solution with concentration of 1 mg/L, reaction temperature of 30 °C, without the addition of sodium hydroxide. Results are shown in Figure 4.

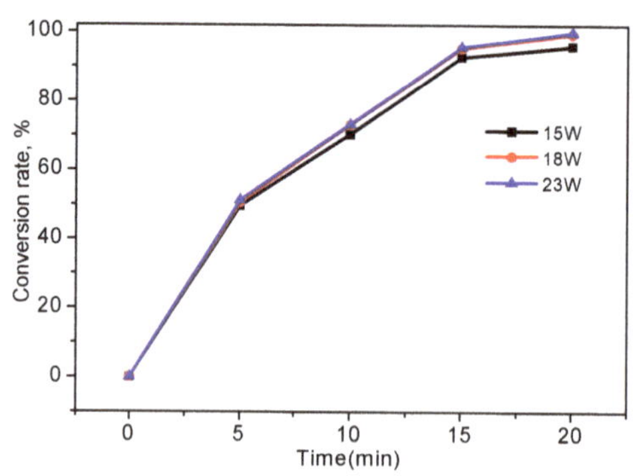

Figure 4. Effect of ultraviolet (UV) intensity on digestion efficiency.

As shown in Figure 4, the digestion efficiency was found to be increasing with the increased power, and no significant changes were observed at power higher than 18 W. Mainly due to the facy that the UV lamp of 18 W has enough energy and photons to induce and motivate O_3 to photolyze and generate •OH, while the 15 W UV lamp is relatively weaker and lower. By using 18 W or 23 W UV lamp, the carbamide solution with concentration of 1 mg/L could be completely digested in 20 min. Consequently, the 18 W UV lamp was selected to be used in the subsequent experiments.

For the UV/O_3 synergistic digestion scheme, pH has a great influence on the conversion of •OH in the reaction system. Under acidic conditions, only about 20% of the ozone in aqueous solution decomposes directly to •OH, whereas under alkaline conditions, the photolysis of ozone and hydroxyl ions (OH^-) can motivate the formation of •OH, and the conversion rate of •OH is much higher [43,44]. Experimental results are shown in Figure 5.

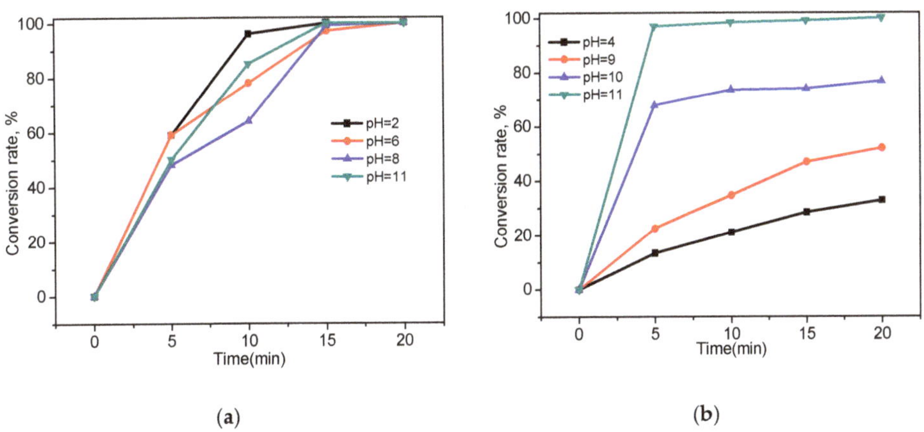

Figure 5. Effect of pH on digestion efficiency. (**a**) Digestion of carbamide solution with concentration of 1 mg/L; (**b**) Digestion of ammonia chloride solution with concentration of 1 mg/L.

It can be seen from Figure 5a that the carbamide can be quickly digested under both alkaline and acidic conditions, and the carbamide has the highest digestion efficiency with the pH at 2, which mainly due to the higher oxidation-reduction potential of ozone under acidic conditions (E^0 = 2.07 V) than that under alkaline conditions (E^0 = 1.24 V). Whereas, OH^- can promote the generation of •OH under alkaline conditions, so that the synergistic effect of the reaction process is enhanced [45] when pH at 11. When the pH is at 6, reactions remained the same conversion rate as that of pH at 2 in the first 5 min, but it declined quicker in the rest of reaction time. It may be explained that: under acidic condition and UV radiation, O_3 dissolved in aqueous solution can form the intermediate products of H_2O_2, which can be motivated by UV to generate •OH, while the ionization of H_2O_2 was greatly influenced by pH of solution, the higher pH of solution is, the more H_2O_2 will be ionized, and the less •OH will be generated by photolysis of H_2O_2.

As shown in Figure 5b, the digestion process of ammonia chloride is greatly influenced by pH of solution. When the pH at 11, the reaction rate constant of •OH with NH_3 is about 1.7–8.7 × 10^7 $M^{-1} \cdot S^{-1}$, so that the digestion efficiency is very high and the digestion time need only 5 min. However, when the pH is at 10, the reaction rate in first 5 min is high, and after that it declines rapidly and remains basically unchanged, which is mainly due to the consumption of OH^- and its rapid decline of concentration. As the ozone and •OH are difficult to react with NH_4^+ [46], the weak conversion rate of ammonia chloride was obtained with the pH at 4. Considering the ammonia nitrogen and carbamide nitrogen coexist in the natural waters or waste waters, high pH value (pH = 11) was selected to be used in this study.

To further verify the effectiveness of UV/O_3 digestion method under alkaline conditions, comparative experiments were carried out based on the above results obtained, and the results are shown in Figure 6.

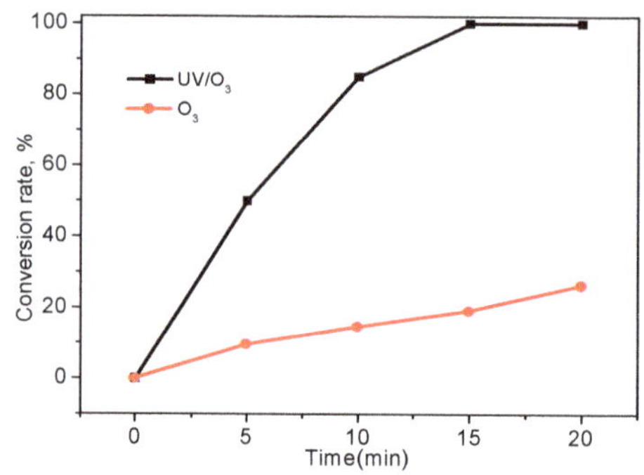

Figure 6. Comparison of digestion efficiency between UV/O_3 and O_3.

Compared with UV/O_3 digestion method, weak percent conversion was obtained by sole O_3 digestion method for carbamide digestion with concentration of 1 mg/L and the UV radiation had excellent intensification effect on digestion process of O_3, which can effectively induce and promote the radical reaction of O_3 in aqueous solution to generate •OH with super oxidation capacity.

3.2. US/O_3 Digestion

In the US/O_3 synergistic digestion scheme, the intensification mechanisms of US are mainly reflected in two aspects, one is to intensify the mass transfer of ozone, the other is to promote the generation of •OH. Under certain frequency and sound intensity, US can crush the lager ozone bubbles into microbubbles with a diameter of about 0.2–0.3 μm, and the total contact area of gas-liquid will be greatly improved. Moreover, the gas-liquid mass transfer coefficient will be greatly enhanced by strong mechanical stirring effect of US. On the other hand, the local high temperature and high pressure environment caused by ultrasound cavitation effect will also promote the generation of •OH [47–49].

In this study, the ultrasound frequency of 20 kHz and power of 75 W were chosen as optimizing parameters for digestion of TDN as suggested by the previous literature [50]. In order to verify the effectiveness of US/O_3 digestion method, comparative experiments were conducted based on the reaction conditions obtained from above results, and carbamide solution with concentration of 100 mg/L. Results are shown in Figure 7.

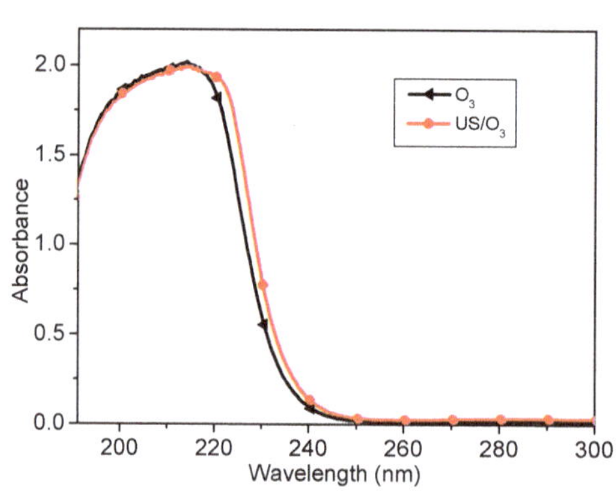

Figure 7. Comparison of digestion efficiency between US/O_3 and O_3.

Compared with sole O_3 digestion method, the intensification effect of US on digestion process of O_3 is not obvious, which is mainly due to the low conversion rate of •OH, meanwhile, the •O generated by US could be consumed by O_3. On the other hand, although the US can promote the diffusion and dissolution of O_3 in water, the lower level (32 mm) of digested solution in the reactor of US/O_3 synergistic digestion experimental device led to the shorter residence time of ozone in aqueous solution, and the reaction time with N-compounds was shortened too.

3.3. UV/US/O_3 Digestion

Combining the advantages of UV, US and O_3, comparative experiments of UV/US/O_3 and UV/O_3 digestion methods were conducted based on the optimal conditions, in which the concentration of carbamide solution is 4 mg/L. Results are displayed in Figure 7.

As shown in Figure 8, the UV/US/O_3 synergistic digestion method had an excellent digestion efficiency, and the carbamide solution with concentration of 4 mg/L could be completely digested in 25 min. Meanwhile, it can be seen that ultrasound can effectively promote the diffusion and dissolution of O_3 in aqueous solution and intensify the gas-liquid mass transfer for reaction system.

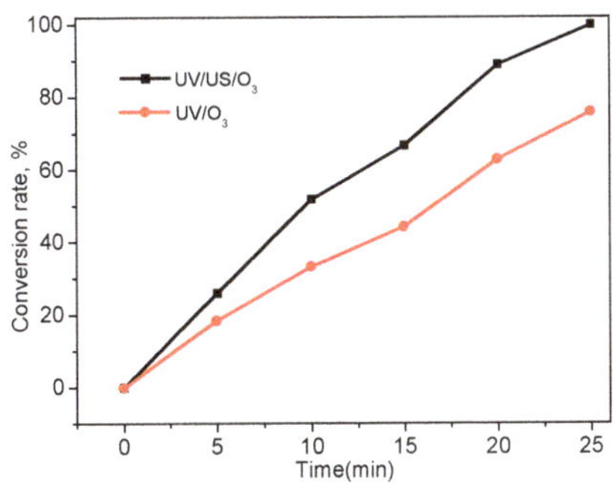

Figure 8. Comparison of digestion efficiency between UV/US/O_3 and UV/O_3.

3.4. Optimization of UV/US/O_3 Process

The optimization of the UV/US/O_3 digestion process was conducted in order to obtain feasible, economical and maximum parameters for percent conversion of TDN. In our work, the orthogonal experimental design was applied, in which a six-factor (four real factors and two dummy factors) and five-level L_{25} (5^6) orthogonal table was employed to investigate the effects of four independent factors (distance between UV lamp and reactor (A), US power (B), water sample volume (C), digestion time (D)) on conversion rate (CR) based on the preliminary experimental results. The orthogonal experiment design and the results of orthogonal experiment are shown in Tables 1 and 2, respectively.

Table 1. Orthogonal experimental design.

Factor	A (cm)	B (W)	C (mL)	D (min)
Level 1	1	50	10	10
Level 2	2	62.5	15	15
Level 3	3	75	20	20
Level 4	4	87.5	25	25
Level 5	5	100	30	30

As shown in Table 2, twenty five experiments were run under the conditions of the carbamide solution with concentration of 4 mg/L, O_3 mass flow rate of 3200 mg/h, reaction temperature of 30 °C, pH at 11, UV lamp power of 18 W and US frequency of 20 kHz, which was fixed from the previous study.

Where Ki refers to the average value of five conversion rates at the i level of each factor, which reflects the influence of each level of each factor on the conversion rate, the larger Ki value, the higher CR at the i level. While Range is the difference between the maximum and minimum Ki values, which reflects the degree of influence of various factors on CR, the larger R value, the more important this factor.

Therefore, the main influencing factors of the UV/US/O_3 digestion process were ranked D > B > C > A, and the recommended optimal combination was $A_2B_3C_1D_4$. Consequently, the optimized parameters were obtained by range analysis, viz.: distance between UV lamp and reactor, 2 cm; US power, 75 W; water sample volume, 10 mL; digestion time, 25 min, according to the actual values of different factor corresponding to each level displayed in Table 1.

Table 2. Results of orthogonal experiment.

Experiment No.	A	B	C	D	Dummy	Dummy	CR (%)
1	1	1	1	1	1	1	26.5
2	1	2	2	2	2	2	40.0
3	1	3	3	3	3	3	64.2
4	1	4	4	4	4	4	80.3
5	1	5	5	5	5	5	75.0
6	2	1	2	3	4	5	57.2
7	2	2	3	4	5	1	71.9
8	2	3	4	5	1	2	93.7
9	2	4	5	1	2	3	12.4
10	2	5	1	2	3	4	50.9
11	3	1	3	5	2	4	69.6
12	3	2	4	1	3	5	21.4
13	3	3	5	2	4	1	26.1
14	3	4	1	3	5	2	77.0
15	3	5	2	4	1	3	77.4
16	4	1	4	2	5	3	37.8
17	4	2	5	3	1	4	40.4
18	4	3	1	4	2	5	97.4
19	4	4	2	5	3	1	74.8
20	4	5	3	1	4	2	18.9
21	5	1	5	4	3	2	89.5
22	5	2	1	5	4	3	52.6
23	5	3	2	1	5	4	18.5
24	5	4	3	2	1	5	35.8
25	5	5	4	3	2	1	46.6
K_1	57.20	56.12	60.88	19.54			
K_2	57.22	45.26	53.58	38.12			
K_3	54.30	59.98	52.08	57.08			
K_4	53.86	56.06	55.92	83.30			
K_5	48.60	53.76	48.68	73.14			
Range	8.62	14.72	12.20	63.76			
Rank	D > B > C > A						
Optimal combination	$A_2B_3C_1D_4$						

3.5. Validation of Digestion Efficiency

In order to further evaluate the digestion efficiency of the proposed method under the optimum conditions obtained from orthogonal experiment. The carbamide solution with concentration ranging from 1.0 to 4.0 mg/L were digested and analyzed, each group was run in triplicate, results are displayed in Table 3. As shown in Table 3, where A_s represents the final absorbance of standard solution of sodium nitrate digested under optimum conditions, A_1, A_2 and A_3 represent the final absorbance of carbamide solution digested under the same conditions in triplicate, respectively, and A_m represents the mean absorbance.

Table 3. Validation of digestion efficiency.

TDN (mg/L)	A_s	A_1	A_2	A_3	A_m	CR (%)
1.0	0.282	0.290	0.287	0.291	0.289	102.48
2.0	0.543	0.540	0.536	0.537	0.538	99.08
3.0	0.806	0.799	0.796	0.803	0.799	99.13
4.0	1.073	1.091	1.052	1.037	1.060	98.79

It can be seen in Table 3 that the percent conversion of TDN ranged from 102.48% to 98.79% with an increase of TDN from 1.0 mg/L to 4.0 mg/L, and the proposed digestion method had a good digestion efficiency, by which the carbamide solution with concentration of 4.0 mg/L could be completely digested in 25 min under the optimum conditions obtained in this study.

However, some interference factors during the experiments caused a certain degree of deviations in experimental results. First of all, the N blank value was the mean value of 14 blank tests in making the calibration curve of TN, while the actual N blank values deviated from the mean value, which may be positive or negative. Secondly, after the reaction, the pH of the digested solution was weakly alkaline, and the OH^- in the solution had an effect on the absorbance at 220 nm. Moreover, the residual ozone dissolved in water could not be completely removed by heating the solution, which also had an effect on the absorbance at 220 nm. Finally, in the detection process, the noise of the instrument also affected the experimental results. Therefore, with the same interference of absorbance, the lower concentration of TDN in water, the higher deviation of the conversion rate.

3.6. Application of Digestion Method

The proposed UV/US/O_3 synergistic digestion method was applied to determine the TDN in a synthetic wastewater samples, which refer to the wastewater produced from urea production process [51]. Analysis results of the actual urea wastewater are shown in Table 4.

Table 4. Analysis results of the actual urea wastewater.

Analysis Contents	Analysis Data
pH	9
alkalinity	19.6 mmol/L
total iron	0.49 mg/L
chloridion	16 mg/L
ammonia chloride	0.034 wt%
urea	0.57 wt%
formic acid	a little

According to the contents of actual wastewater, synthetic wastewater samples were prepared by the ratio 19:1 of urea to ammonia chloride, in which the concentration of TDN was in the range of 1.0~4.0 mg/L, other contents and its concentrations remained unchanged, and then each group was run in 6 times, the concentration of TDN was determined by UV spectrophotometric method. Determination results are displayed in Table 5.

Table 5. Determination results of the synthetic wastewater samples.

TDN (mg/L)	Test 1 (mg/L)	Test 2 (mg/L)	Test 3 (mg/L)	Test 4 (mg/L)	Test 5 (mg/L)	Test 6 (mg/L)	Mean (mg/L)	RSD (%)	CR (%)
1.0	1.03	1.01	1.04	0.98	1.01	0.99	1.014	2.09	101.4
2.0	2.04	1.97	2.02	2.03	1.99	1.96	2.010	1.50	100.5
3.0	2.96	3.06	2.96	3.01	2.95	3.02	2.988	1.33	99.6
4.0	4.03	4.05	3.97	4.03	4.09	3.97	4.034	1.08	100.9

In this case, it can be seen from Table 5 that the proposed UV/US/O_3 synergistic digestion method combined with common UV spectrophotometric detection method had good performance, relative standard deviations (RSD) and the conversion rate (CR) were in the range 1.08~2.09% and 99.6~101.4%, respectively.

4. Conclusions

This study is the first time that the combination of both UV radiation and US simultaneously with O_3 oxidation was used to digest the N-compounds for determination of total dissolved nitrogen in waters. In this work, the advantage of UV/US/O_3 synergistic digestion method combined with common UV spectrophotometric detection method has been clearly shown by the digestion efficient and fast determination of total dissolved nitrogen. Results of experimental investigation showed that UV can effectively induce and promote the decomposition and photolysis of O_3 in aqueous solution to generate •OH, while US can intensify the gas-liquid mass transfer for reaction system, by which the carbamide solution with concentration of 4.0 mg/L could be completely digested under the optimum conditions: water sample volume, 10 mL; pH of water sample, 11; O_3 mass flow rate, 3200 mg/h; reaction temperature, 30 °C; digestion time, 25 min; UV lamp power, 18 W; distance between UV lamp and reactor, 2 cm; US frequency, 20 kHz; US power, 75 W. The conversion rate (CR) of synthetic wastewater samples varied from 99.6% to 101.4% for TDN in the range of 1.0~4.0 mg/L.

However, only three species of standard N-compounds (i.e., nitrate, ammonia chloride and carbamide) were investigated in our work. Natural waters may contain more species of TDN, especially for DON, the composition of which is unknown, but which usually include proteins, amino acid, acylamide, fulvic acid and humic acid, etc. In this case, more species of DON would need to be investigated. Meanwhile, in order to simulate and optimize the digestion progresses of N-compounds by UV/US/O_3 method, the degradation mechanism and reaction kinetics of N-compounds need for further study. Furthermore, due to the limitation of pressure and quantity of ozone generated by self-made ozone generation system, the ozone mass flow rate could not be adjusted conveniently, a novel ozone generation system based on pure oxygen would be established. Finally, a new jacketed digestion reactor need to be designed and developed for further studies and applications. Future work will focus on these issues.

Author Contributions: Conceptualization, X.S., L.X. and M.Z.; Methodology, X.S. and L.X.; Software, H.C.; Validation, H.C. and Z.L.; Formal analysis, Z.L.; Investigation, H.C.; Data curation, Z.L.; Writing—original draft preparation, X.S. and H.C.; Writing—review and editing, L.X.; Supervision, M.Z.; Project administration, Y.C. and L.X.; Funding acquisition, H.P. All authors have read and agreed to the published version of the manuscript.

Funding: This research was supported by Application Research of Public Welfare Technology in Zhejiang Province, China, Grant No. LGF19B060003 and LGF20E090005; National Natural Science Foundation of China, Grant No. 21676251.

Acknowledgments: The authors would also like to acknowledge everyone who has provided helpful guidance and would also like to thank the anonymous reviewers for their useful comments.

Conflicts of Interest: The authors declare no conflict of interest.

References

1. Gruber, N.; Galloway, J.N. An Earth-system perspective of the global nitrogen cycle. *Nature* **2008**, *451*, 293–296. [CrossRef] [PubMed]
2. Ma, J.; Yuan, D.; Lin, K.; Feng, S.; Zhou, T.; Li, Q. Applications of flow techniques in seawater analysis: A review. *Trends. Environ. Anal.* **2016**, *10*, 1–10. [CrossRef]
3. Water Quality-Determination of Total Nitrogen-Alkaline Potassium Persulfate Digestion UV Spectrophoto-Metric Method. Available online: http://www.mee.gov.cn/ywgz/fgbz/bz/bzwb/jcffbz/201203/t20120307_224383.shtml (accessed on 7 March 2012).
4. Oms, M.T.; Cerda, A.; Cerda, V.; Nollet, L.M. (Eds.) *Handbook of Water Analysis*; Marcel Dekker Inc.: New York, NY, USA, 2000; p. 201.

5. Wang, Q.H.; Yu, L.J.; Liu, Y.; Lin, L.; Lu, R.G.; Zhu, J.P.; He, L. Methods for the detection and determination of nitrite and nitrate: A review. *Talanta* **2017**, *165*, 709–720. [CrossRef]
6. Moorcroft, M.J.; Davis, J.; Compton, R.G. Detection and determination of nitrate and nitrite: A review. *Talanta* **2001**, *54*, 785–803. [CrossRef]
7. Neue, K.J. Methode zur Bestimmung des Stickstoffs in organischen Körpern. *Fresenius Z. Anal. Chemie.* **1883**, *22*, 366–382.
8. Armstrong, F.A.J.; Williams, P.M.; Strickland, J.D.H. Photo-oxidation of organic matter in seawater by ultra-violet radiation analytical and other applications. *Nature* **1966**, *221*, 481–483.
9. Koroleff, F. Determination of total nitrogen in natural waters by means of persulfate oxidation. International Council for the Exploration of the Sea (ICES). *Pap. C. M.* **1969**, *8*, 73–78.
10. Patton, C.J.; Kryskalla, J.R. *Methods of Analysis by the U.S. Geological Survey National Water Quality Laboratory: Evaluation of Alkaline Persulfate Digestion as an Alternative to Kjeldahl Digestion for Determination of Total and Dissolved Nitrogen and Phosphorus in Water*; U.S. Geological Survey: Reston, VI, USA, 2003.
11. Walsh, T.W. Total dissolved nitrogen in seawater: A new-high-temperature combustion method and a comparison with photo-oxidation. *Mar. Chem.* **1989**, *26*, 295–311. [CrossRef]
12. Frankovich, T.A.; Jones, R.D. A rapid, precise and sensitive method for the determination of total nitrogen in natural waters. *Mar. Chem.* **1998**, *60*, 227–234. [CrossRef]
13. Rogora, M.; Minella, M.; Orrù, A.; Tartari, G.A. A comparison between high-temperature catalytic oxidation and persulphate oxidation for the determination of total nitrogen in freshwater. *Int. J. Environ. Anal. Chem.* **2006**, *86*, 1065–1078. [CrossRef]
14. D'Elia, C.F.; Steudler, P.A.; Corwin, N. Determination of total nitrogen in aqueous samples using persulfate digestion. *Limnol. Oceanogr.* **1977**, *22*, 760–764. [CrossRef]
15. Nydahl, F. On the peroxodisulphate oxidation of total nitrogen in waters to nitrate. *Water Res.* **1978**, *12*, 1123–1130. [CrossRef]
16. Rossi, G.; Savarese, M. On-line determination of total nitrogen in natural sea water samples by alkaline persulphate oxidation. *Mar. Pollut. Bull.* **1997**, *35*, 174–175. [CrossRef]
17. Ferree, M.A.; Shannon, R.D. Evaluation of a second derivative UV/visible spectroscopy technique for nitrate and total nitrogen analysis of wastewater samples. *Water Res.* **2001**, *35*, 327–332. [CrossRef]
18. Gross, A.; Boyd, C.E. A digestion procedure for the simultaneous determination of total nitrogen and total phosphorus in pond water. *J. World Aquac. Soc.* **1998**, *29*, 300–303. [CrossRef]
19. De Borba, B.M.; Jack, R.F.; Rohrer, J.S.; Wirt, J.; Wang, D. Simultaneous determination of total nitrogen and total phosphorus in environmental waters using alkaline persulfate digestion and ion chromatography. *J. Chromatogr.* **2014**, *A1369*, 131–137. [CrossRef]
20. Lin, K.N.; Pei, J.X.; Li, P.C.; Ma, J.; Li, Q.L.; Yuan, D.X. Simultaneous determination of total dissolved nitrogen and total dissolved phosphorus in natural waters with an on-line UV and thermal digestion. *Talanta* **2018**, *185*, 419–426. [CrossRef]
21. International Organization for Standardization 11905-1:1997. *Water Qualit-Determination of Nitrogen-Part. 1: Method Using Oxidative Digestion with Peroxodisulfate*; IOS: Geneva, Switzerland, 1997.
22. McKelvie, I.D.; Mitri, M.; Hart, B.T.; Hamilton, I.C.; Stuart, A.D. Analysis of total dissolved nitrogen in natural waters by on-line photooxidation and flow injection. *Anal. Chim. Acta* **1994**, *293*, 155–162. [CrossRef]
23. Cerdà, A.; Oms, M.T.; Forteza, R.; Cerdà, V. Speciation of nitrogen in wastewater by flow injection. *Analyst* **1996**, *121*, 13–17. [CrossRef]
24. Gentle, B.S.; Ellis, P.S.; Grace, M.R.; McKelvie, I.D. Flow analysis methods for the direct ultra-violet spectrophotometric measurement of nitrate and total nitrogen in freshwaters. *Anal. Chim. Acta* **2011**, *704*, 116–122. [CrossRef]
25. Worsfold, P.J.; Clough, R.; Lohan, M.C.; Monbet, P.; Ellis, P.S.; Quétel, C.R.; Floor, G.H.; McKelvie, I.D. Flow injection analysis as a tool for enhancing oceanographic nutrient measurements-a review. *Anal. Chim. Acta* **2013**, *803*, 15–40. [CrossRef] [PubMed]
26. Kroon, H. Determination of nitrogen in water: Comparison of a continuous-flow method with on-line UV digestion with the original Kjeldahl method. *Anal. Chim. Acta* **1993**, *276*, 287–293. [CrossRef]
27. Oms, M.T.; Cerda, A.; Cerda, V. Sequential injection system for on-line analysis of total nitrogen with UV- mineralization. *Talanta* **2003**, *59*, 319–326. [CrossRef]

28. Enea, P.; Beatrice, C.; Lisa, S.; Marie, P.T.; Massimo, O.; Steven, C.; Jeremy, E.M.; Zoltán, M. Determination of total dissolved nitrogen in seawater by isotope dilution gas chromatogrphy mas spectrometry following digestion with persulfate and derivatization with aqueous triethyloxonium. *J. Chromatogr.* **2018**, *A1569*, 193–199.
29. Bronk, D.A.; Lomas, M.W.; Glibert, P.M.; Schukert, K.J.; Sanderson, M.P. Total dissolved nitrogen analysis: Comparisons between the persulfate, UV and high temperature oxidation methods. *Mar. Chem.* **2000**, *69*, 163–178. [CrossRef]
30. Sun, Y.; Yu, L.L.; Huang, H.B.; Yang, J.W.; Cheng, S.A. Research trend and practical development of advanced oxidation process on degradation of recalcitrant organic wastewater. *CIESC J.* **2017**, *68*, 1743–1756.
31. Antonio, G.-C.J.; Belén, E.-G.; Ana, A.; Antonio, S.-P.J.; Francisco, M.-A. Wastewater treatment by advanced oxidation process and their worldwide research trends. *Int. J. Environ. Res. Public Health* **2019**, *17*, 1–19.
32. Zhang, X.; Wang, Q.S. Application of advanced oxidation technologies in wastewater treatment. *Tech. Water. Treat.* **2009**, *3*, 24–28.
33. Luo, D. Study on Process Intensification of Ozonation of Dye Wastewater. Master's Thesis, Chongqing University of Technology, Chongqing, China, 25 March 2017.
34. Deng, Y.; Zhao, R. Advanced Oxidation Processes (AOPs) in Wastewater Treatment. *Curr. Pollut. Rep.* **2015**, *1*, 167–176. [CrossRef]
35. Domini, C.; Vidal, L.; Cravotto, G.; Canals, A. A simultaneous, direct microwave/ultrasound-assisted digestion procedure for the determination of total Kjeldahl nitrogen. *Ultrason. Sonochem.* **2009**, *16*, 564–569. [CrossRef]
36. Roth, J.A.; Sullivan, D.E. Solubility of ozone in water. *Ind. Eng. Chem. Fundam.* **1981**, *20*, 137–140. [CrossRef]
37. Katsoyiannis, I.A.; Canonica, S.; Gunten, U.V. Efficiency and energy requirements for the transformation of organic micropollutants by ozone, O_3/H_2O_2 and UV/H_2O_2. *Water Res.* **2011**, *45*, 3811–3822. [CrossRef] [PubMed]
38. Wardenier, N.; Liu, Z.; Nikiforov, A.; Van Hulle, S.W.H.; Leys, C. Micropollutant elimination by O_3, UV and plasma-based AOPs: An evaluation of treatment and energy costs. *Chemosphere* **2019**, *234*, 715–724. [CrossRef] [PubMed]
39. Wang, Y.; Li, H.Y.; Yi, P.; Zhang, H. Degradation of clofibric acid by UV, O_3 and UV/O_3 processes: Performance comparison and degradation pathways. *J. Hazard. Mater.* **2019**, *379*, 120771. [CrossRef] [PubMed]
40. García-Ballesteros, S.; Mora, M.; Vicente, R.; Vercher, R.F.; Sabater, C.; Castillo, M.A.; Amata, A.M.; Arques, A. A new methodology to assess the performance of AOPs in complex samples-Application to the degradation of phenolic compounds by O_3 and UV-A-Vis. *Chemosphere* **2019**, *222*, 114–123. [CrossRef]
41. Peyton, G.R.; Glaze, W.H. Destruction of pollutants in water with ozone in combination with ultraviolet radiation. 3. Photolysis of aqueous ozone. *Environ. Sci. Technol.* **1988**, *22*, 761–767. [CrossRef]
42. Mishchuk, N.A.; Goncharuk, V.V.; Vakulenko, V.F. Theoretical analysis of physicochemical processes occurring during water treatment by ozone and ultraviolet radiation. *Adv. Colloid Interface Sci.* **2008**, *139*, 62–73. [CrossRef]
43. Ince, N.H. Light-enhanced chemical oxidation for tertiary treatment of municipal landfill leachate. *Water Environ. Res.* **1998**, *70*, 1161–1169. [CrossRef]
44. Neta, P.; Maruthamuthu, P.; Carton, P.M.; Fessenden, R.W. Formation and reactivity of the amino radical. *J. Phys. Chem.* **1978**, *82*, 1875–1878. [CrossRef]
45. Glaze, W.H.; Kang, J.-W.; Chapin, D.H. The chemistry of water treatment processes involving ozone, hydrogen peroxide and ultraviolet radiation. *Ozone Sci. Eng.* **1987**, *9*, 335–352. [CrossRef]
46. Ku, Y.; Wen, W.; Yung-Shuen, S. Reaction behaviors of decomposition of monocrotophos in aqueous solution by UV and UV/O_3 processes. *J. Hazard. Mater.* **2000**, *72*, 25–37. [CrossRef]
47. Raymond Sierka, A.; Gary Amy, L. Catalytic effects of ultraviolet light and/or ultrasound on the ozone oxidation of humic acid and trihalomethane precursors. *Ozone Sci. Eng.* **1985**, *7*, 47–62. [CrossRef]
48. Guo, W.Q.; Yin, R.-L.; Zhou, X.-J.; Du, J.-S.; Cao, H.-O.; Yang, S.-S.; Ren, N.-Q. Sulfamethoxazole degradation by ultrasound/ozone oxidation process in water: Kinetics, mechanisms, and pathways. *Ultrason. Sonochem.* **2015**, *22*, 182–187. [CrossRef] [PubMed]
49. Ghanbari, F.; Khatebasreh, M.; Mahdavianpour, M.; Lin, K.A. Oxidative removal of benzotriazole using peroxymonosulfate/ozone/ultrasound: Synergy, optimization, degradation intermediates and utilizing for real wastewater. *Chemosphere* **2020**, *244*, 1–11. [CrossRef] [PubMed]

50. Zhang, A.P.; Yang, Z.; Xia, R.T.; Ding, Q.; Zhang, Y.Y. Study on the effect of ultrasound frequency and casting on the effect of ultrasound cavitation. *Chem. Eng. Mach.* **2016**, *4*, 517–521. [CrossRef]
51. Yu, Z.Y.; Zhang, J.; Pei, C.Z. Application of urea analytical wastewater recycling technology. *Small N. Fertilizer.* **2006**, *2*, 6–7.

MDPI
St. Alban-Anlage 66
4052 Basel
Switzerland
Tel. +41 61 683 77 34
Fax +41 61 302 89 18
www.mdpi.com

Processes Editorial Office
E-mail: processes@mdpi.com
www.mdpi.com/journal/processes

www.ingramcontent.com/pod-product-compliance
Lightning Source LLC
Chambersburg PA
CBHW061459210926
51850CB00008B/1763
9783039368884